STATE OF THE WORLD

2004

Other Norton/Worldwatch Books

State of the World 1984 through *2003*
(an annual report on progress toward a sustainable society)

Vital Signs 1992 through *2003*
(an annual report on the trends that are shaping our future)

Saving the Planet
Lester R. Brown
Christopher Flavin
Sandra Postel

How Much Is Enough?
Alan Thein Durning

Last Oasis
Sandra Postel

Full House
Lester R. Brown
Hal Kane

Power Surge
Christopher Flavin
Nicholas Lenssen

Who Will Feed China?
Lester R. Brown

Tough Choices
Lester R. Brown

Fighting for Survival
Michael Renner

The Natural Wealth of Nations
David Malin Roodman

Life Out of Bounds
Chris Bright

Beyond Malthus
Lester R. Brown
Gary Gardner
Brian Halweil

Pillar of Sand
Sandra Postel

Vanishing Borders
Hilary French

STATE OF THE WORLD 2004

A Worldwatch Institute Report on
Progress Toward a Sustainable Society

Brian Halweil and Lisa Mastny,
Project Directors

Erik Assadourian
Christopher Flavin
Hilary French
Gary Gardner
Danielle Nierenberg
Sandra Postel
Michael Renner
Radhika Sarin
Janet Sawin
Amy Vickers

Linda Starke, *Editor*

W·W·NORTON & COMPANY

NEW YORK   LONDON

The STATE OF THE WORLD and WORLDWATCH INSTITUTE trademarks are registered in the U.S. Patent and
Trademark Office.

The views expressed are those of the authors and do not necessarily represent those of the Worldwatch Institute;
of its directors, officers, or staff; or of its funders.

The text of this book is composed in Galliard, with the display set in Gill Sans. Book design by Elizabeth Doherty;
cover design by Lyle Rosbotham; composition by Worldwatch Institute; manufacturing by Phoenix Color Corp.

First Edition
ISBN 0-393-05860-3
ISBN 0-393-32539-3 (pbk)

W.W. Norton & Company, Inc., 500 Fifth Avenue, New York, N.Y. 10110
www.wwnorton.com

W.W. Norton & Company Ltd., Castle House, 75/76 Wells Street, London W1T 3QT

2 3 4 5 6 7 8 9 0

⊕ This book is printed on recycled paper.

Worldwatch Institute Staff

Acknowledgments

This twenty-first edition of *State of the World* draws on the dedication and hard work of every member of the talented Worldwatch staff. Backed by the generous support of Board members, funders, and friends, the Institute's researchers, editors, marketers, communications specialists, librarians, interns, and administrative staff all deserve our sincere thanks for their contributions to this year's special report on the consumer society.

We begin by acknowledging the foundation community, whose faithful backing sustains and encourages the Institute's work. The Overbrook Foundation and the Merck Family Fund awarded grants specifically for our consumption efforts. We also would like to acknowledge several other funders who generously support Worldwatch: the Aria Foundation, the Richard & Rhoda Goldman Fund, The William and Flora Hewlett Foundation, The Frances Lear Foundation, the NIB Foundation, the V. Kann Rasmussen Foundation, the A. Frank and Dorothy B. Rothschild Fund, The Shared Earth Foundation, The Shenandoah Foundation, Turner Foundation, Inc., the U.N. Environment Programme, the Wallace Global Fund, the Weeden Foundation, and The Winslow Foundation.

In addition, we give thanks to the Institute's individual donors, including the 3,500+ Friends of Worldwatch, who with their enthusiasm have demonstrated their strong commitment to Worldwatch and its efforts to contribute to a sustainable world. We are particularly indebted to the Worldwatch Council of Sponsors—Adam and Rachel Albright, Tom and Cathy Crain, John and Laurie McBride, and Wren and Tim Wirth—who have consistently shown their confidence and support of our work with generous annual contributions of $50,000 or more.

For this 2004 edition of *State of the World*, the Institute drew on the talents of an unprecedented number of outside authors, putting into action our strategic goal of strengthening ties with leading thinkers and practitioners in the field of sustainability. Sandra Postel, a Worldwatch Senior Fellow and director of the Global Water Policy Project, and Amy Vickers, award-winning author, engineer, and water conservation specialist, wrote the chapter on water productivity. We are also delighted to include contributions from consumption experts Isabella Marras, Solange Montillaud-Joyel, and Guido Sonnemann of the U.N. Environment Programme; William McDonough and Michael Braungart of McDonough Braungart Design Chemistry; Paul McRandle and Mindy Pennybacker of the Green Guide Institute; Juliet Schor of Boston College; and David Tilford of the Center for a New American Dream.

Chapter authors are grateful too for the enthusiasm and dedication of the 2003 team of interns, who doggedly pursued elusive facts and produced graphs, tables, and text boxes. Clayton Adams cheerfully extended his stay with us to locate information for Chapters 4, 6, and 7; Zoë Chafe contributed invaluable research assistance for both Chapter 7 and the "Year in Review" timeline; Claudia Meulenberg graciously juggled endless requests for help with Chapter 1; Shawn Powers tenaciously gathered information for Chapters 1, 3, and 5; and Anand Rao and Tawni Tidwell enriched Chapter 2 with their solid knowledge of global energy trends. Highly capable and good-natured, each was a stimulating addition to our staff this past summer and autumn.

The immense job of tracking down articles, journals, and books from all around the world fell to Research Librarian Lori Brown. In addition to keeping researchers up to date on the latest issues in their fields, she drew on her remarkable knack for gathering information to assemble this year's "Year in Review" timeline of significant global events.

After the initial research and writing were completed, an internal review process by Worldwatch staff members helped ensure that we would communicate our findings as clearly and accurately as possible. At August's day-long meeting, chapter authors were challenged, complimented, and critiqued by interns, magazine staff, and other reviewers. The authors greatly appreciate the excellent advice they received, including the input of research colleagues Chris Bright and Molly O'Meara Sheehan, who were busy working on other projects this year, although they did find time to contribute one "Behind the Scenes" feature each. The magazine staff of Ed Ayres and Tom Prugh also turned their editing skills to improving the drafts.

Reviews from outside experts, who generously gave us their time, were also indispensable to this year's final product. For their thoughtful comments and suggestions, as well as for the information many people provided, we are particularly grateful to: Abe Agulto, Mark Anielski, Michael Appelby, Matt Bentley, William Browning, David Brubaker, William Cain, Scot Case, Maurie Cohen, Dwight Collins, John de Graaf, Bas de Leeuw, Ed Diener, Chad Dobson, John Ehrenfeld, Andrea Fava, Tom Ferguson, Bette Fishbein, David Fridley, Bruce Friedrich, Sakiko Fukuda-Parr, Howard Geller, Gerard Gleason, Edward Groth III, Kirsty Hamilton, Marlene Hendrickson, Rich Howarth, Bobby Inocencio, Daniel Katz, Jonathan Louie, Philip Lymbery, Mia Macdonald, Michael Marx, Jim Mason, William Moomaw, Rosa Moreno, Nick Parrott, Enrique Peñalosa, David Pimentel, Robert Prescott-Allen, Lynn Price, Howard Rappaport, Richard Reynnells, John Rodwan, Hiroyuki Sato, Hans Schiere, Lee Schipper, Robert Schubert, Paul Shapiro, Ted Smith, Freyr Sverrisson, Joel Swisher, Ted Trainer, Arthur Weissman, Eric Williams, Paul Willis, and David Wood. In addition, we thank Norman Myers of Oxford University, coauthor of the forthcoming book *The New Consumers: Rich Lives or Richer Lifestyles?*, for steering us toward a new understanding of consumers in the world.

Further refinement of each chapter took place under the careful eye of independent editor Linda Starke, whose 27 years of experience with Worldwatch publications ensured that we conveyed our messages in the clearest way possible and met our deadlines. After the edits and rewrites were complete, Art Director Lyle Rosbotham skillfully crafted the design of each chapter, the timeline, and this year's new "Behind the Scenes" features. Ritch Pope finalized the production by preparing the index.

Writing is only the beginning of getting *State of the World* to readers. The task then passes to our excellent communications department, which works on multiple fronts to ensure that the *State of the World* message circulates widely beyond our Washington offices. Director of Communications Leanne Mitchell, Media Coordinator Susan Finkelpearl, and Communications Associate Susanne Martikke worked closely with researchers to craft our messages for the press, public, and decisionmakers around the world. Webmaster Steve Conklin used his technical expertise to convey our information via our newly redesigned Web site, and Information Technology Manager Patrick Settle ensured that the lines of communication ran smoothly both within and outside the office.

Vice President for Business Development Elizabeth Nolan coordinated cooperation with our global publishing partners and brought creativity and zest to our marketing efforts. Executive Assistant Katherine Dirks, with her willingness to tackle any project, was a tremendous asset to business development and to Worldwatch President Christopher Flavin, and ensured that meetings and travel ran according to schedule. Director of Finance and Administration Barbara Fallin ably handled the Institute's finances and kept the rest of the office sane during an office renovation this past summer. Joseph Gravely continued his reign as czar of Worldwatch's mail room.

Sadly, we had to say goodbye to Adrianne Greenlees, the Institute's Vice President for Development, in July, but were fortunate to find an excellent replacement in John Holman. He and Development Associate Cyndi Cramer work tirelessly to spread the Worldwatch word and to welcome new supporters into the Worldwatch family. The Institute's foundation fundraising activities continue under the able leadership of Kevin Parker, our Director of Foundation Relations, with assistance from Development Associate Mary Redfern. Both have worked closely with our foundation supporters to cultivate new relationships that will sustain the Institute's work for years to come.

We give special thanks to our partners around the globe for their extraordinary efforts to spread the message of sustainable development. *State of the World* is regularly published in 31 languages and 39 different editions, thanks largely to the dedication of a host of publishers, nongovernmental organizations, civil society groups, and individuals who provide advice as well as translation, outreach, publishing, and distribution assistance for our research. We would particularly like to acknowledge the help we receive from Øystein Dahle, Magnar Norderhaug, and Helen Eie in Norway; Brigitte Kunze, Christoph Bals, Klaus Milke, Bernd Rheinberg, and Gerhard Fischer in Germany; Soki Oda in Japan; Anna Bruno Ventre and Gianfranco Bologna in Italy; Lluis Garcia Petit and Marisa Mercado in Spain; Jung Yu Jin in Korea; George Cheng in Taiwan; Yesim Erkan in Turkey; Viktor Vovk in Ukraine; Tuomas Seppa in Finland; Zsuzsa Foltanyi in Hungary; Ioana Vasilescu in Romania; Hamid Taravaty in Iran; Eduardo Athayde in Brazil; and Jonathan Sinclair Wilson in the United Kingdom. For a complete listing of our global partners, see www.worldwatch.org/partners.

We would also like to express our gratitude to our long-time U.S. publisher, W.W. Norton & Company. Thanks to the dedication of its staff—especially Amy Cherry, Leo Wiegman, Andrew Marasia, Julia Druskin, and Lucinda Bartley—Worldwatch publications are available everywhere, from university campuses to small-town bookstores.

We are particularly grateful for the hard work and loyal support of the members of the Institute's Board of Directors, who have provided key input on strategic planning, organizational development, and fundraising over the last year. We are pleased that three distinguished new members joined the Board:

ACKNOWLEDGMENTS

Akio Morishima, Board Chairman of the Institute for Global Environmental Strategies in Japan; Geeta Aiyer, President of Boston Common Asset Management in the United States; and Satu Hassi, former Minister of Environment in Finland and current member of the Finnish Parliament.

We are dedicating this edition of *State of the World* to Tom and Cathy Crain, members of both the Institute's Board of Directors and the Council of Sponsors. In the late 1990s, the Crains had an "epiphany" while working in the financial industry that led them to embrace the cause of sustainability. Today, Tom and Cathy are as well informed on the economics of sustainability and globalization, and as dedicated to reforming the global economic system, as anyone on our staff.

Since joining the Worldwatch Board in 1998, Tom and Cathy Crain have played a major role in the Institute's strategic planning, helped to build both the membership and operations of the Board of Directors, and contributed strongly to the Institute's financial base. Tom is now the Institute's Vice Chairman and Treasurer, and Cathy chairs the Nominating Committee. We cannot thank them enough for the strength of their commitment to a sustainable future.

Finally, we would like to extend a warm welcome to the youngest member of the Worldwatch family, Elizabeth Rose McGinn, who was born in November 2002 to Senior Researcher Anne Platt McGinn. It is for Elizabeth and her generation that we strive ceaselessly to find ways to make our planet more livable.

Brian Halweil and Lisa Mastny
Project Directors

Contents

List of Boxes, Tables, and Figures

Boxes

Tables

Figures

Foreword

A skeptical reader might ask whether the world needs yet another report that details the extent and urgency of global challenges. But *State of the World 2004* is different. It focuses on consumption—one of the most central and also one of the most neglected elements in the global search for a sustainable future.

The Plan of Implementation that emerged from the World Summit on Sustainable Development in Johannesburg, South Africa, in 2002 states: "Fundamental changes in the way societies produce and consume are indispensable for achieving global sustainable development." Upholding this solemn commitment is now in part my responsibility since I was elected as the 2004 chair of the Commission on Sustainable Development (CSD), the United Nations' follow-up body for both the Rio and the Johannesburg agreements.

Consumption is of course essential to human well-being, but consuming too much or consuming the wrong things undermines both our personal health and the health of the natural environment on which we depend. The CSD offers a unique opportunity for a diverse range of communities—from indigenous peoples to educators, farmers, and business executives—to share ideas for addressing the challenges posed by our consumer society. At the annual CSD gatherings, people learn about what has worked and what has not worked in various parts of the world, and they gain strength from the experience of others. Governments learn how to apply the politically difficult polluter pays principle and how to eliminate harmful subsidies.

The twelfth meeting of the CSD, which I will chair at the United Nations in April 2004, will set in motion the 10-year program for sustainable production and consumption that was called for in Johannesburg. In addition, this meeting will have an important focus on water, sanitation, and human settlements, all essential elements of sustainable consumption and also key to achieving another central priority of the United Nations—eliminating poverty around the globe.

As Worldwatch Institute's expert researchers show in rigorous detail in this edition of *State of the World*, new patterns of consumption will be required in order to lift billions of people out of poverty in a manner that is consistent with global sustainability. We all make important daily decisions that affect not only our own communities but also the world as a whole—both its current and its future inhabitants.

As recognized in Johannesburg, much of the responsibility for bringing our consumer society into balance with the planet falls on richer nations, not simply because they are responsible for most global consumption but because they can help developing countries "leapfrog" some of the unsustainable choices that industrial nations are now exporting. In the end, sustainable consumption is a common concern that requires our common effort.

One example of a new approach to consumption involving both producers and consumers is the work of the nonprofit Rainforest Alliance. It has signed landmark accords with a leading coffee company as well as a banana producer and other large food companies to monitor the environmental practices on farms where their raw materials are produced. Environmentally conscious consumers will be able to buy the ecolabeled products that will be the fruits of this project.

We need more examples like this to show that we have the means—if we choose to use them—to apply the cleaner production concept, to let consumers make informed choices, and to demand and provide environmental information. We also have the means to apply the polluter pays principle, to eliminate harmful subsidies, and to create new markets. When we use these tools, we will change our patterns of consumption and production, making them more sustainable.

I am committed to fulfilling one of the CSD's unique capabilities when it comes to consumption and other issues: forging alliances between the corporate sector, citizens' groups, and others in order to achieve positive change for the world. I hope that by pointing to the accomplishments of these alliances and by highlighting innovative solutions, *State of the World 2004* will contribute to the evolution of a sustainable society.

I have not been able to read every word of this book and therefore cannot endorse every idea that it contains. But based on what I have seen so far, I am convinced that *State of the World*, as it has in the past, will provide a powerful array of innovative ideas that are sure to be useful to me and to others involved in the 2004 deliberations of the CSD. As the report's authors point out, what is lacking now is determined action. This is what we must all commit ourselves to achieving.

The challenge is formidable, but the alternative is unthinkable. More than ever before, what becomes clear in *State of the World 2004* is that each of us plays a daily role in changing the world. And while this realization might appear intimidating, it can also be a source of collective strength.

Børge Brende
Norwegian Minister of the Environment
Chairman, Commission on
Sustainable Development

Preface

In his book *An All-Consuming Century*, history professor Gary Cross argues that "consumerism" won the ideological wars of the twentieth century. Although most histories of recent economic and political developments suggest that it was capitalism or democracy that triumphed over communism, Cross makes a persuasive case that consumerism is what defines our age and is the lens through which most people view our times.[1]

The consumer society's long reach can be measured by the vast increases in purchases of automobiles, fast food, electronic devices, and other emblems of modern lifestyles. But the argument that consumerism defines our age runs much deeper: the drive to acquire and consume now dominates many peoples' psyches, filling the place once occupied by religion, family, and community. Consumption has given hundreds of millions of people a new sense of independence and has become a common benchmark to measure personal accomplishment. Time spent at church is now dwarfed by time spent at the "mall"—and consumption's connection to broader economic goals such as employment is a touchstone for politicians. In the aftermath of September 11, 2001, President Bush advised his fellow Americans that it was their patriotic duty to go to the malls and "buy."

Although Gary Cross's book is focused on the United States, his analysis can be applied to a rapidly growing share of the world's population. According to a recent study, 1.7 billion people—27 percent of humanity—have now entered the consumer society. Of that group, roughly 270 million are in the United States and Canada, 350 million in Western Europe, and 120 million in Japan. But nearly half of global consumers now live in developing countries, including 240 million in China and 120 million in India—numbers that have surged dramatically in the past two decades as globalization has introduced millions of people to consumer goods while also providing the technology and capital needed to build and disseminate them.[2]

On Worldwatch Institute's thirtieth anniversary, this edition of *State of the World* examines how we consume, why we consume, and what impact our consumption choices have on our fellow human beings and the planet. With chapters on food, water, energy, governance, economics, the power of purchasing, and redefining the good life, Worldwatch's award-winning research team asks whether a less-consumptive society is possible—and then argues that it is essential.

Consumption is of course necessary for human life and well-being, and if the choice is between being part of the consumer society or being among the 2.8 billion people who barely survive on less than $2 per day, the answer is easy. Massive increases in calorie intake, housing quality, household appliances, and scores of other amenities of the past half-century have helped lift hundreds of millions of people out of poverty.

But consumption among the world's wealthy elites, and increasingly among the middle class, has in recent decades gone well beyond satiating needs or even fulfilling dreams to become an end in its own right. It is as if much of the world is now following the exhortation of post–World War II American retailing analyst Victor Lebow, who said "our enormously productive economy...demands that we make consumption our way of life, that we convert the buying and use of goods into rituals, that we seek our spiritual satisfaction, our ego satisfaction, in consumption.... We need things consumed, burned up, worn out, replaced, and discarded at an ever increasing rate." This model, while rarely described so nakedly, helped fuel the unparalleled growth in the global economy over the past five decades, creating incomes and jobs for hundreds of millions of people.[3]

The unbounded pursuit of consumption has also exacted a heavy cost, however, a cost that is now growing at least as fast as consumption itself. Consumption today absorbs vast quantities of resources, many of which are now being used far beyond sustainable levels. In just the last 50 years, global use of fresh water has grown threefold, while fossil fuel use has risen fivefold. Renewable resources are particularly under threat, from the falling water tables of northern China to the depleted fisheries of the North Atlantic. Over time, the efficiency of human resource use has improved, and depleted resources have been replaced by others, but the pattern of the past half-century is clear: the pollution and resource degradation that flow from ever-growing consumption continue to worsen, and the toll is being measured not just in damaged ecosystems but in human disease and misery—particularly for the poorest among us. The billions of tons of carbon

dioxide building up in the atmosphere as a result of rising levels of fossil fuel consumption are now taking these burdens to the global level in the form of climate change.[4]

The real challenge lies ahead. The global consumption bandwagon has acquired an extraordinary momentum that will place rapidly growing demands on human society and the natural world in the decades to come. This momentum is, for example, seen in China, which went from having virtually no private cars in 1980 to having 5 million in 2000—and which is likely to have 24 million cars in 2005, leaving still more than 1 billion prospective car buyers in China.[5]

Not only will hundreds of millions of people in the developing world enter the consumer society in the near future, but the per capita consumption levels of those who are already in it continue to surge as automobiles and houses get larger and as new gadgets proliferate. And even as per person consumption expands, the absolute number of people also continues to grow—close to 3 billion human beings are likely to be added by mid-century. The combined effect of consumption and population is particularly alarming, but between the two, consumption is the more intractable. Most projections show world population leveling off in the second half of the century as fertility rates decline. But consumption just keeps growing.

It is this daunting prospect that drew the Worldwatch research team to focus most of its work over the past year on consumption, following in the path-breaking footsteps of our former colleague Alan Durning, who wrote *How Much is Enough?* in 1992. Durning pointed out a conundrum that has become even more apparent today: the single-minded pursuit of consumption not only will undermine the quality of life of those in the consumer society, it will diminish the ability of those outside the consumption class to meet their basic needs.[6]

In pursuing the theme of consumption throughout this year's *State of the World*, we have sought to do more than describe the dilemmas posed by consumption, going on to explore ways that consumption can be restrained and redirected in order to improve prospects for human well-being and sustainability. In the pages that follow, the authors have shown how in everything from our use of energy and water to our consumption of food we can make choices that will improve our health, create jobs, and reduce pressures on the world's natural ecosystems.

To accomplish this goal, we have interspersed the chapters in *State of the World 2004* with short articles on a variety of everyday products—from computers to chickens and cans of soda—in order to allow readers to see common goods in a new light. In addition, we have pointed to many cases in which consumers are banding together to purchase goods such as sustainably grown wood products, organic cocoa, and "fair-trade" coffee. Although most of these movements are tiny compared with the larger consumer economy, they are growing rapidly and could soon become a powerful force in many markets.

Our goal in *State of the World 2004* is not only to address one of the most important issues of our time in a way that informs and motivates our readers, but to work with our partners around the world to provide concrete ideas for those who want to get off the consumer treadmill. Consumption is of course in part a societal challenge that will require effective use of government regulation and fiscal policy to achieve the common good. But more so than most issues, changes in consumption practices will require millions of individual decisions that can only begin at the grassroots.

To help in this process, we will soon be launching a consumption portal on the Worldwatch Web site that will contain selected material from *State of the World*, links to organizations working actively on consumption campaigns, and tips for becoming a more informed consumer. This Web portal will also include a companion guide to *State of the World*, which will have dozens of vignettes about commonly used items as well as suggestions for finding more sustainable alternatives. In addition, the portal will provide contact information for the many partner organizations around the world that helped us gather the information for this book and that are working to change consumption habits, including the Green Guide, the Silicon Valley Toxics Coalition, and the Center for a New American Dream.

It would be foolish to underestimate the challenge of checking the consumption juggernaut. Few forces are as powerful or widespread. But as the costs of unbridled consumption become clear, we believe that the innovative responses described in these pages will also catch on at an accelerating pace. In the long run, it will become apparent that achieving generally accepted goals—meeting basic human needs, improving human health, and supporting a natural world that can sustain us—will require that we control consumption rather than allow consumption to control us.

We hope that you will read, analyze, and question the information and ideas developed in these pages. We look forward to hearing your suggestions for strengthening future editions of *State of the World*.

Christopher Flavin

President
Worldwatch Institute
November 2003

State of the World:
A Year in Review

Following the positive responses to last year's innovation of "State of the World: A Year in Review," we are continuing to document the many developments and challenges along the road to sustainable development. This year's edition was compiled by Research Librarian Lori Brown, with contributions from all the staff. Zöe Chafe, Lisa Mastny, and Radhika Sarin were particularly helpful in putting together the new timeline.

The timeline this year covers significant announcements and reports from October 2002 until *State of the World 2004* went to press in October 2003. It is once again a mix of progress, setbacks, and missed steps around the world that are affecting society's environmental and social goals.

There is no attempt to be comprehensive, but instead to highlight events that increase awareness of the connections between people and the world's environment. We always welcome your feedback, and we hope to continue building the timeline as we move forward in the new millennium.

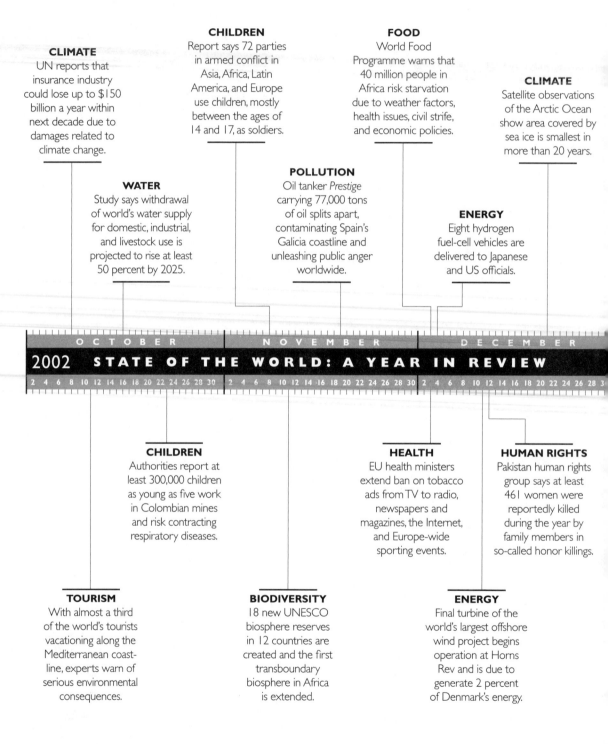

CLIMATE
UN reports that insurance industry could lose up to $150 billion a year within next decade due to damages related to climate change.

CHILDREN
Report says 72 parties in armed conflict in Asia, Africa, Latin America, and Europe use children, mostly between the ages of 14 and 17, as soldiers.

FOOD
World Food Programme warns that 40 million people in Africa risk starvation due to weather factors, health issues, civil strife, and economic policies.

CLIMATE
Satellite observations of the Arctic Ocean show area covered by sea ice is smallest in more than 20 years.

WATER
Study says withdrawal of world's water supply for domestic, industrial, and livestock use is projected to rise at least 50 percent by 2025.

POLLUTION
Oil tanker *Prestige* carrying 77,000 tons of oil splits apart, contaminating Spain's Galicia coastline and unleashing public anger worldwide.

ENERGY
Eight hydrogen fuel-cell vehicles are delivered to Japanese and US officials.

OCTOBER   NOVEMBER   DECEMBER

2002   STATE OF THE WORLD: A YEAR IN REVIEW

CHILDREN
Authorities report at least 300,000 children as young as five work in Colombian mines and risk contracting respiratory diseases.

HEALTH
EU health ministers extend ban on tobacco ads from TV to radio, newspapers and magazines, the Internet, and Europe-wide sporting events.

HUMAN RIGHTS
Pakistan human rights group says at least 461 women were reportedly killed during the year by family members in so-called honor killings.

TOURISM
With almost a third of the world's tourists vacationing along the Mediterranean coastline, experts warn of serious environmental consequences.

BIODIVERSITY
18 new UNESCO biosphere reserves in 12 countries are created and the first transboundary biosphere in Africa is extended.

ENERGY
Final turbine of the world's largest offshore wind project begins operation at Horns Rev and is due to generate 2 percent of Denmark's energy.

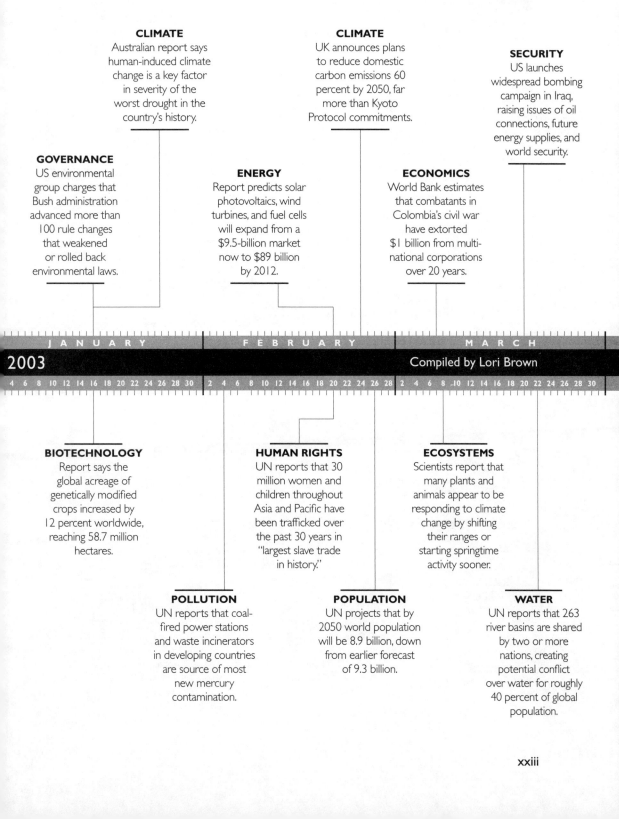

CLIMATE
Australian report says human-induced climate change is a key factor in severity of the worst drought in the country's history.

CLIMATE
UK announces plans to reduce domestic carbon emissions 60 percent by 2050, far more than Kyoto Protocol commitments.

SECURITY
US launches widespread bombing campaign in Iraq, raising issues of oil connections, future energy supplies, and world security.

GOVERNANCE
US environmental group charges that Bush administration advanced more than 100 rule changes that weakened or rolled back environmental laws.

ENERGY
Report predicts solar photovoltaics, wind turbines, and fuel cells will expand from a $9.5-billion market now to $89 billion by 2012.

ECONOMICS
World Bank estimates that combatants in Colombia's civil war have extorted $1 billion from multinational corporations over 20 years.

JANUARY — **FEBRUARY** — **MARCH**

2003   Compiled by Lori Brown

BIOTECHNOLOGY
Report says the global acreage of genetically modified crops increased by 12 percent worldwide, reaching 58.7 million hectares.

HUMAN RIGHTS
UN reports that 30 million women and children throughout Asia and Pacific have been trafficked over the past 30 years in "largest slave trade in history."

ECOSYSTEMS
Scientists report that many plants and animals appear to be responding to climate change by shifting their ranges or starting springtime activity sooner.

POLLUTION
UN reports that coal-fired power stations and waste incinerators in developing countries are source of most new mercury contamination.

POPULATION
UN projects that by 2050 world population will be 8.9 billion, down from earlier forecast of 9.3 billion.

WATER
UN reports that 263 river basins are shared by two or more nations, creating potential conflict over water for roughly 40 percent of global population.

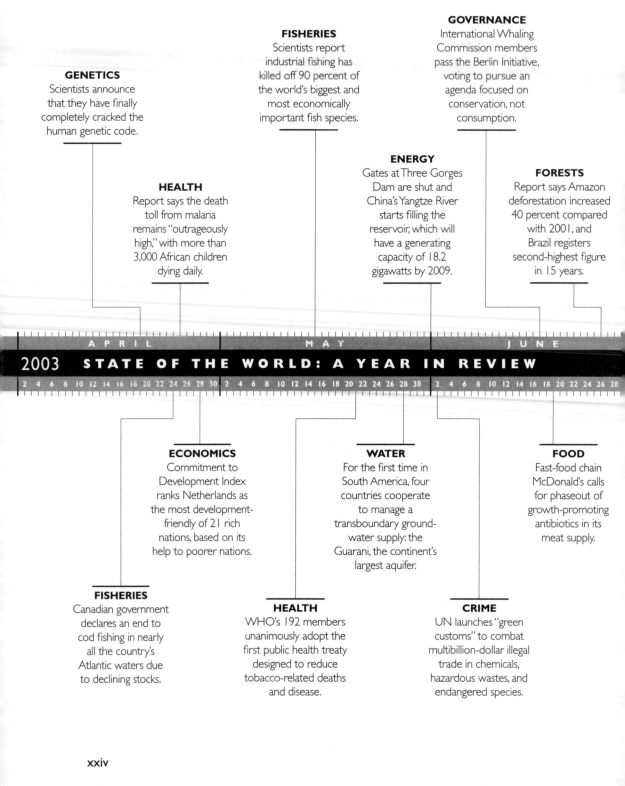

GENETICS
Scientists announce that they have finally completely cracked the human genetic code.

FISHERIES
Scientists report industrial fishing has killed off 90 percent of the world's biggest and most economically important fish species.

GOVERNANCE
International Whaling Commission members pass the Berlin Initiative, voting to pursue an agenda focused on conservation, not consumption.

HEALTH
Report says the death toll from malaria remains "outrageously high," with more than 3,000 African children dying daily.

ENERGY
Gates at Three Gorges Dam are shut and China's Yangtze River starts filling the reservoir, which will have a generating capacity of 18.2 gigawatts by 2009.

FORESTS
Report says Amazon deforestation increased 40 percent compared with 2001, and Brazil registers second-highest figure in 15 years.

APRIL   MAY   JUNE

2003   STATE OF THE WORLD: A YEAR IN REVIEW

2 4 6 8 10 12 14 16 18 20 22 24 26 28 30   2 4 6 8 10 12 14 16 18 20 22 24 26 28 30   2 4 6 8 10 12 14 16 18 20 22 24 26 28

ECONOMICS
Commitment to Development Index ranks Netherlands as the most development-friendly of 21 rich nations, based on its help to poorer nations.

WATER
For the first time in South America, four countries cooperate to manage a transboundary ground-water supply: the Guarani, the continent's largest aquifer.

FOOD
Fast-food chain McDonald's calls for phaseout of growth-promoting antibiotics in its meat supply.

FISHERIES
Canadian government declares an end to cod fishing in nearly all the country's Atlantic waters due to declining stocks.

HEALTH
WHO's 192 members unanimously adopt the first public health treaty designed to reduce tobacco-related deaths and disease.

CRIME
UN launches "green customs" to combat multibillion-dollar illegal trade in chemicals, hazardous wastes, and endangered species.

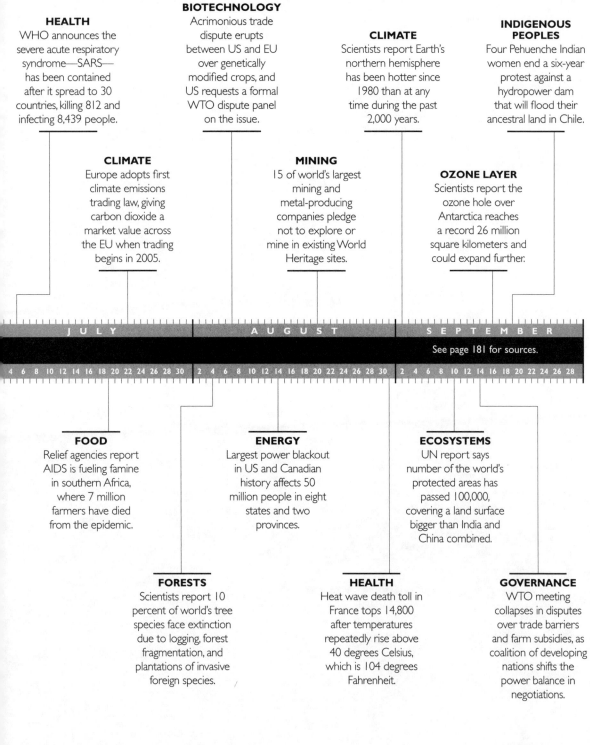

HEALTH
WHO announces the severe acute respiratory syndrome—SARS—has been contained after it spread to 30 countries, killing 812 and infecting 8,439 people.

BIOTECHNOLOGY
Acrimonious trade dispute erupts between US and EU over genetically modified crops, and US requests a formal WTO dispute panel on the issue.

CLIMATE
Scientists report Earth's northern hemisphere has been hotter since 1980 than at any time during the past 2,000 years.

INDIGENOUS PEOPLES
Four Pehuenche Indian women end a six-year protest against a hydropower dam that will flood their ancestral land in Chile.

CLIMATE
Europe adopts first climate emissions trading law, giving carbon dioxide a market value across the EU when trading begins in 2005.

MINING
15 of world's largest mining and metal-producing companies pledge not to explore or mine in existing World Heritage sites.

OZONE LAYER
Scientists report the ozone hole over Antarctica reaches a record 26 million square kilometers and could expand further.

JULY — AUGUST — SEPTEMBER

See page 181 for sources.

4 6 8 10 12 14 16 18 20 22 24 26 28 30 | 2 4 6 8 10 12 14 16 18 20 22 24 26 28 30 | 2 4 6 8 10 12 14 16 18 20 22 24 26 28

FOOD
Relief agencies report AIDS is fueling famine in southern Africa, where 7 million farmers have died from the epidemic.

ENERGY
Largest power blackout in US and Canadian history affects 50 million people in eight states and two provinces.

ECOSYSTEMS
UN report says number of the world's protected areas has passed 100,000, covering a land surface bigger than India and China combined.

FORESTS
Scientists report 10 percent of world's tree species face extinction due to logging, forest fragmentation, and plantations of invasive foreign species.

HEALTH
Heat wave death toll in France tops 14,800 after temperatures repeatedly rise above 40 degrees Celsius, which is 104 degrees Fahrenheit.

GOVERNANCE
WTO meeting collapses in disputes over trade barriers and farm subsidies, as coalition of developing nations shifts the power balance in negotiations.

STATE OF THE WORLD

2004

CHAPTER 1

The State of Consumption Today

Gary Gardner, Erik Assadourian,
and Radhika Sarin

China has a well-deserved reputation as the land of the bicycle. Throughout the twentieth century, the streets of her cities were filled with literally millions of bikes, not only providing personal transportation but also serving as delivery vehicles—carrying everything from construction materials to chickens on their way to market. As recently as the early 1980s, few private cars were found on China's streets.[1]

A visitor from the 1980s who returns to Beijing, Shanghai, or other Chinese cities today will hardly recognize them. By 2002 there were 10 million private cars, and growth in ownership was accelerating: every day in 2003 some 11,000 more cars merged into the traffic on Chinese roads—4 million new private cars during the year. Auto sales increased by 60 percent in 2002 and by more than 80 percent in the first half of 2003. By 2015, if growth continues apace, industry analysts expect 150 million cars to be jamming

China's streets—18 million more than were driven on U.S. streets and highways in 1999. The emerging class of Chinese consumers is enthusiastically embracing the increased mobility and higher social status that the automobile now represents—millions wait months and take on significant debt in order to become pioneer members of China's new automobile culture.[2]

The advantages of this development path are clear to the government officials who are encouraging it. Each new Chinese-made car provides two new jobs to Chinese workers, and the income they receive then stimulates other sectors of the Chinese economy. Moreover, the rush to meet demand is attracting massive investments by foreign companies—General Motors has spent $1.5 billion on a new factory in Shanghai, while Volkswagen has committed $7 billion over the next five years to increase its production capacity.[3]

China is of course following a well-blazed trail, albeit roughly eight decades after widespread use of the automobile first caught on in the United States. Yet China's automobile

Units of measure throughout this book are metric unless common usage dictates otherwise.

story is tied to neither the Chinese nor the automobile. From fast food to disposable cameras and from Mexico to South Africa, a good deal of the world is now entering the consumer society at a mind-numbing pace. By one calculation, there are now more than 1.7 billion members of "the consumer class" today—nearly half of them in the "developing" world. A lifestyle and culture that became common in Europe, North America, Japan, and a few other pockets of the world in the twentieth century is going global in the twenty-first.[4]

The consumer society clearly has a strong allure, and carries with it many economic benefits. And it would certainly be unfair to argue that advantages gained by an earlier generation of consumers should not be shared by those who come later. Yet the headlong growth of consumption in the last decade—and the staggering projections that flow logically from that growth—suggest that the world as a whole will soon run smack into a stark dilemma. If the levels of consumption that several hundred million of the most affluent people enjoy today were replicated across even half of the roughly 9 billion people projected to be on the planet in 2050, the impact on our water supply, air quality, forests, climate, biological diversity, and human health would be severe.[5]

Despite the dangers ahead, there is little evidence that the consumption locomotive is braking—not even in countries like the United States, where most people are amply supplied with the goods and services needed to lead a dignified life. As of 2003 the United States had more private cars than licensed drivers, and gas-guzzling sport-utility vehicles were one of the best-selling vehicles. New houses were 38 percent bigger in 2002 than in 1975, despite having fewer people in each household on average. Americans themselves are larger as well—so much bigger, in fact, that a multi-billion-dollar industry has emerged to cater to the needs of large Americans, supplying them with oversized clothing, sturdier furniture, even supersized caskets. If the consumption aspirations of the wealthiest of nations cannot be satiated, the prospects for corralling consumption everywhere before it strips and degrades our planet beyond recognition would appear to be bleak.[6]

Yet there are many reasons to be hopeful. Consumer advocates, economists, policy-makers, and environmentalists have developed creative options for meeting people's needs while dampening the environmental and social costs associated with mass consumption. In addition to helping individuals find the balance between too much and too little consumption, they stress placing more emphasis on publicly provided goods and services, on services in place of goods, on goods with high levels of recycled content, and on genuine choice for consumers. Together, these measures can help deliver a high quality of life with a minimum of environmental abuse and social inequity. The key is to look critically not only at the "how much" of consumption, but also the "how." (See Chapters 5 and 8.)

Consumption is not a bad thing. People must consume to survive, and the world's poorest will need to consume more if they are to lead lives of dignity and opportunity. But consumption threatens the well-being of people and the environment when it becomes an end in itself—when it is an individual's primary goal in life, for example, or the ultimate measure of the success of a government's economic policies. The economies of mass consumption that produced a world of abundance for many in the twentieth century face a different challenge in the twenty-first: to focus not on the indefinite accumulation of goods but instead on a better quality of life for all, with minimal environmental harm.

Consumption by the Numbers

By virtually any measure—household expenditures, number of consumers, extraction of raw materials—consumption of goods and services has risen steadily in industrial nations for decades, and it is growing rapidly in many developing countries. The numbers tell the story of a world being transformed by a consumption revolution.

Private consumption expenditures—the amount spent on goods and services at the household level—topped $20 trillion in 2000, up from $4.8 trillion in 1960 (in 1995 dollars). Some of this fourfold increase occurred because of population growth (see Box 1–1), but much of it was due to advancing prosperity in many parts of the globe. These overall numbers mask enormous disparities in spending. The 12 percent of the world living in North America and Western Europe account for 60 percent of global private con-

---

BOX 1–1. WHAT ABOUT POPULATION?

The United Nations Population Division projects that world population will increase 41 percent by 2050, to 8.9 billion people. Just as growing acquisition of appliances and cars can eliminate energy savings achieved by efficiency improvements, this increase in human numbers threatens to offset any progress in reducing the amount of goods that each person consumes. For example, even if the average American eats 20 percent less meat in 2050 than in 2000, total meat consumption in the United States will be roughly 5 million tons greater in 2050 due to population growth alone.

With 99 percent of global population growth projected to occur in developing nations, these countries need to consider carefully the twin goals of population stabilization and increased consumption for human development. The industrial world can help developing countries stabilize their populations by supporting family planning, education, and the improvement of women's status. And it can lower the impact of increased consumption by assisting with the adoption of cleaner, more efficient technologies.

But it would be a mistake to think of population growth as a challenge facing only poor nations. When population growth and high levels of consumption mix, as they do in the United States, the significance of the former balloons. For example, although the U.S. popula-

tion increases by roughly 3 million a year, whereas India's increases by nearly 16 million, the additional Americans have greater environmental impact. They are responsible for 15.7 million tons of additional carbon to the atmosphere, compared with only 4.9 million tons in India. Wealthy countries with expanding populations need to look at the impact of both their consumption and their population policies.

Other less discussed demographic trends mix with consumption in surprising ways as well. For instance, as a result of rising incomes, urbanization, and smaller families, the number of people living under one roof fell between 1970 and 2000 from 5.1 to 4.4 in developing countries and from 3.2 to 2.5 in industrial countries, while the total number of households increased. Each new house requires space and materials, of course. In addition, savings gained from having more people share energy, appliances, and home furnishings are lost when fewer people live in the same house. Thus a one-person household in the United States uses 17 percent more energy per person than a two-person household does. So even in some European nations and Japan, where total population is not growing much if at all, changing household dynamics should be examined as drivers of increased consumption.

SOURCE: See endnote 7.

---

sumer spending, while the one third living in South Asia and sub-Saharan Africa account for only 3.2 percent. (See Table 1–1.)[7]

In 1999, some 2.8 billion people—two of every five humans on the planet—were living on less than $2 a day, which the United Nations and the World Bank say is the minimum for meeting basic needs. Roughly 1.2 billion people were living in "extreme poverty," measured by an average daily income of less than $1. Among the poorest are hundreds of millions of subsistence farmers, who by definition do not earn wages and who seldom engage in money-based market transactions. For them, and for all of the world's poor, consumption expenditures are focused almost entirely on meeting basic needs.[8]

Although most consumer spending occurs in the wealthier regions of the world, the number of consumers is spread a bit more evenly between industrial and developing regions. This is clear from research done by former U.N. Environment Programme (UNEP) consultant Matthew Bentley, who describes the existence of a global "consumer class." These people have incomes over $7,000 of purchasing power parity (an income measure adjusted for the buying power in local currency), which is roughly the level of the official poverty line in Western Europe. The global consumer class itself ranges widely in levels of wealth, but members are typically users of televisions, telephones, and the Internet, along with the culture and ideas that these products transmit. This global consumer class totals some 1.7 billion people—more than a quarter of the world. (See Table 1–2.)[9]

Almost half of this global consumer class lives in developing nations, with China and India alone claiming more than 20 percent of the global total. (See Table 1–3.) In fact, these two countries' combined consumer class of 362 million people is larger than this class

Table 1–1. Consumer Spending and Population, by Region, 2000

Region	Share of World Private Consumption Expenditures	Share of World Population
	(percent)	
United States and Canada	31.5	5.2
Western Europe	28.7	6.4
East Asia and Pacific	21.4	32.9
Latin America and the Caribbean	6.7	8.5
Eastern Europe and Central Asia	3.3	7.9
South Asia	2.0	22.4
Australia and New Zealand	1.5	0.4
Middle East and North Africa	1.4	4.1
Sub-Saharan Africa	1.2	10.9

SOURCE: See endnote 7.

in all of Western Europe (although the average Chinese or Indian member of course consumes substantially less than the average European). Much of the rest of the developing world is not represented in this surge of new consumption, however: sub-Saharan Africa's consumer class, the smallest, has just 34 million people. Indeed, the region has essentially been a bystander to the prosperity experienced in most of the world in recent decades. Measured in terms of private consumption expenditures per person, sub-Saharan Africa in 2001 was 20 percent worse off than two decades earlier, creating a yawning gap between that region and the industrial world.[10]

In addition to having large consumer blocs, developing countries tend to have the greatest potential to expand the ranks of consumers. For example, China and India's large consumer set constitutes only 16 percent of

Table 1–2. Consumer Class, by Region, 2002

Region	Number of People Belonging to the Consumer Class	Consumer Class as Share of Regional Population	Consumer Class as Share of Global Consumer Class[1]
	(million)	(percent)	(percent)
United States and Canada	271.4	85	16
Western Europe	348.9	89	20
East Asia and Pacific	494.0	27	29
Latin America and the Caribbean	167.8	32	10
Eastern Europe and Central Asia	173.2	36	10
South Asia	140.7	10	8
Australia and New Zealand	19.8	84	1
Middle East and North Africa	78.0	25	4
Sub-Saharan Africa	34.2	5	2
Industrial Countries	912	80	53
Developing Countries	816	17	47
World	1,728	28	100

[1]Total does not add to 100 due to rounding.
SOURCE: See endnote 9.

the region's population, whereas in Europe the figure is 89 percent. Indeed, in most developing countries the consumer class accounts for less than half of the population—often much less—suggesting considerable room to grow. Based on population projections alone, the global consumer class is conservatively projected to hold at least 2 billion people by 2015.[11]

These numbers suggest that the story of consumption in the twenty-first century could be as much about emerging consumer nations as about traditional ones. In a 2003 Background Paper, the U.N. Environment Programme noted that boosting Asian car ownership rates to the world average would add 200 million cars to the global fleet—one and a half times the number of cars currently found in the United States. Concerns about the impact of developments like these suggest the urgency of pursuing alternative, sustainable paths to prosperity in the region. At the same time, worries about potential

increases in Asian consumption are misplaced if they obscure the need for reform in wealthy countries, where high levels of consumption have been the norm for decades. The early industrializing countries in Europe

Table 1–3. Top 10 National Consumer Class Populations, 2002

Country	Number of People in Consumer Class, 2002	Share of National Population
	(million)	(percent)
United States	242.5	84
China	239.8	19
India	121.9	12
Japan	120.7	95
Germany	76.3	92
Russian Federation	61.3	43
Brazil	57.8	33
France	53.1	89
Italy	52.8	91
United Kingdom	50.4	86

SOURCE: See endnote 10.

and North America, along with Japan and Australia, are responsible for the bulk of global environmental degradation associated with consumption.[12]

Consumption trends cover virtually every conceivable good and service, and these can be categorized in many ways. Of particular interest are fundamentals such as food and water; trends for these give a sense of whether basic needs are being met. Other consumer items indicate the degree to which life options are expanding for people, and how much more comfortable life is becoming.

In terms of basic needs, trends are mixed. Daily intake of calories has increased in both the industrial and the developing worlds since 1961 as food supplies have become bountiful, at least at the global level. Yet the U.N. Food and Agriculture Organization (FAO) reports that 825 million people are still undernourished and that the average person in the industrial world took in 10 percent more calories daily in 1961 (2,947 calories) than the average person in the developing world consumes today (2,675 calories). The existence of hunger in the face of record food supplies reflects the fact that food remains expensive for much of the world's poor relative to their meager incomes. In Tanzania, for instance, where per capita household expenditures were $375 in 1998, 67 percent of household spending went to food. In Japan, per capita household expenditures stood at $13,568 that year, but only 12 percent of that was spent on food. (See Table 1–4.)[13]

Not only do the world's wealthy take in more calories than the poor, but those calories are likelier to come from more resource-intensive foods, such as meat and dairy products, which are produced using large quantities of grain, water, and energy. (See Chapters 3 and 4.) People in industrial countries get 856 of their daily calories from ani-

Table 1–4. Share of Household Expenditures Spent on Food

Country	Per Capita Household Expenditures, 1998	Share Spent on Food
	(dollars)[1]	(percent)
Tanzania	375	67
Madagascar	608	61
Tajikistan	660	48
Lebanon	6,135	31
Hong Kong	12,468	10
Japan	13,568	12
Denmark	16,385	16
United States	21,515	13

[1]Purchasing power parity.
SOURCE: See endnote 13.

mal products, while in developing countries the figure is 350. Still, meat consumption is rising in the more prosperous regions of the developing world as incomes and urbanization rates increase. Half of the world's pork is eaten in China, for example, while Brazil is the second largest consumer of beef, after the United States. And meat is increasingly consumed as fast food, which is often more energy-intensive to produce. According to a recent marketing research study, the fast-food industry in India is growing by 40 percent a year and is expected to generate over a billion dollars in sales by 2005. Meanwhile, a quarter of India's population remains undernourished—a number virtually unchanged over the past decade.[14]

Clean water and adequate sanitation, which are instrumental in preventing the spread of infectious disease, are also basic consumption needs. As with most goods, access to water and sanitation is skewed in favor of wealthier populations, although this situation has improved for poorer people somewhat in the past decade. In 2000, 1.1 billion people did not have access to safe drinking water, defined as the availability of at least 20 liters per person per day from a source within one kilometer of the user's dwelling. And two out of every five people did not have adequate sanitation facilities, such as a connection to a sewer or septic tank, or even a simple pit latrine. People in rural areas suffer the most. In 2000, only 40 percent of people living in rural areas were using adequate sanitation facilities, compared with 85 percent of urban inhabitants.[15]

As incomes rise, people gain access to non-food consumer items that indicate greater prosperity. Paper use, for example, tends to increase as people become more literate and as communications links increase. Globally, paper use increased more than sixfold between 1950 and 1997 and has doubled since the mid-1970s; the average Briton used 16 times more paper at the end of the twentieth century than at its start. Indeed, most of the world's paper is produced and consumed in industrial countries: the United States alone produces and uses a third of the world's paper, and Americans use more than 300 kilograms each annually. In developing nations as a whole, in contrast, people use 18 kilograms of paper each year. In India, the

annual figure is 4 kilos, and in 20 nations in Africa, it is less than 1 kilo. UNEP estimates that 30–40 kilos of paper are the minimum needed to meet basic literacy and communication needs.[16]

Rising prosperity also opens access to goods that promise new levels of comfort, convenience, and entertainment to millions. (See Table 1–5.) In 2002, 1.12 billion households, about three quarters of the world's people, owned at least one television set. Watching TV has become a leading form of leisure, with the average person in the industrial world spending three hours—half of their daily leisure time—in front of a television each day. The TV offers viewers access to local news and entertainment, but also exposure to countless consumer products that are shown in advertisements and during programs. And the view emerging from the screen is increasingly global in scope. Of the 1.12 billion households with TVs, 31 percent subscribed to a cable television service, often exposing them to a global entertainment culture.[17]

Many of these conveniences were considered luxuries when first introduced but are

Table 1–5. Household Consumption, Selected Countries, Circa 2000

Country	Household Consumption Expenditure	Electric Power	Television Sets	Telephone Mainlines	Mobile Phones	Personal Computers
	(1995 dollars per person)	(kilowatt-hours per person)	(per thousand population)			
Nigeria	194	81	68	6	4	7
India	294	355	83	40	6	6
Ukraine	558	2,293	456	212	44	18
Egypt	1,013	976	217	104	43	16
Brazil	2,779	1,878	349	223	167	75
South Korea	6,907	5,607	363	489	621	556
Germany	18,580	5,963	586	650	682	435
United States	21,707	12,331	835	659	451	625

SOURCE: See endnote 17.

now perceived to be necessities. Indeed, where societal infrastructures have developed around them, some of these consumer goods have become integral to day-to-day life. Telephones, for example, have become an essential tool of communication—in 2002, there were 1.1 billion fixed-lines and another 1.1 billion mobile lines. A significant percentage of the world's people, including the vast majority of the world's global consumers, now has at least basic access to telephones. Communications have also advanced with the introduction of the Internet. This most recent addition to modern communications now connects about 600 million users.[18]

A large share of consumer spending focuses on goods that are arguably unnecessary for comfort or survival but that make life more enjoyable. These purchases include everything from seemingly minor daily indulgences, such as sweets and soda, to major purchases, such as ocean cruises, jewelry, and sports cars. Expenditures on these products are not necessarily an indictment of the global consumer class, since reasonable people can disagree on what constitutes excessive consumption. But the sums spent on them are an indication of the surplus wealth that exists in many countries. Indeed, figures on consumer spending at the extreme undercut the perception

that many of the unmet basic needs of the world's poor are too costly to address. Providing adequate food, clean water, and basic education for the world's poorest could all be achieved for less than people spend annually on makeup, ice cream, and pet food. (See Table 1–6.)[19]

The growing frenzy of consumption during the twentieth century led to greater use of raw materials, which complements household expenditures and numbers of consumers as a measure of consumption. Between 1960 and 1995, world use of minerals rose 2.5-fold, metals use increased 2.1-fold, wood products 2.3-fold, and synthetics, such as plastics, 5.6-fold. This growth outpaced the increase in global population and occurred even as the global economy shifted to include more service industries such as telecommunications and finance, which are not as materials-intensive as manufacturing, transportation, and other once-dominant industries. The doubling of metals use, for example, happened even as metals became less critical to generating wealth: in 2000, the global economy used 45 percent fewer metals than three decades earlier to generate a dollar's worth of economic output.[20]

Fuel and materials consumption reflects the same pattern of global inequity found in

Table 1–6. Annual Expenditure on Luxury Items Compared with Funding Needed to Meet Selected Basic Needs

Product	Annual Expenditure	Social or Economic Goal	Additional Annual Investment Needed to Achieve Goal
Makeup	$18 billion	Reproductive health care for all women	$12 billion
Pet food in Europe and United States	$17 billion	Elimination of hunger and malnutrition	$19 billion
Perfumes	$15 billion	Universal literacy	$5 billion
Ocean cruises	$14 billion	Clean drinking water for all	$10 billion
Ice cream in Europe	$11 billion	Immunizing every child	$1.3 billion

SOURCE: See endnote 19.

final goods consumption. The United States alone, with less than 5 percent of the global population, uses about a quarter of the world's fossil fuel resources—burning up nearly 25 percent of the coal, 26 percent of the oil, and 27 percent of the world's natural gas. Add consumption by other wealthy nations, and the dominance of just a few countries in global materials use is clear. In terms of metals use, the United States, Canada, Australia, Japan, and Western Europe—with among them 15 percent of the world's population—use 61 percent of the aluminum produced each year, 60 percent of lead, 59 percent of copper, and 49 percent of steel. Use is high on a per person basis as well, especially relative to use in poorer nations. The average American uses 22 kilograms of aluminum a year, while the average Indian uses 2 kilos and the average African, less than 1 kilo.[21]

Meanwhile, the world's growing appetite for paper makes increasing demands on the world's forests. Virgin wood stocks destined for paper production, for instance, account for approximately 19 percent of the world's total wood harvest and 42 percent of wood harvested for "industrial" uses (everything but fuelwood). By 2050, pulp and paper manufacture could account for over half of the world's industrial wood demand.[22]

Consumption of raw materials such as metals and wood could, in principle, be largely independent of the consumption of goods and services, since many products could be remanufactured or made from recycled materials. But materials in most economies in the twentieth century did not circulate for even a second or third use. Even today, recycling provides only a small share of the materials used in economies worldwide. About half of the lead used today comes from recycled sources, as does a third of the aluminum, steel, and gold. Only 13 percent of copper is

from recycled sources, down from 20 percent in 1980. Meanwhile, recycling of municipal waste generally remains low, even in nations that can afford recycling infrastructure. The 24 nations in the Organisation for Economic Co-operation and Development (OECD) that provide data on this, for example, have an average recycling rate of only 16 percent for municipal waste; half of them recycle less than 10 percent of their waste.[23]

Meanwhile, the share of total paper fiber supply coming from recycled fiber has grown only modestly, from 20 percent in 1921 to 38 percent today. This small increase, in the face of far greater increases in paper consumption, means that the amount of paper not recycled is higher than ever. In light of FAO projections that global paper consumption will increase by nearly 30 percent between 2000 and 2010, the share of paper that is recycled is especially critical, and it will have a large impact on the health of the world's forests in coming years.[24]

Disparate Drivers, Common Result

The global appetite for goods and services is driven by a set of largely independent influences, from technological advances and cheap energy to new business structures, powerful communications media, population growth, and even the social needs of human beings. These disparate drivers—some are natural endowments, others accidents of history, still others human innovations—have interacted to send production and demand to record levels. In the process, they have created an economic system of unprecedented bounty and unparalleled environmental and social impact.

The story starts with the consumer. Mainstream economists since Adam Smith have claimed that consumers are "sovereign" actors who make rational choices in order to maximize their gratification. Instead, consumers make imperfect decisions using a set of judgments that are shaped by incomplete and biased information. Their decisions are primarily driven by advertising, cultural norms, social influences, physiological impulses, and psychological associations, each of which can boost consumption.[25]

Physiological drives play a central role in stimulating consumption. The innate desire for pleasurable stimulation and the alleviation of discomfort are powerful motivators that have evolved over millennia to facilitate survival, as when hunger leads a person to search for food. These impulses are reinforced by consumers' experiences. Products that have satisfied us in the past are remembered as pleasurable, bolstering the desire to consume them again. In consumer societies where food and other goods are abundant, these impulses are leading to unhealthful levels of consumption, in part because they are further stimulated by advertising. Indeed, recent psychological studies have revealed that these impulses can even be primed subconsciously, arousing a desire for increased consumption, as for a thirst-quenching beverage after a feeling of thirst is aroused.[26]

Consumption habits also have social roots. Consumption is in part a social act through which people express their personal and group identity—choosing the newspaper of a particular political party, for instance, or the fashions favored by a peer group. Social motivators can be insatiable drivers of consumption, in contrast to the desire for food, water, or other goods, which is circumscribed by capacity limits. In 1954, the average Briton, for example, could count on an ample material base—enough food, clothing, shelter, and access to transportation to live a dignified life. So the increased spending that accompanied a doubling of wealth by 1994 was likely an attempt to satisfy social and psychological needs.

Beyond the first pair of shoes, for instance, shoe ownership may not be about protecting a person's feet but about comfort, style, or status. Such desires can be boundless and therefore have the potential of driving consumption ever upward.[27]

Cornucopian stocks of goods, the product of huge increases in production efficiency since the Industrial Revolution, further stimulate humans' social and psychological proclivity to consume. Modern industrial workers now produce in a week what took their eighteenth-century counterparts four years. Innovations such as Henry Ford's assembly line slashed production time per automobile chassis from 12.5 hours in 1913 to 1.5 hours in 1914—and have been greatly improved since then. Today, a Toyota plant in Japan rolls out 300 completed Lexuses per day, using only 66 workers and 310 robots. Efficiency increases like these have reduced costs dramatically and fueled sales. This is perhaps most evident in the semiconductor industry, where production efficiencies helped to drive the cost per megabit of computing power from roughly $20,000 in 1970 to about 2¢ in 2001. Such order-of-magnitude increases in computing power at greatly reduced prices have spurred the modern computer revolution.[28]

Globalization has also lowered prices and stimulated consumption. Since 1950, successive rounds of trade negotiations have driven tariffs on many products steadily lower, with real consequences for individual consumers: Australians, for example, pay on average A$2,900 less for a car today because of tariff reductions that took effect after 1998. And the World Trade Organization's 1996 Information Technology Agreement has eliminated tariffs entirely for most computers and other information technologies, often reducing a product's cost by 20–30 percent. The eight rounds of global trade negotiations since 1950 are credited as a major spur to economic expansion worldwide.[29]

A globalizing world has also allowed large corporations to look across national borders for cheaper labor—and to pay workers as little as pennies per hour. (See Chapter 5.) Export processing zones (EPZs), minimally regulated manufacturing areas that produce goods for global commerce, have multiplied in the past three decades in response to the demand for inexpensive labor and a desire to boost exports. From 79 EPZs in 25 countries in 1975, the number expanded to some 3,000 in 116 nations by 2002, with the zones employing some 43 million workers who assemble clothing, sneakers, toys, and other goods for far less than it would cost in industrial nations. The zones boost the availability of inexpensive goods for global consumers, but are often criticized for fostering abuses of labor and human rights.[30]

Meanwhile, technological innovations of all kinds have increased production efficiency, often by raising the capacity of people and machinery to extract resources. Today's "supertrawler" fishing vessels, for example, can process hundreds of tons of fish per day. They are part of the reason that communities of many oceanic fish have suffered declines on the order of 80 percent within 15 years of the start of commercial exploitation. Mining equipment is also more muscular: in the United States, mining companies now engage in "mountaintop removal," which can leave a mountain dozens of meters shorter than its original height. In addition, the capacity of hauling trucks increased some eightfold, from 32 tons to 240 tons, between 1960 and the early 1990s. And output per U.S. miner more than tripled in the same period. Finally, chip mills—facilities that grind whole trees into wood chips for paper and pressed lumber products—can convert more than 100 truckloads of trees into chips every day. These

advances in humanity's ability to exploit vast swaths of resources, and at lower cost, help supply markets with inexpensive goods—a prod to greater consumption.[31]

In consumer societies where food and other goods are abundant, impulses are leading to unhealthful levels of consumption.

Cheap energy and improved transportation have also fueled production, lowering costs and facilitating increased distribution. Despite a spike in oil prices in the 1970s, the inflation-adjusted price of oil was only 7 percent higher in the 1997–2001 period than in 1970–74, for example. And reductions in transportation costs have helped make goods affordable to more people. Air freight rates dropped by nearly 3 percent annually for most international routes between 1980 and 1993, which helps to explain why perishables such as apples from New Zealand or grapes from Chile are now commonly found in European and North American supermarkets. Expanded markets also allow companies to increase the division of labor used in producing and delivering goods and services and to achieve greater economies of scale, each of which further lowers the costs of production.[32]

The unparalleled pace of these technological and transportation developments in the twentieth century led to increasingly rapid adoption of new products. In the United States, it took 38 years for the radio to reach an audience of 50 million people, 13 years for television to reach the same number, and only 4 years for the Internet to do the same. This has kept production lines humming in the information technologies industries, where Moore's Law—the rule of thumb that microprocessor capacity will double every 18 months—has prompted regular introductions of ever-more powerful computers and other digital products. The regular supply of new products, in turn, has prompted rapid turnover of these products in the last two decades—increasing consumption even further.[33]

Forces driving consumption are even found in the economic realities facing modern corporations. Most companies have substantial fixed costs—for heavy machinery, factory buildings, and delivery vehicles needed to produce and sell their products. Today's state-of-the-art semiconductor manufacturing plant, for instance, now costs around $3 billion, a huge investment that must be paid for even when sales are poor. Fixed costs therefore represent financial risk. This danger can be reduced by increasing output and sales so that fixed costs are spread over a greater volume of products and a greater diversity of markets. Thus the ongoing pressure to cover fixed costs creates an urgency to expand production—and to find new customers to buy the steady output of goods.[34]

The need for new customers gives business a strong incentive to develop a host of new tools designed to stimulate consumer demand, many of which play on the physiological, psychological, and social needs of human beings. Advertising has perhaps been the most powerful of these tools. Today advertising pervades nearly all aspects of the media, including commercial broadcasting, print media, and the Internet. Global spending on advertising reached $446 billion in 2002 (in 2001 dollars), an almost ninefold increase over 1950. (See Figure 1–1.) More than half of this is spent in the United States, where ads account for about two thirds of the space in the average newspaper, almost half of the mail that Americans receive, and

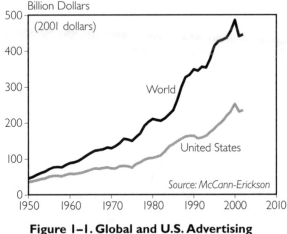

Figure 1–1. Global and U.S. Advertising Expenditures, 1950–2002

about a quarter of network television programming. But advertising is surging in the rest of the world as well. Non-U.S. advertising expenditures have risen three and a half times over 20 years, with emerging markets showing particularly rapid growth. In China, ad spending increased by 22 percent in 2002 alone.[35]

Advertising is increasingly targeted and sophisticated, as seen in efforts to place products in movies and television programs. In recent studies, more than half of the cases of new smoking among teenagers could be traced to their exposure to smoking in movies, for example. And despite a 1990 voluntary "ban" on product placements by the tobacco industry, in the United States actual placements have almost doubled, with 85 percent of the top 250 movies between 1988 and 1997 containing smoking. Indeed, smoking is three times more prevalent in the movies than in the actual U.S. population. With Hollywood earning perhaps half of its revenues from movie sales outside the United States, smoking in movies continues to shape global smoking patterns as well. And non-American studios increasingly serve as vehicles for tobacco advertising. Some three fourths of the films produced between 1991 and 2002 by Bollywood (India's equivalent of Hollywood) included scenes with smoking.[36]

Innovative business practices have also helped boost consumer demand. The introduction of the credit card in the United States in the 1940s helped to increase total consumer credit almost elevenfold between 1945 and 1960. Today, heavy use of credit cards is promoted vigorously, since the profits of companies issuing the cards depend on having consumers maintain large monthly balances. In 2002, 61 percent of American credit card users carried an outstanding monthly balance, which on average was $12,000 at an interest rate of 16 percent. (See Chapter 5.) At this rate, a cardholder would pay about $1,900 a year in finance charges—more than the average per capita income (in purchasing power parity) in at least 35 countries.[37]

Credit is also spurring spending in Asia, Latin America, and Eastern Europe. In East Asia, the household share of total bank lending increased from 27 percent in 1997 to 40 percent in 2000. In several countries, major automobile manufacturers are expanding their product lines because of this explosion in credit lending. General Motors official Philip Murtaugh underlines the importance of credit in China: "Once we establish the type of comprehensive GM financing systems we have in the U.S., we expect to see a huge jump in purchases."[38]

Finally, government policies are sometimes responsible for priming the consumption pump. Economic subsidies, now totaling around $1 trillion globally each year, can ripple throughout an economy, stimulating consumption along the way. The U.S. government, for instance, has subsidized suburban homebuilding since World War II with tax benefits and other enticements. Roomy suburban homes helped spur the consump-

tion of a wide array of consumer durables, including refrigerators, televisions, furniture, washing machines, and automobiles. Cars, in turn, require vast quantitites of raw materials: a third of U.S. iron and steel, a fifth of the aluminum, and two thirds of the lead and rubber. And the spread of suburbs led to greater public spending for new roads, firehouses, police stations, and schools. The Center for Neighborhood Technology in Chicago found in the late 1990s that low-intensity development is about 2.5 times more materials-intensive than high-intensity development. Thus the decision to subsidize suburban homebuilding had a profound effect on U.S. consumption patterns in the last half of the twentieth century.[39]

Problems in Paradise

In *Natural Capitalism*, their 1999 analysis of industrial economies, Paul Hawkins, Amory Lovins, and Hunter Lovins suggested that the United States generates a gargantuan amount of what the authors called "waste"—any expenditure for which no value is received. These outlays pay for a host of unintended byproducts of the American economic system, including air and water pollution, time spent idle in traffic, obesity, and crime, among many others. By the authors' calculations, this waste cost the United States at least $2 trillion in the mid-1990s—some 22 percent of the value of the economy. The estimate can only be a rough one, but the analysis is useful in calling attention in a comprehensive way to the little-noticed underbelly of modern industrial economies. The environmental and social toll of industrial economies is becoming difficult to ignore.[40]

Indeed, the very existence of waste in the more traditional sense—from households, mines, construction sites, and factories—shows that industrial economies are defective in their design. In contrast to the goods and services produced by the millions of other species on our planet, which generate useful byproducts but not worthless waste, human economies are designed with little attention to the residuals of production and consumption. The impact of this design flaw is enormous, starting with the extraction process. For every usable ton of copper, for example, 110 tons of waste rock and ore are discarded. As metals become rarer, the wastes tend to increase: roughly 3 tons of

Nearly all the world's ecosystems are shrinking to make way for humans and their homes, farms, malls, and factories.

toxic mining waste are produced in mining the amount of gold needed in a single wedding ring.[41]

Consumer waste is equally sobering, especially in wealthy countries. The average resident of an OECD country generates 560 kilos of municipal waste per year, and all but three of the 27 reporting countries generated more per person in 2000 than in 1995. Even in nations considered leaders in environmental policy, such as Norway, reducing waste flows is a continuing challenge. In 2002, the average Norwegian generated 354 kilograms of waste, 7 percent more than the previous year. The share of waste recycled also grew, but it has stalled at less than half of total waste generated. Meanwhile, Americans remain the world's waste champions, producing 51 percent more municipal waste per person than the average resident of any other OECD country. The glimmer of good news from the United States is that per person rates appear to have plateaued in the 1990s. Still, the high waste levels per American, coupled with continuing growth of the U.S.

population, adds up to a lot of trash.[42]

Trends in resource use and ecosystem health indicate that natural areas are also under stress from growing consumption pressures. (See Table 1–7.) An international team of ecologists, economists, and conservation biologists published a study in *Science* in 2002 indicating that nearly all the world's ecosystems are shrinking to make way for humans and their homes, farms, malls, and factories. Seagrass and algae beds, the study says, are declining by 0.01–0.02 percent each year, tropical forests by 0.8 percent, marine fisheries by 1.5 percent, freshwater ecosystems (swamps, floodplains, lakes, and rivers) by 2.4 percent, and mangroves by a staggering 2.5 percent. It also cited large but harder to quantify annual losses of coral reefs, rangeland, and cropland. Only temperate and boreal forests showed a resurgence, increasing by 0.1 percent annually after decades of decline. Consistent findings of global environmental decline are found in the Living Planet Index, a tool developed by WWF International to measure the health of forests, oceans, freshwater systems, and other natural systems. The Index shows a 35-percent decline in the planet's ecological health since 1970. (See Figure 1–2.)[43]

One measure of the impact of human consumption on global ecosystems is provided by the "ecological footprint" accounting system, which measures the amount of productive land an economy requires to produce the resources it needs and to assimilate its wastes. Calculations done by the California-based group Redefining Progress show that Earth has 1.9 hectares of biologically productive land per person to supply resources and absorb wastes. Yet the environmental demands of the world's economies are so large that the average person today uses 2.3 hectares worth of productive land. This overall number masks, of course, a tremendous

Table 1–7. Global Natural Resource and Environmental Trends

Environmental Indicator	Trend
Fossil fuels and atmosphere	Global use of coal, oil, and natural gas was 4.7 times higher in 2002 than in 1950. Carbon dioxide levels in 2002 were 18 percent higher than in 1960, and estimated to be 31 percent higher since the onset of the Industrial Revolution in 1750. Scientists have linked the warming trend during the twentieth century to the buildup of carbon dioxide and other heat-trapping gases.
Ecosystem degradation	More than half of Earth's wetlands, from coastal swamps to inland floodplains, have been lost, largely due to draining or filling for human settlements or agriculture. About half of the world's original forest cover is also gone, while another 30 percent of it is degraded or fragmented. In 1999, global use of wood for fuel, lumber, paper, and other wood products was more than double that in 1950.
Sea level	Sea level rose 10–20 centimeters in the twentieth century, an average of 1–2 millimeters per year, as a result of melting continental ice masses and the expansion of oceans due to climate change. Small island nations, though accounting for less than 1 percent of global greenhouse emissions, are at risk of being inundated by rising sea levels.
Soil/land	Some 10–20 percent of the world's cropland suffers from some form of degradation, while over 70 percent of the world's rangelands are degraded. Over the past half-century, land degradation has reduced food production by an estimated 13 percent on cropland and 4 percent from pasture.
Fisheries	In 1999, total fish catch was 4.8 times the amount in 1950. In just the past 50 years, industrial fleets have fished out at least 90 percent of all large ocean predators—tuna, marlin, swordfish, sharks, cod, halibut, skate, and flounder.
Water	Overpumping of groundwater is causing water tables to decline in key agricultural regions in Asia, North Africa, the Middle East, and the United States. The quality of groundwater is also deteriorating as a result of runoff of fertilizers and pesticides, petrochemicals that leak out of storage tanks, chlorinated solvents and heavy metals discarded by industries, and radioactive wastes from nuclear facilities.

SOURCE: See endnote 43.

range of ecological footprints—from the 9.7 hectares claimed by the average American to the 0.47 hectares used by the average Mozambican. Footprint analysis shows that total consumption levels had already exceeded the planet's ecological capacity by the late 1970s or early 1980s. Such overconsumption is possible only by drawing down stocks of resource reserves, as when wellwater is pumped to the point that groundwater levels decrease.[44]

Aggressive pursuit of a mass consump-

tion society also correlates with a decline in health indicators in many countries. "Diseases of consumption" continue to surge. Smoking, for example, a consumer habit fueled by tens of billions of dollars in advertising, contributes to around 5 million deaths worldwide each year. In 1999, tobacco-related medical expenditures and productivity losses cost the United States more than $150 billion—almost 1.5 times the revenue of the five largest multinational tobacco companies that year. Similarly, overweight

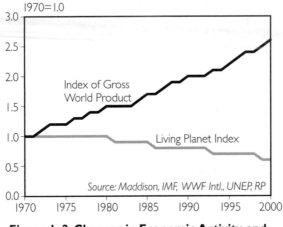

1970=1.0

Figure 1–2. Changes in Economic Activity and Ecosystem Health, 1970–2000

and obesity, generally the result of poor diet and an increasingly sedentary lifestyle, affect more than a billion people, lowering day-to-day life quality, costing societies billions in health care, and contributing to rapid increases in diabetes. In the United States, an estimated 65 percent of adults are overweight or obese, leading to an annual loss of 300,000 lives and to at least $117 billion in health care costs in 1999.[45]

The failure of additional wealth and consumption to help people have satisfying lives may be the most eloquent argument for reevaluating our approach to consumption.

Overall "social health" has declined in the United States in the past 30 years as well, according to Fordham University's Index of Social Health. This documents increases in poverty, teenage suicide, lack of health insurance coverage, and income inequity since 1970. And despite higher levels of consumption than in most other indus-

trial nations, the United States scores worse on numerous indices of development: it ranks last among 17 OECD countries measured in the U.N. Development Programme's Human Poverty Index for industrial countries, for instance, which compiles indicators of poverty, functional illiteracy, longevity, and social inclusion.[46]

An OECD study has also documented disengagement from civic involvement in some industrial nations, especially the United States and Australia. In both countries, rates of membership in formal organizations have fallen, as has the intensity of participation in terms of meeting attendance and willingness to take on leadership roles. Meanwhile, informal social interactions—playing cards with neighbors, going on picnics, and the like—have also declined markedly in both countries, as have levels of trust among people and in institutions. The data on other prosperous countries are more encouraging, although early signs of social disengagement are evident. Organizational membership remains high in many European nations, but the level of involvement and of personal interaction has shown declines in some, and membership is often more transient than in the past. Even in Sweden, which appears to have strong social and community networks, signs of concern are appearing: political engagement is increasingly passive, and levels of trust in institutions are declining.[47]

Harvard Professor of Public Policy Robert Putnam has identified time limitations, residential sprawl, and high rates of television viewing as three features of American society that may explain a decline in civic engagement, together accounting for about half of the situation. All three are linked to high consumption: time pressures are often linked to the need to work long hours to support consumption habits, sprawl is a function of car dependence and the desire for larger homes and properties, and heavy television viewing helps promote consumption through exposure to advertising and programming that often romanticizes the consumer lifestyle.[48]

Perhaps the most damning evidence that continued consumption is generating diminishing benefits is found in studies that compare the ever-rising level of personal wealth in rich countries with the stagnant share of people in these nations who claim to be "very happy." Although self-reported happiness among the poor tends to rise with increased income, studies show that the linkage between happiness and rising income is broken once modest levels of income are reached. The failure of additional wealth and consumption to help people have satisfying lives may be the most eloquent argument for reevaluating our current approach to consumption.[49]

Disappointment in the ability of consumption to deliver lives of fulfillment is producing discontent among scholars, policymakers, and the public. A slew of books published in the 1990s documented dissatisfaction with societies organized around consumption. The titles tell the story: *The Overspent American, The Overworked American, An All-Consuming Century, Confronting Consumption,* and *The High Price of Materialism,* among others. Although the analyses differ, all these authors express the view that consumption-oriented societies are not sustainable, for environmental or social reasons.

Discontent with a commitment to high consumption was evident at the policy and grassroots levels as well. Several European governments are implementing or considering reforms to working hours and family leave benefits, for example. And some people in Europe and the United States are starting to adopt simpler lifestyles. Slowly but steadily, people's interest in recasting consumption in a supporting rather than the leading role is now evident.[50]

A New Role for Consumption?

Despite the problems associated with a consumer society, and notwithstanding the tentative steps taken to shift societies to a less damaging path, most people in industrial countries are still on an upward consumption track, and many in developing countries remain mired in poverty. In order to advance the tentative interest in a new role for consumption, any vision will need to include responses to four key questions:

- Is the global consumer class experiencing a higher quality of life from its growing levels of consumption?
- Can societies pursue consumption in a balanced way, especially in putting consumption in harmony with the natural environment?
- Can societies reshape consumer options to offer genuine choice?
- Can societies make a priority of meeting the basic needs of all?

All things considered, are consumers benefiting from the global consumer culture? Individuals, the important arbiters of this question, might consider the personal costs associated with heavy levels of consumption: the financial debt; the time and stress associated with working to support high consumption; the time required to clean, upgrade, store, or otherwise maintain possessions; and the ways in which consumption replaces time with family and friends.

Individuals as well as policymakers might consider the seeming paradox that quality of life is often improved by operating within clear limits on consumption. Forests, for example, can be available to all indefinitely if they are harvested no faster than the rate of regrowth. Similarly, someone who adopts clear parameters of personal well-being—exercising daily and eating well, for example—is likely to have a higher quality of life than a person who consumes in an open-ended and unrestrained way. Indeed, the underlying premise of mass consumption economics—that unlimited consumption is acceptable, even desirable—is fundamentally at odds with life patterns of the natural world and with the teaching on moderation that is common to philosophers and religious leaders across many cultures and throughout much of human history.

The underlying premise of mass consumption economics—that unlimited consumption is acceptable, even desirable—is fundamentally at odds with life patterns of the natural world.

Second, is our consumption in balance economically, socially, and environmentally? In societies of mass consumption, laws and economic incentives often encourage people to cross key economic, environmental, and social thresholds. Banks and credit agencies urge consumers to take on heavy burdens of debt; businesses and individuals use forests, groundwater, and other renewable resources beyond their rates of renewal; and employers often reward workers for spending long hours on the job. Each of these excesses exacts a price in personal or societal well-being. Numerous imaginative ways for bringing consumption choices in better harmony with social and environmental needs—from legislation mandating levels of recycled content to product "take back" laws that make producers responsible for the products and waste

they create—are available.

Third, are consumers given genuine choices that help them to meet their needs? Clearly, mass consumption societies offer more products and services than any other economic system in human history. Yet consumers do not always find what they need. Consider transportation: safe and convenient access to just five transportation alternatives—walking, cycling, mass transit, car-sharing, or private cars—may offer more real options for getting people where they want to go than a choice of 100 models at a car dealership would. And where genuine choice is present, the most desirable choice may not be affordable, as happens with organic food in some countries. Governments need to reshape economic incentives and regulations to ensure that businesses offer affordable options that meet consumers' needs. They also have a role in curbing consumption excess, primarily by removing incentives to consume—from subsidized energy to promotion of low-density development.

Last, can societies create a consumption ethic that gives priority to meeting the basic needs of all? Physical well-being—including sufficient access to healthy food, clean water and sanitation, education, health care, and physical security—is the foundation of all individual and societal achievement. Neglecting these basics will inevitably limit the capacity of many to realize their personal potential—and their ability to make meaningful contributions to society. In a world in which there are more people living on less than $2 per day than there are in the global consumer class, the continued pursuit of greater wealth by the rich when there is little evidence that it increases happiness raises serious ethical questions.

Beyond the ethical imperative to care for all is a self-serving motive. Lack of attention to the needs of the poorest can result in greater insecurity for the prosperous and in increased spending on defensive measures. The need to spend billions of dollars on wars, border security, and peacekeeping arguably is linked to a disregard for the world's pressing social and environmental problems. The same is true at the community level. Expenditures for private education, gated communities, and home alarm systems are just a few of the ways that failing to invest in the poorest comes back to haunt the wealthy. Meeting the basic needs of all, it seems, is both right and smart.

Addressing these four questions would give consumption a less central place in our lives and would free up time for community building and strengthening interpersonal relationships—factors that psychologists tell us are essential for a satisfying life. By reorienting societal priorities toward improving people's well-being rather than merely accumulating goods, consumption can act not as the engine that drives the economy but as a tool that delivers an improved quality of life.

Plastic Bags

Plastic shopping bags could be the most ubiquitous consumer item on Earth. Their light weight, low cost, and water resistance make them so convenient for carrying groceries, clothing, or any other routine purchase that it is hard to imagine life without them.

The first plastic "baggies" for bread, sandwiches, fruits, and vegetables were introduced in the United States in 1957. Plastic trash bags were appearing in homes and along curbsides around the world by the late 1960s. But these items really took off in the mid-1970s, when a new process for cheaply manufacturing separate plastic bags made it possible for major retailers and supermarkets to offer their customers an alternative to paper sacks. Today, four out of five bags used in grocery stores are the plastic "T-shirt" variety with two handles that look like shirt-sleeves.[1]

These bags start as crude oil, natural gas, or other petrochemical derivatives, which are transformed in plastics factories into chains of hydrogen and carbon molecules known as polymers or polymer resin. (High-density polyethylene resin is the industry standard for plastic bags.) The polyethylene is superheated and the molten resin is extruded as a tube, sort of like the process of making pasta. After the desired shape is achieved, the resin is cooled, hardens, and can be flattened, sealed, gusseted, punched, or printed on.[2]

The typical plastic bag that weighs just a few grams and is a few millimeters thick might seem thoroughly innocuous were it not for the sheer volume of global production. Factories around the world churned out roughly 4–5 trillion plastic bags—from large trash bags to thick shopping bags to thin grocery bags—in 2002, according to estimates from the Chemical Market Associates, a consulting firm for the petrochemical industry. North America and Western Europe account for nearly 80 percent of the use of these products. Americans throw away 100 billion plastic grocery bags each year. These are becoming more and more common in poorer nations as well. And bags produced in Asia now account for one quarter of those used in wealthy nations.[3]

Producing plastic bags uses about 20–40 percent less energy and water than paper sack production does, and generates less air pollution and solid waste, according to lifecycle assessments by both industry and nonindustry groups. Officials from the plastics industry also note that plastic bags take up less space in a landfill, and that neither product decomposes under the prevailing conditions in most landfills. (Given the proper conditions, the paper sack would decompose rapidly, while the plastic bag would not.)[4]

But many mischievous bags do not find their way to landfills. Instead they go airborne after they are discarded. In Kenya, farmers and conservationists complain about the bags getting caught in fences, trees, and even the throats of birds. In Beijing, the government was spending so much money cleaning plastic bags out of gutters, sewers, and ancient temples that it launched a propaganda campaign to encourage people to tie knots in the bags so they wouldn't fly away. The Irish apparently call the ever-present bags their "national flag"; South Africans have dubbed them the "national flower." [5]

Some manufacturers have recently introduced biodegradable or compostable plastic bags, made from starches, polymers or polylactic acid, and no polyethylene. So far, these account for less than 1 percent of the market and are prohibitively expensive, according to the Biodegradable Products Institute, an association that promotes the use of biodegradable polymeric materials. Nonetheless, the organizers of the 2000 Olympic Games in Sydney, Australia, were able to collect 76 percent of the food waste generated at the sports venues and the athletes' village by using biodegradable food utensils and plastic bags that composted as easily as the food and that eliminated the need to separate the garbage. (The following spring, the compost nourished city gardens.) [6]

Elsewhere, governments and individuals are suggesting a more permanent solution that does not depend on new technology. The Ladakh Women's Alliance and other citizens groups led a successful campaign in the early 1990s to ban plastic bags in the Indian province, where the first of May is now celebrated as "Plastic Ban Day." Bangladesh began enforcing its own ban after discovering that discarded bags were clogging drainage and sewage lines, which increased flooding and the incidence of waterborne diseases. [7]

In January 2002, the government of South Africa took action by requiring industry to make bags more durable and more expensive, to discourage their disposal—prompting a 90-percent reduction in use. Ireland instituted a 15¢-per-bag tax in March 2002, which led to a 95-percent reduction in use. Australia, Canada, India, New Zealand, the Philippines, Taiwan, and the United Kingdom also have plans to ban or tax plastic bags. [8]

Supermarkets around the world are voluntarily encouraging shoppers to forgo bags—or to bring their own—by giving a small per-bag refund or charging extra for plastic bags. Weaver Street Market, a community-owned grocery in North Carolina, has gone a step further by selling canvas bags at a discounted price. Sales of these durable alternatives have grown fivefold, said store manager James Watts, and usage of plastic bags has plummeted. "It's good for business but also for the environment," he adds. Yet the idea of bringing reusable bags whenever you go shopping is so simple and obvious that most people may not realize the big impact it could have. [9]

—*Brian Halweil*

CHAPTER 2

Making Better Energy Choices

Janet L. Sawin

Nestled among the rolling green hills of southern West Virginia lie old towns like Clear Creek, Duncan Fork, Superior Bottom, and White Oak. The Appalachian Mountains of this region are home to some of the poorest people in the United States. For generations, residents have relied on coal mining for jobs and as a way of life. But many believe that "Appalachia is under assault" and that the industry that has sustained them for generations is now impoverishing them. Mountaintops are being blasted off to reach coal that lies within. In the process, mountains become wasteland, hardwood forests are lost, streams fill with toxic sludge, wells dry up, and entire communities are being driven away.[1]

Miles away from these barren mountain tops, someone arrives home and flips a switch, wanting only light in the darkness and not thinking about what this involves beyond the walls of the house. For most people, electricity is an invisible force that flows in magically and silently to brighten a room, cool a refrigerator, heat a stovetop, or bring a television to life. Between monthly utility bills, most people give it little thought.

Yet the moment someone flips a light switch or turns on a computer, a chain reaction is set in motion. Current flows into the building from transmission lines that stretch across open land and city streets to bring electricity from distant power plants. Along the way, much of this energy is lost to resistance in the transmission lines and dissipates as heat.

To create electricity, in much of the world enormous piles of coal move by conveyor belt to be pulverized into a fine powder and then fed into a blazing fire in the power plant's furnace. The fire produces steam from water, which turns a generator to produce an electrical current. In the process, the plant emits pollutants that cause acid rain and smog, as well as mercury and carbon dioxide (CO_2), a global warming gas. At most, 35 percent of the coal's energy converts to electricity, meaning that nearly two thirds of it is lost as waste heat, benefiting no one and often harming surrounding ecosystems. And all this coal must be transported to the power plant, by rail or barge, from places like the Appalachian Mountains of southern West Virginia.[2]

Everything we consume or use—our homes, their contents, our cars and the roads we travel, the clothes we wear, and the food we eat—requires energy to produce and package, to distribute to shops or front doors, to operate, and then to get rid of. We rarely consider where this energy comes from or how much of it we use—or how much we truly need.

Whether in the form of gasoline to fuel a car or uranium to generate electricity, the energy required to support our economies and lifestyles provides tremendous convenience and benefits. But it also exacts enormous costs on human health, ecosystems, and even security. Energy consumption affects everything from a nation's foreign debt (due to fuel imports) to the stability of the Middle East. From the air we breathe to the water we drink, our energy use affects the health of current and future generations. Inefficient and unsustainable use of biomass in poor nations leads to deforestation and desertification, while unsustainable use of fossil fuels is altering the global climate. And as we seek out more-remote sources of fuel, we endanger the culture and way of life of indigenous peoples from the Amazon to the Arctic.

The energy intensity—that is, the energy input per dollar of output—of the global economy is declining, and recent decades have seen continuing improvements in energy efficiency. Yet these encouraging developments are being offset by an ever-increasing level of consumption worldwide. Of course, it is not surprising that energy use is rising in developing countries, where most people have never driven a car, turned on an air conditioner, or cooked with anything other than wood or animal dung. As their lives improve, their use of energy increases, and vice versa.

More surprising is the dramatic surge in energy use in many industrial countries. Compared with just 10 years ago, for example, Americans are driving larger and less efficient cars and buying bigger homes and more appliances. As a result, U.S. oil use has increased over the decade by nearly 2.7 million barrels a day—more oil than is used daily in total in India and Pakistan, which together contain more than four times as many people as the United States does. Is this ever-growing demand sustainable? And is a fundamental shift required in the way we produce and use energy?[3]

The type and amount of energy that people use is influenced by a number of factors, including income, climate, available resources, and corporate and government policies. Through taxes and subsidies, regulations and standards, and investments in infrastructure, governments influence how, where, how much, and what form of energy we use. But we as consumers are not powerless bystanders. We are the ones who buy vehicles, appliances, clothing, homes, and other goods and services, based on features such as price, fashion, and values. Ultimately, within the limits of availability and affordability, it is consumers who choose what to buy and how to use it, and thus it is consumers who can drive change.

Global Energy Use Trends

Between 1850 and 1970, the number of people living on Earth more than tripled and the energy they consumed rose 12-fold. By 2002, our numbers had grown another 68 percent and fossil fuel consumption was up another 73 percent. Energy use has driven economic growth and vice versa, but they are not as closely linked as once believed. Before the first global oil crisis, many economists thought that using more energy was a prerequisite for economic growth. But when oil prices suddenly leaped skyward in the early 1970s, governments and consumers reacted by setting efficiency standards and conserving fuel. Between 1970 and 1997, global energy intensity declined 28 percent as economic output continued to rise.[4]

The more efficiently we produce and use energy, the less energy we require for the same services. If the United States used as much energy per dollar of gross domestic product in 2000 as it did in 1970, energy consumption would have totaled 177 quads rather than the 98.5 quads actually used. According to energy analyst Amory Lovins, energy efficiency measures enacted since the mid-1970s saved the United States an estimated $365 billion in 2000 alone.[5]

With only 2 percent of global reserves and 4.5 percent of total population, the United States remains the world's largest oil consumer.

The potential for future savings in the United States and elsewhere remains enormous. We still waste huge amounts of energy. Consider the path from coal mine to light switch, and imagine those energy losses throughout the entire economy, and in every country. In the United States, for example, for every 100 units of energy fed into power plants, buildings, vehicles, and factories, no more than 37 units emerge as useful services such as heat, electricity, and mobility. Globally, the average efficiency of converting primary energy into useful energy is 28 percent. And losses vary greatly from one use or country to the next: for example, Lovins estimates that only 14 percent of oil at the wellhead reaches the wheels of a typical modern car.[6]

Consumption of energy, particularly oil, has increased steadily, with only a brief slow-down during the oil crises of the 1970s. Only Eastern Europe and the former Soviet states have experienced declines in energy consumption. Industrial countries continue to use the largest share of global oil—62 percent. U.S. oil use has doubled since 1960. Although the nation's share of global consumption has declined considerably since 1960, during the 1990s it began to inch up again. With only 2 percent of global reserves and 4.5 percent of total population, the United States remains the world's largest oil consumer.[7]

Today the world's richest people use on average 25 times more energy per person than the poorest ones do. In fact, almost a third of the people in the world have no access to electricity or other modern energy services, while another third have only limited access. About 2.5 billion people, most in Asia, have only wood or other biomass for energy. The average American consumes five times more energy than the average global citizen, 10 times more than the average Chinese, and nearly 20 times more than the average Indian. (See Table 2–1.)[8]

Extreme inequalities exist within the developing world as well, where energy consumption is rising fastest and petroleum use alone has quadrupled since 1970. For example, India has a rapidly growing consumer class that is accumulating cars and home appliances, while 48 percent of Indian families live without permanent housing. The same can be said for countries from Ghana to Viet Nam.[9]

More and more people in the South are using as much energy as people in the North do on average, and studies suggest that their incomes are rising faster than anything experienced by the industrial world. (See Box 2–1.) China is already the world's number one coal consumer and the third largest oil user, while Brazil is the sixth largest oil consumer, India ranks eighth, and Mexico tenth.[10]

Table 2–1. Annual Energy Consumption and Carbon Dioxide Emissions, Selected Countries

Country	Commercial Energy	Oil	Electricity	Carbon Dioxide Emissions
	(tons of oil equivalent per person)	(barrels per day per thousand population)	(kilowatt-hours per person)	(tons per person)
United States	8.1	70.2	12,331	19.7
Japan	4.1	42.0	7,628	9.1
Germany	4.1	32.5	5,963	9.7
Poland	2.4	10.9	2,511	8.1
Brazil	1.1	10.5	1,878	1.8
China[1]	0.9	4.2	827	2.3
India	0.5	2.0	355	1.1
Ethiopia	0.3	0.3	22	0.1

[1]Excluding Hong Kong.
SOURCE: See endnote 8.

BOX 2–1. SURGING ENERGY DEMAND IN CHINA AND INDIA

Although more than a third of the world's people live in China and India, they now account for only 13 percent of global energy consumption. But their energy use is rising rapidly, and these two nations both rely heavily on coal—China for more than 70 percent of its commercial energy and India for over 50 percent. The International Energy Agency projects that rising energy demand in China and India will account for more than two thirds of the expected global increase in coal use between now and 2030. These population giants will thus have enormous impacts on the global energy market and the environment in the decades ahead.

Income levels have risen rapidly in both countries thanks to declining population growth rates and surging economic growth. China's economy has more than quadrupled in size since 1980. During the 1980s, electricity demand in China increased more than 400 percent because of appliance purchases. In India, the number of "affluent" households—those earning $220 per month or more—grew sixfold in just five years, while the number of low-income families declined significantly. Such trends promise to accelerate, feeding a growing consumer class that wants access to the conveniences of home appliances, lighting, gas-powered cooking, and increased mobility.

Demand for oil will grow rapidly too, as more and more people obtain private cars. Domestic oil currently meets about two thirds of China's needs, but consumers will soon rely far more on imports if demand doubles by 2025 as expected—causing China to overtake Japan as the world's second largest oil consumer, behind the United States. Car sales in China grew by 82 percent during the first half of 2003 relative to the same period the previous year. At projected growth rates, China's private vehicle fleet could swell from just over 5 million in 2000 to nearly 24 million by the end of 2005, adding substantially to already congested streets and polluted air. Growth in purchases of sport-utility vehicles (SUVs), which have poor fuel economy, has exceeded even manufacturers' expectations. In India, SUV sales now represent 10 percent of vehicle purchases, and they could soon surpass car sales.

—*Tawni Tidwell*

SOURCE: See endnote 10.

To date, energy use in the South has been limited primarily by income and infrastructure—lack of access to roads and electricity has restrained use of cars and appliances, for example, even among the growing middle and upper classes. In the future, however, people in the developing world will be more constrained by depleting resources and environmental realities. Earth cannot provide enough for today's global population to live like the average American, or even the average European. (See Chapter 1.) For example, if the average Chinese consumer used as much oil as the average American uses, China would require 90 million barrels per day—11 million more than the entire world produced each day in 2001. In the future, population growth, climate change, and other environmental challenges could stress natural systems to their limits, while conventional fuels cannot meet projected energy demand growth. In fact, many analysts predict that even at current consumption rates world oil production will peak before 2020. This has enormous implications for the way people live and get around.[11]

Energy That Moves Us

During the twentieth century, humans became a vastly more mobile species. In industrial countries today, it is not unusual for someone to travel 10,000 or even 50,000 kilometers in a year. And a good deal of what we use as consumers, from our personal computers to the food we eat, crosses continents and oceans to reach us. Just 150 years ago, movement was limited to the distance a person or animal could travel on foot. For roughly a third of humanity, of course, this is still true. For the other two thirds, however, increased mobility of people and property has had profound impacts, altering everything from work to family to the nature and design of our cities.

Today transportation accounts for nearly 30 percent of world energy use and 95 percent of global oil consumption. The United States is by far the world's largest consumer of energy for transport, devouring more than a third of the global total. Starting from a low base, however, transport energy use is currently increasing most rapidly in Asia, the Middle East, and North Africa.[12]

In fact, transport is the world's fastest-growing form of energy use, driven in part by a shift of people and freight to more flexible but more energy-intensive transport modes. Although more passengers travel by rail than air, and more freight by ship than other means, even relatively small shifts in transport choices have significant impacts. Only 0.5 percent of the total distance people travel each year is done by air, yet planes use up about 5 percent of transportation energy. And trucks require four to five times more energy than railroads or ships do for the same weight and distance shipped.[13]

But the most significant driver of rising energy consumption for transportation is growing reliance on the private car. Some 40.6 million passenger vehicles rolled off the world's assembly lines in 2002, five times as many as in 1950. The global passenger car fleet now exceeds 531 million, growing by about 11 million vehicles annually.[14]

About one fourth of those cars are found on U.S. roads, where cars and light trucks account for 40 percent of the nation's oil use and contribute about as much to climate change as all economic activity in Japan does. The total distance traveled by Americans exceeds that of all other industrial nations combined—not only because the country is larger, but also because Americans opt to drive when others bike or walk. As a U.S. transportation consultant recently noted, "The automobile has gotten like TV sets:

There are more of them in the house than eyes to look at them." Today there are more cars than Americans licensed to drive them, and most households own two or more vehicles.[15]

On a per capita basis, car ownership in Western Europe and Japan is comparable to ownership levels in the United States in the early 1970s, while Eastern Europe is more like the United States in the 1930s. But rising incomes, lifestyle changes, women entering the work force, national policies that encourage mobility, and declining fuel costs have all spurred significant growth. Per-person car ownership in Japan quadrupled from 1975 to 1990, and Poland saw a 15-fold increase from the early 1970s to 2001. Only about 20 percent of the world's vehicles are in Asia and the Pacific region, but the numbers there are growing by 10–15 percent a year. (See Table 2–2.)[16]

The size and weight of vehicles are expanding as well—a trend that has wiped out more than 20 years of efficiency improvements gained in the United States through federally mandated fuel efficiency standards. In fact, the fuel economy of U.S. vehicles would be one third higher than it is today if weight and performance had remained constant since 1981.

Ironically, Ford Motor Company's Model T got better gas mileage nearly a century ago than the average vehicle Ford puts on the roads today (albeit with a top speed of 45 miles per hour). Nearly half the vehicles that Americans buy now are gas-guzzling SUVs and light trucks. And the yearning for larger vehicles is contagious. If current trends continue, by 2030 half of the world's passenger vehicles will be SUVs or other light trucks.[17]

People are also driving greater distances. Between 1952 and 1992, while the number of people in the United Kingdom increased 15 percent, the distance they drove tripled. And from 1970 to 2000, the kilometers traveled in European Union (EU) countries more than doubled. In the United States, the number of trips per household rose 46 percent between 1983 and 1995, while average trip length increased more than 5 percent.[18]

Although mobility contributes to economic and social well-being, there are high external costs associated with the extent and nature of our travel. Worldwide, nearly a million people—most of them pedestrians—are killed in traffic accidents each year, and the number of deaths from vehicular air pollution is higher. As vehicle use increases, roads

Table 2–2. Private and Commercial Vehicle Fleets, Selected Countries and Total, 1950–99

Country	1950	1960	1970	1980	1990	1999
			(million vehicles)			
United States	49.2	73.9	108.4	155.8	188.8	213.5
Japan	–	1.3	17.3	37.1	56.5	71.7
Germany	–	5.6	15.5	24.6	32.2	45.8
China	–	–	–	1.7	5.8	12.8
India	–	0.5	1.1	1.9	4.2	8.2
Argentina	–	0.9	2.3	4.3	5.9	6.6
South Africa	0.6	1.2	2.1	3.4	5.1	6.6
Czech and Slovak Reps.	0.2	0.4	1.0	2.6	3.7	5.1
World	70.4	126.9	246.4	411.0	583.0	681.8

SOURCE: See endnote 16.

become congested, wasting productive hours and reducing vehicle efficiency. The costs of road transport not covered by drivers—air pollution, noise, congestion, accidents, and road damage—start at 5 percent of gross domestic product in industrial countries and go higher in some developing-country cities. And money we invest in road-based infrastructure means less invested elsewhere, which worsens existing social inequities for those who cannot use the predominant means of transport. Even in the United States, about a third of the population is too poor, too old, or too young to drive a car.[19]

West Europeans now use public transit for 10 percent of all urban trips, and Canadians for 7 percent, compared with Americans at only 2 percent.

The choices people make about how to move around are greatly influenced by government policies, such as vehicle and fuel taxes, land use rules, and support for air and car transport versus public transit and bicycle use. A century ago, the United States led the world in public transit. In 1910, almost 50 times more Americans commuted to work by rail than by car, and a decade later almost every major U.S. city had a rail system. But following World War II, the government emphasized construction of roads and freeways. Today U.S. commuters receive subsidies for parking while those who use public transit receive considerably less, and bicyclists get nothing. Thus it is not surprising that public transit ridership is lower today than it was 50 years ago, despite a doubling in U.S. population. An exception to this trend is New York City, where due to high density and a proliferation of taxis and public transit options, only 25 percent of residents are

licensed to drive. And in cities such as Denver, Colorado, where services are improving or expanding, the use of mass transit is again on the rise.[20]

In contrast, many countries have devoted significant resources to public transport while discouraging the use of private vehicles through traffic policies and user fees. In Japan and Europe, much of the investment in transportation infrastructure after World War II focused on passenger trains and transit systems. Today nearly 92 percent of downtown Tokyo travelers commute by rail, and the Japanese do only 55 percent of their traveling by car. West Europeans now use public transit for 10 percent of all urban trips, and Canadians for 7 percent, compared with Americans at only 2 percent. This is significant because for every kilometer people drive by private vehicle, they consume two to three times as much fuel as they would by public transit.[21]

Differences in transport trends are also explained by prices. The most rapid growth in private vehicle ownership and use typically occurs in countries with the lowest fuel and car prices. Cars and gas are cheaper in the United States than in Europe, for example, because they are not taxed as heavily. In fact, some of the worst gas guzzlers are subsidized: in 2003 the U.S. Congress passed legislation that effectively tripled a federal business tax credit for SUV purchases, to $75,000 each, compared with a $2,000 deduction for hybrid-electric vehicles.[22]

Despite U.S. policies and low fuel prices, some Americans are opting to pay more in order to consume less. While today's internal combustion engine is only about 20 percent efficient, hybrid cars can go much farther on a liter of fuel. By January 2003, some 150,000 drivers around the world had bought a hybrid car; many of these new owners are in the United States, where monthly sales of the

Toyota Prius were quadruple the figure in Japan. (See Box 2–2.)[23]

Growing awareness of air pollution, safety, and congestion problems associated with cars have motivated strong measures to reduce traffic growth, particularly in developing countries. In 1999, a group of citizens in Santiago, Chile, joined with the environmental group Greenpeace on a year-long campaign to upgrade that city's public transit system. As a result, Santiago now has special corridors for buses, the largest streets are restricted to public transit on high pollution days, and public transit usage has risen considerably.[24]

The municipal government in Bogotá, Colombia, began shifting roadways from cars to bicycles in the late 1980s and plans to ban private car use during peak hours by 2015. Former Mayor Enrique Peñalosa, the driving force behind this movement, believes that cars are "the most powerful instrument of social differentiation and alienation that we have in society" as they divert money away from education and other social services. Today, Bogotá has a good public transit system, pollution levels have declined, and commuting times during rush hours have been cut in half. Numerous other cities, from Zurich in Switzerland to Portland in Oregon, have lowered pollution levels while increasing public transport use by reorganizing urban areas and improving transport efficiencies.[25]

"Congestion charges" on vehicles entering city centers, combined with investments in public transit, have also reduced car use and pollution. London drivers used to spend half their time stuck in traffic, traveling the same speed as Londoners a century ago. But in response to a toll enacted in early 2003, traffic levels dropped by an average of 16 percent in the first few months, and most former car users began commuting by public transit. Central-city congestion charges were estab-

---

BOX 2–2. EFFICIENCY IS NOT ENOUGH

Numerous studies have concluded that over the next 10–15 years, the fuel economy of new U.S. cars and light trucks can increase by as much as one third with existing technologies. In the longer term, the use of light but strong space-age composite material, based on carbon fibers, advanced design, and hybrid or fuel-cell technology, could at least triple fuel economy. However, improvements in efficiency will only begin to resolve the problems associated with our transportation choices. And efficiency advances, in isolation, can actually encourage people to use more energy, increasing their travel and vehicle purchases because energy costs represent a smaller share of total expenses.

SOURCE: See endnote 23.

---

lished years ago in Singapore and in Trondheim, Norway, and more recently in Toronto and Melbourne, with similar results.[26]

Elsewhere people have chosen to share fleets of vehicles rather than owning them, and in some places people are moving away from cars completely. A joint EUROCITIES-European Commission network is now promoting a "new mobility culture" throughout the EU, aiming to improve quality of life and shift reliance from cars to public transit, cycling, and walking. Zermatt, Switzerland, uses its long-time car-free status as a selling point for tourists, and 280 households in Freiburg, Germany, were the first of more than 40 German communities to commit to living without cars. It seems that once people have alternatives that are safe, comfortable, and reliable, more of them choose to live car-free.[27]

Energy Where We Live and Work

Worldwide, people use about a third of all energy in buildings—for heating, cooling, cooking, lighting, and running appliances. Building-related energy demand is rising rapidly, particularly within our homes.[28]

But there are large differences in household energy use from one country to the next. People in the United States and Canada consume 2.4 times as much energy at home as those in Western Europe. The average person in the developing world uses about one ninth as much energy in buildings as the average person in the industrial world, even accounting for noncommercial fuels. Yet a much higher share of total energy in developing countries is used at home because available fuels and technologies are so inefficient. In China, households use about 40 percent of the nation's energy, in India the figure is 50 percent, and in most of Africa it is even higher, compared with 15–25 percent in the industrial world.[29]

Although perhaps one fourth of the world's inhabitants have inadequate shelter or no house at all, for many other people home sizes are expanding even as the number of people per household shrinks. The United States represents the extreme case, where average new homes grew nearly 38 percent between 1975 and 2000, to 210 square meters (2,265 square feet)—twice the size of typical homes in Europe or Japan and 26 times the living space of the average person in Africa.[30]

As homes become bigger, largely due to low energy prices, each individual house has more space to heat, cool, and light, as well as room for bigger appliances and more of them. And as the number of people per household contracts, due to a variety of social trends, the number of houses needed for a given population rises. Each additional home requires construction materials, lighting, heating and cooling, appliances, and often more cars and roads—all of which require energy to produce and to operate. Between 1973 and 1992, shrinking household size in industrial countries alone accounted for a 20-percent increase in energy use per person.[31]

Home appliances are the world's fastest-growing energy consumers after automobiles, accounting for 30 percent of industrial countries' electricity consumption and 12 percent of their greenhouse gas emissions. Saturation in the ownership of large appliances in these nations is continually offset by the diffusion of new ones, including computers and other forms of information technology (IT), while efficiency gains made since the 1970s are being squandered in exchange for more and larger amenities. (See Box 2–3.) The average size of refrigerators in U.S. households, for example, increased by 10 percent between 1972 and 2001, and the number per home rose as well. Air conditioning has taken a similar path: in 1978, 56 percent of American homes had cooling systems, most of which were small window units; 20 years later, three quarters of U.S. homes had air conditioners, and nearly half were large central systems.[32]

Between 2000 and 2020, electricity use for appliances in the industrial world could rise 25 percent. Standby power—the electricity consumed when televisions, computers, fax machines, stereos, and many other appliances are turned "off" but are not unplugged—will likely be the fastest-growing consumer. By 2020, it could represent 10 percent of total electricity use in these countries, requiring almost 400 additional 500-megawatt power plants that will emit more than 600 million tons of carbon dioxide annually.[33]

In developing countries, most building-related energy needs are for cooking, water

BOX 2–3. THE BANE AND BOON OF INFORMATION TECHNOLOGY

The information technology age held the promise of a clearer path to sustainable development and a paperless economy. Yet it appears to have delivered neither. To date the impacts of IT on global energy consumption present a complex picture.

Americans use 2 percent of their electricity just to run computers and Internet equipment, while lifecycle consumption of the entire Internet infrastructure in Germany takes 3–4 percent of national energy. Much of the increase in energy use is attributable to new industries (such as Internet service providers), methods of communication (such as cell phones), and new forms of information management created through the use of IT. Rather than reducing paper consumption, electronic mail has actually increased it by 40 percent, with dramatic impacts on associated energy use—from running printers to supporting one of the most energy-intensive industries in the world, the paper industry. And although ordering products electronically would intuitively seem to require less energy than driving to different stores, this has turned out not to be true. One study of "tele-shopping" showed no net transport savings, while another found it took 55 percent more fuel to deliver groceries purchased online.

By some accounts, however, consumers in industrial nations are reducing their energy use with IT through changes in product inventories and through telecommuting, which cuts the amount of energy used to construct and occupy new buildings. IT has also been credited with driving much of the dramatic economic growth that some countries experienced in the late 1990s—a trend not matched by similar increases in energy consumption due to expansion of less energy-intensive industries like banking and financing.

Other benefits of IT include:
- A new computer chip and equipment design can reduce standby power requirements up to 90 percent.
- Toyota's newest hybrid-electric vehicle has an electronic control and software system that continuously optimizes the operation of key components, ensuring that the car always performs in its most efficient mode.
- Researchers at the Pacific Northwest National Laboratory are developing a "smart grid" that will prompt consumers to vary their energy loads as electricity prices change. This will increase the operating efficiency of existing plants by reducing the need for additional grid capacity and by enabling advanced controls and sensors to improve the efficiency of appliances. It will also provide renewable energy with easier access to the grid and energy markets.

—*Tawni Tidwell*

SOURCE: See endnote 32.

heating, and space heating—life's essentials—and most of this energy is derived from non-commercial, traditional fuels. For example, nearly three fourths of India's population relies on traditional biomass for cooking. Even so, demand for modern appliances is rising across the developing world as well. (See Table 2–3.)[34]

In fact, most of the growth in electricity demand since 1990 has occurred in the developing world, where per capita consumption has risen faster than income and where energy use for buildings tripled between 1971 and 1996. Television ownership increased fivefold in East Asia and the Pacific between 1985 and 1997. But appliance penetration rates are still relatively low in developing countries, so the potential for growth is enormous. In India, sales of frost-free refrigerators are projected to grow nearly 14 percent annu-

Table 2–3. Appliance Ownership in Industrial and Developing Countries, Selected Years

Country	Year	Refrigerator	Clothes Washer	Dishwasher	Air Conditioner
			(number per 100 households)		
United States	1973	100	70	25	47
	1992	118	77	45	69
	1998	115[1]	77	50	72
Japan	1973	104	96	0	16
	1992	117	99	0	131
Western Europe[2]	1973	91	69	5	0
	1992	111	89	24	1
India	1994	7	2	–	–
	1996	9	4	–	–
	1999	12	6	–	–
China rural/urban	1981	0/0	0/6	–	0/0
	1992	2/53	12/83	–	0/0
	1998	9/76	23/91	–	1/20

[1]Other sources show continued increases to 2000. [2]United Kingdom, West Germany, France, and Italy.
SOURCE: See endnote 34.

ally. The International Energy Agency expects that, based on current policies, world electricity demand will double between 2000 and 2030, with the greatest demand increase in the developing world and the most rapid growth in people's homes.[35]

Yet the same needs could be met with far less energy. Efficiency programs have proved highly effective to date, and continued improvements could take us a long way toward meeting this rising demand. The United States established national standards in 1987 after a proliferation of state-level programs. In response, manufacturers achieved major savings in appliance energy use, nearly tripling the efficiency of new refrigerators between 1972 and 1999 while saving consumers money. In Europe, higher energy prices combined with standards and labels, such as the German Blue Angel, have influenced consumer choice and have led manufacturers to produce more-efficient products in order to compete, thereby transforming

entire markets. (See Chapter 5.) Still, much more could be done. Technologies available today could advance appliance efficiency by at least an additional 33 percent over the next decade, and further improvements in dryers, televisions, lighting, and standby power consumption could avoid more than half of projected consumption growth in the industrial world by 2030.[36]

In developing countries, people could save as much as 75 percent of their energy through improvements in building insulation, cooking, heating, lighting, and electrical appliances. Unfortunately, diffusion of more-efficient technologies is extremely slow due to high initial costs, the lack of modern fuels such as piped gas, and failures of existing distribution systems. Experiences in Thailand and Brazil show what is possible, however. In the early 1990s, facing a 14-percent annual increase in demand for electricity, the Thai government initiated a partnership with manufacturers to improve the efficiency of buildings, lighting, and cold appliances (such as refrigerators and air conditioners). By 2000, Thailand had exceeded its energy savings and CO_2 reduction targets by at least 200 percent. Between 1996 and 1998 alone, the market share of efficient refrigerators skyrocketed from 12 to 96 percent. And in Brazil, thanks greatly to labeling and voluntary standards, consumers lowered refrigerator-related energy use by 15 percent from 1985 to 1993.[37]

Improvements in the design and construction of buildings could also yield significant energy savings. According to energy analyst Donald Aitken, "Buildings remain the most underrated aspect of energy economics, and the most unexploited opportunity for improving efficiency." Potential savings in existing buildings are enormous, and consumers have already begun making improvements. In the EU, building construction is responsible for more than 12 percent of economic activity, and over half of this is for retrofitting existing buildings. But new buildings offer the greatest potential for savings, and the numbers are not insignificant—Americans alone erected 1.7 million new private homes in 2002.[38]

Because many buildings stand for at least 50–100 years—and some last for centuries—it is essential to get them right the first time. Even in cold climates, people can reduce the heating needs of new buildings by 90 percent through a combination of design and material improvements, while lighting and other energy needs can be lowered as well. As most people do not build their own homes or offices, efficiency of building design and materials generally rests on government regulations. California's buildings are far more efficient than the U.S. average because the state's building code is updated regularly, based on current technologies. Perhaps most telling is the fact that while per capita elec-

tricity use has doubled over the past 30 years in the rest of the country, it has remained constant in California.[39]

The energy demands of buildings can be dramatically reduced without increasing construction costs by using an integrated approach to the building "envelope" (the walls, roofs, foundations, and so on) and the mechanical and electrical systems. Many new buildings in Europe and the Asia-Pacific region were built incorporating this approach, including the European Parliament building in Strasbourg in France, Potsdamer Platz in Berlin, and Aurora Place in Sydney, Australia.[40]

Around the world, people are erecting "green" buildings that include additional energy-saving features such as daylighting, natural cooling, high performance windows, superior insulation, and photovoltaics (PVs). According to the Rocky Mountain Institute, lighting consumes up to 34 percent of U.S. electricity, including energy requirements for offsetting waste heat. Full use of advanced lighting technologies alone could eliminate the need for 120 1,000-megawatt power plants in the United States, saving money and improving human health and productivity as well.[41]

Encouraged by the U.S. Leadership in Energy and Environmental Design (LEED) program—a voluntary standard for green buildings—developers built the world's first green "high-rise" in New York City. These Battery Park City apartments will use 35 percent less energy and 65 percent less electricity than an average building during peak hours, with PVs meeting at least 5 percent of the demand. And on Earth Day in May 2003, Toyota opened a new California building complex constructed with steel from recycled automobiles, including efficient design and lighting and boasting one of North America's largest commercial PV systems.[42]

From California to Kenya to Germany, consumers are installing minute to megawatt-sized PV systems on the rooftops of houses and businesses. In 2002, more than 40,000 Japanese homeowners added 140 megawatts of PV installations, thanks to supportive government policies. Generating power locally with solar or wind energy is not only cleaner than conventional electricity, it also reduces or eliminates transmission and distribution losses, which range from 4–7 percent in industrial nations to more than 40 percent in parts of the developing world.[43]

Elsewhere people are painting roofs and planting rooftop gardens to reduce their energy consumption by 10–50 percent. Germans developed modern green roof technology, inspired by the sod roofs and walls in Iceland. Replacing dark, heat-absorbing surfaces of rooftops with plants lowers ambient temperatures and reduces energy use for heating and cooling. Examples of green roofs can be seen around the world, from Chicago's City Hall to Amsterdam's Schiphol Airport and Ford Motor Company's Rouge River Plant in Michigan—all 4.2 hectares (10.4 acres) of it.[44]

Energy in Everything We Buy

Everything we use has associated and compounding energy inputs, and the largest share of global energy consumption goes into producing our vehicles, appliances, buildings, and even our clothes and food. In the 1970s, manufacturing such products required 25–70 percent of total energy (varying widely by country). This share has declined steadily in all countries as the transportation and building sectors have grown even more rapidly, but energy use in manufacturing is still rising as we buy and use more and more stuff.[45]

Many manufactured goods cross borders and oceans to reach us, where the energy required to make and move them is omitted from national accounts. As a result, some experts argue that energy intensity is actually increasing in some nations, because they effectively import energy inputs from overseas. For example, by one estimate, the energy embodied in Australia's imports exceeds that of its exports.[46]

The energy invested in a particular thing during its life, from cradle to grave, is called the "embodied energy" of that object. The amount of embodied energy that an item contains depends greatly on the technology used to create it, the degree of automation, the fuel used by and the efficiency of a particular machine or power plant, and the distance the item travels from inception to purchase. The value differs considerably from place to place, and even from house to house.

By some estimates, people can live in a typical house for 10 years before the energy they use in it exceeds what went into its components—steel beams, cement foundation, window glass and frames, tile floors and carpeting, drywall, wood paneling or stairs—and its construction. And the embodied energy in the structure is rarely static. As people remove old materials and install new ones, add another room or a new deck, the embodied energy in the house increases.[47]

As with houses, large amounts of energy are required to assemble our automobiles, to construct and operate the manufacturing plants, and to fabricate the various inputs that make up a car. Most of the energy use associated with making a vehicle is for the manufacture of steel, plastic, glass, rubber, and other material inputs. The larger a vehicle, the more energy required, adding further significance to the trend toward larger cars and SUVs. And once we take a car on the road, its requirements extend to all the energy needed to construct and maintain the highways and bridges we travel, the parking lots, the auto dealers and parts stores, and the many fueling stations needed to keep it running. In total, the energy use associated with a car can be 50–63 percent higher than the direct fuel consumption of the vehicle over its lifetime, and the environmental impacts are also enormous.[48]

But the largest share of energy use associated with vehicles is driving them. To run our vehicles we extract petroleum from the earth, transport it to convenient locations, and refine it into useful fuel. Petroleum refining is one of the world's most energy-demanding industries—and the most energy-intensive in the United States. In 1998, petroleum refining accounted for 8 percent of total U.S. energy consumption.[49]

In many countries, an increasing share of fuel use for household transport is to reach enormous out-of-town "hypermarkets" that are replacing neighborhood food stores. Today many people use almost as much energy to collect some foods as producers do to get them to market. And the farther food travels, the higher its embodied energy, not only because it requires more transport fuel but also because it needs more preservatives and additives, refrigeration, and packaging. In much of the world, food transportation to local stores and then to our homes is among the largest and fastest-growing sources of greenhouse gas emissions.[50]

Producing our food also requires massive amounts of energy. While much comes from the sun, nearly 21 percent of the fossil energy we use goes into the global food system. David Pimentel of Cornell University estimates that the United States devotes about 17 percent of its fossil fuel consumption to the production and consumption of food: 6 percent for crop and livestock production, 6 percent for processing and packaging, and 5 percent for distribution and cooking.[51]

The question of whether embodied energy consumption is rising in some nations is open for debate, and depends greatly on the energy intensity of countries manufacturing the goods consumed. For example, South Korea has the most energy-efficient steel industry in the world, and transporting by ship is relatively energy-efficient. So exporting Korean cars to countries that manufacture steel less efficiently, such as the United States, can actually reduce the embodied energy of vehicles on U.S. roads.[52]

In 1998, petroleum refining accounted for 8 percent of total U.S. energy consumption.

In fact, there are extreme differences in the energy intensity of manufacturing industries from one country to another. In the early 1990s, the Japanese and Germans used less than half as much energy per unit of output in their heavy industries as Canadians and Americans did, due primarily to differences in energy prices. Japan, South Korea, and countries in Western Europe have the most efficient manufacturing sectors, whereas developing countries, the former Soviet bloc, and a few industrial countries—particularly the United States and Australia—have the least efficient. Yet some developing countries have taken the opportunity to leapfrog to modern technologies, rivaling Japan and Europe in manufacturing efficiency.[53]

By supporting items and processes that have lower embodied energy, as well as the companies that produce them, consumers can significantly reduce society's energy use. Unfortunately, labeling programs so far report only direct energy consumption of products, not their full embodied energy, making it difficult to compare one product to another. In spite of this, many consumers have already saved large amounts of energy by recycling and by purchasing recycled materials rather than relying on virgin resources. Producing aluminum from recycled material, for example, requires 95 percent less energy than manufacturing it from raw materials would.[54]

Policy and Choice

We are constantly making choices that affect our energy use. In fact, the amount and type of energy we consume is a result of two kinds of choices: those we make as a society and those we make as individuals and families. Society's decisions to tax or subsidize activities—such as driving and road building, for example—encourage people to adopt certain lifestyles, both extending and limiting their choices. But as individual consumers we still have important choices to make, from how much we drive to whether we insulate our homes. Two individuals with the same incomes, living in the same societies and in similar climates, often use very different amounts of energy as a result of personal choices.

If we are to meet the energy needs of the billions of people who now lack modern energy services, and at the same time bring energy use into balance with the natural world, new energy choices will be required—both at the individual level and at the societal level. Government policies are one way that societies make energy choices, and policies that affect the price of energy are among the most important.

As economies develop, factors such as climate, population density, and rate of urbanization become less important, and energy prices become the fundamental factors determining a nation's energy intensity. In fact, countries with higher energy prices—like Japan and Germany—also have lower energy intensities, while those with lower prices are generally quite energy-intensive, such as the United States for gas and oil, Australia for coal, and Scandinavia for electricity.[55]

When prices are low, energy use for individuals represents a smaller share of the cost of doing business, manufacturing a product, or running a home; consequently, investments in energy savings are low. Over the long term, prices affect what we choose to own, the size of our homes, how much we use our cars and appliances, and even the goods and technologies available to us. However, government policy plays a large role, sometimes propping up or undercutting energy prices. Through subsidies, taxes, standards, and other measures, government policies have a direct impact on energy supplies, demand, and the efficiency of our homes, appliances, cars, and factories.[56]

Auto and fuel taxes in many countries, in conjunction with investments in public transit and bicycle infrastructure, affect ownership trends and distance traveled by car and even the characteristics of the vehicle fleet, and they can encourage the use of public transportation, bikes, and rail. Where governments or companies subsidize public transit, people are more apt to commute by bus or subway than by car. In Denmark, where the tax on auto registrations exceeds a car's retail price, and where rail and bike infrastructure are well developed, more than 30 percent of families do not even own cars—mostly because they do not want them rather than because they cannot afford them. Congestion charges, such as those recently introduced in London, can also encourage commuters to make more-efficient energy choices.[57]

Even choice of home size and location is influenced by taxes, housing policies, and standards. The United States offers a full tax deduction on home mortgage interest, which enables people to buy homes of all sizes but also encourages large homes in sprawling communities. Sweden's tax policy also favors home ownership, but because housing policy has focused for decades on apartments, most people choose apartment living, and cities are more compact as a result. Homes and their contents are more efficient in places like California and Japan, where building codes and appliance standards are becoming increasingly stringent as technologies improve.[58]

And governments can influence the amount of embodied energy in the products people use and the waste left behind. At least 28 countries—from Brazil and Uruguay to China and throughout Europe—now require manufacturers to take back products for reuse and recycling. As a result, companies become more interested and invested in the disassembly and recycling of their goods, increase the quality and lifetime of their products, and thus reduce the amount of energy that goes into making them.[59]

Still, most countries favor auto and air travel over less energy-intensive alternatives, and they are biased toward conventional energy over renewable energy and toward new supplies over efficiency measures. In the mid-1990s, governments worldwide were handing out $250–300 billion annually to subsidize fossil fuels and nuclear power. Since then, several transitioning and developing countries have reduced energy subsidies significantly, but global subsidies for conventional energy remain many magnitudes higher than those for alternatives such as renewables and efficiency. And countries the world over invest enormous amounts of money in large transport infrastructure and energy-intensive manufacturing instead of less intensive, less damaging alternatives.[60]

Because subsidies artificially reduce the price of energy, they can lead to overconsumption. South Korean policies have suppressed electricity prices, undermining

national objectives to improve efficiency. By the late 1990s, per capita household energy consumption there exceeded average levels in Europe. And subsidies are often of greatest benefit to those who do not need them. Up until 2003, for example, the Nigerian government provided annually more than $2 billion in fuel subsidies that benefited the rich at least as much as the poor. The subsidies also encouraged the smuggling of cheap fuel out of the country, requiring Nigeria to import fuel at higher cost.[61]

At least 28 countries now require manufacturers to take back products for reuse and recycling.

Combined with billions of dollars provided each year by the World Bank and export-credit agencies for carbon-intensive fossil fuel projects, national subsidies also forestall possible alternatives such as efficiency and renewable energy technologies, encourage energy-intensive industries to move to developing countries, and amount to lost opportunities for those nations to leapfrog to new technologies.[62]

Failure to internalize the full costs of energy acts as a subsidy as well because consumers do not pay directly for the environmental, social, or security impacts of their energy choices—whether the choice is for the source of the energy or for the amount they decide to use. For decades, government attempts to resolve energy problems and associated challenges have focused almost entirely on reducing intensity of production rather than tackling the motivations and problems associated with our ever-rising consumption. Unfortunately, the energy efficiency improvements made on the production end have been more than offset by rising levels of energy use on the consumer end.

But several countries have begun promoting sustainable consumption through green taxes, shifting the tax burden from labor to energy and other resources, often due to concerns about environmental problems such as climate change. As part of its effort to dramatically reduce national greenhouse gas emissions, for instance, Germany enacted new taxes on conventional energy in 1999, providing financial incentives for energy conservation and renewable energy technologies and a reduction in payroll taxes.[63]

While government policy acts to influence energy consumption in many ways, consumers' own decisions also have a major impact. Around the globe, consumers are making a difference, for better or worse. Whether people buy a new hybrid vehicle or a "Hummer," whether they travel by plane, train, or bicycle, or whether they decide not to go somewhere at all are choices that make a difference. Unfortunately, more and more consumers are choosing larger appliances, bigger homes, and tank-like vehicles for single-person trips on urban roads to malls or hypermarkets. As a counterbalance, though not a large one, other consumers are purchasing efficient hybrid cars, choosing locally grown produce, installing PVs, and buying green power. (See also Chapter 6.)

In much of Germany and Denmark, individuals singly and as part of cooperatives have installed wind turbines to provide local power for their communities. Elsewhere, people are tapping into renewable energy through green power markets. By the end of 2002, more than 980 megawatts of new renewable energy capacity had been added to meet the demand of green power customers in the United States, and another 430 megawatts were planned or under construction. And as the result of a student-run campaign calling for national leadership in

environmental stewardship, the University of California campuses and Los Angeles College District committed to reducing energy consumption, purchasing green power, and installing photovoltaics on campus buildings. These two university systems together could increase U.S. grid-connected PV installations by 30 percent.[64]

Some consumers are going even further. Local authorities and representatives of municipalities from all over Europe have signed the Brussels Declaration for a Sustainable Energy Policy in Cities, committing to work for sustainable energy use in Europe and encouraging the creation of a legal framework to support the effort. In 1992, people in more than 30 Dutch municipalities voted to eliminate cars from their inner cities, while all over the Netherlands parking for bicycles far exceeds spaces for cars at railway stations as a result of customer demand. The Germans and Swiss started car sharing in the 1980s, and the concept has since spread to more than 550 communities in eight European countries with at least 70,000 members. Car sharing is now catching on in North America as well, with programs in more than 40 U.S. cities, from Seattle to Washington, D.C.[65]

Communities of people in more than 40 countries have created "ecovillages," working to achieve sustainable lifestyles through ecological design and construction, renewable and passive energy use, community building spaces, and local, organic agriculture. But it is not necessary to live in an ecovillage to reduce overall energy use and impact on the natural environment. Californians proved this when the energy crisis of 2001 led them, through behavioral and technological changes, to use 7.5 percent less electricity than in the previous summer. And in London a new community—Zero Emissions Development (ZED), whose housing units were sold out before its completion—

was built to minimize pollution and energy use through a combination of green technologies and designs, proximity to public transit, a shared fleet of electric vehicles, and prominently displayed meters that enable residents to track their resource consumption. Architect Bill Dunster, inspired by the fact that most energy is wasted through everyday choices, asserts that "you can get a better quality of life through making these changes, so why not do it?"[66]

It is widely assumed that quality of life and energy consumption are inextricably linked. Energy can improve lives by providing services that meet basic needs and lift people out of sickness, hunger, cold, and poverty. And the desire for a "better quality of life"— still too often defined as a larger home and more vehicles, appliances, and possessions— drives further energy consumption. But does there come a point beyond which more energy use provides only small marginal benefits? How much do we really need to achieve a good quality of life?

To answer these questions, it is useful to look at the relationship between perceived quality of life in various nations and their use of energy. The Human Development Index (HDI) was created by the United Nations to emphasize people rather than economic growth alone as the focus of development. It measures knowledge, longevity, and living standards. Energy analyst Carlos Suárez has mapped out the correlation between HDI and energy consumption. For the poorest people, even small increases in energy use can bring about dramatic improvements in the quality of their lives, both directly and indirectly. For example, electric lighting reduces eye strain and lengthens hours available for education, modern fuels for cooking lower health risks, and powered pumps reduce time spent collecting water. Improvements in energy services can also provide opportuni-

ties for increased income, and thus for further quality-of-life improvements. According to Suárez, the additional benefit per unit of energy drops as energy consumption approaches 1,000 kilograms of oil equivalent (kgoe) per person per year, and at 1,000–3,000 kgoe per person the benefits of additional energy use begin to decline significantly. Beyond this point, even tripling a country's per-person energy consumption does not correlate with an increase in that nation's HDI number. Countries that are nearing 3,000 kgoe per person include Italy, Greece, and South Africa; in contrast, Americans use nearly three times as much energy per person.[67]

Germany enacted new taxes on conventional energy in 1999, providing financial incentives for energy conservation and renewable energy technologies.

In a different attempt to measure quality of life, researcher Robert Prescott-Allen has developed the Wellbeing Index. (See Chapter 8.) This is a numerical ranking of 180 countries based on 87 indicators of human and ecosystem well-being, including health, education, wealth, and individual rights and freedoms, as well as diversity and quality of ecosystems, air and water quality, and resource use. According to the index, Sweden ranks first in well-being in the world while the United Arab Emirates (UAE) is nearly last, and yet the average person in the UAE consumes nearly twice as much energy as the average Swede does. (See Table 2–4.) Austrians, on the other hand, use about 61 percent as much energy as Swedes, yet still rank near the top for well-being. Thus there is no fixed relationship between energy use and

Table 2–4. Energy Use and Well-being, Selected Countries

Country	Well-being Rank[1]	Per Capita Energy Use Rank[2]	Share of Sweden's Per Capita Energy Use
			(percent)
Sweden	1	10	100
Finland	2	6	112
Norway	3	8	104
Austria	5	26	61
Japan	24	19	70
United States	27	4	140
Russian Federation	65	17	71
Kuwait	119	3	162
United Arab Emirates	173	2	190

[1]Out of 180 countries. [2]Based on total primary energy supply.
SOURCE: See endnote 68.

perceived well-being, and there is potential for great advances on the consumption front while improving quality of life.[68]

This is encouraging news because the status quo is not sustainable—socially, economically, or environmentally. There is growing evidence worldwide that current patterns of energy consumption are actually degrading the quality of life for many people—worsening air and water pollution, increasing health problems, raising economic and security costs associated with fuel extraction and use, and weakening the natural systems on which we rely for our very existence, including the global climate. Many developing nations, with huge populations in densely settled areas, are rapidly realizing these limits and starting to address them. For example, severe congestion and pollution problems in Shanghai have forced the city to limit the number of new vehicle registrations each month.[69]

Can Earth sustain our growing energy needs in the twenty-first century, even with a rapid and dramatic shift to more efficient technologies and the heavy use of renewable energy? No one knows for sure, but it certainly will not be easy. Increasing populations and growing levels of per capita consumption—particularly in the developing countries, where 75 percent of the world now lives—have the potential to overwhelm even the most ambitious energy technology efforts.[70]

By 2050, global population is projected to increase more than 40 percent, to 8.9 billion people. If everyone in the developing world were to consume the same amount of energy as the average person in high-income countries does today—a level significantly below per capita consumption in the United States—energy use in the developing world would increase more than eightfold between 2000 and 2050. If everyone on Earth consumed at this rate, total global energy use would increase fivefold over this period.[71]

Although this rate of growth is highly improbable, conventional sources of fossil fuels are unlikely to meet rising demand over the next century. And increasing our use of conventional fuels and technologies will further threaten the natural environment, public health, and international stability, with significant implications for our quality of life. We will be hard-pressed to meet global energy needs even with renewable energy and major improvements in efficiency if current consumption trends continue. Consumption patterns will have to change as well. We will have to find new ways to satisfy the needs of mind and body while reducing consumption of energy for transportation and in our buildings, and while minimizing the energy embodied in all that we buy.

The Secretary-General of the Organisation for Economic Co-operation and Development, which represents the world's wealthiest nations, recently acknowledged that around the world "there is growing consensus that energy use patterns need to be radically altered." Governments can help to shape energy use through measures such as infrastructure investments, regulation, incentives, and energy pricing. Political will and effective, appropriate policies are essential for driving change.[72]

But it is also up to us as individuals—both as consumers and as members of diverse communities—to recognize the links between our consumption choices and the impacts we have on the world around us. We must come to grips with the limits we face and change the way we use energy.

Computers

The post-industrial information economy is often mistakenly characterized as heralding an era of "dematerialization" because tiny semiconductors, the base ingredient for computer chips and electronic devices, yield high value and utility. But semiconductors are in fact more materials-intensive than most "traditional" goods. A single 32-megabyte microchip requires at least 72 grams of chemicals, 700 grams of elemental gases, 32,000 grams of water, and 1,200 grams of fossil fuels. Another 440 grams of fossil fuels are used to operate the chip during its typical life span—four years of operation for three hours a day. The total mass of secondary materials used to produce the 2-gram chip is 630 times that of the final product. For comparison, the resources needed to build a car weigh about twice as much as the final product.[1]

The chips are made in "clean rooms" free of dust and other particles that can spoil the delicate silicone wafers. But workers in these rooms are exposed to a host of chemicals that might be associated with cancers, miscarriages, and birth defects. These facilities also generate huge volumes of chemical waste that have contaminated groundwater at numerous high-tech sites. Santa Clara County in California, the birthplace of the semiconductor industry, contains more toxic waste sites than any other county in the United States.[2]

The number of personal computers in the world rose fivefold from 1988 to 2002—from 105 million to over half a billion. Each of these machines is a toxics trap. A typical monitor with a cathode ray tube (CRT) display contains two to four kilograms of lead, as well as phosphor, barium, and hexavalent chromium. Other toxic ingredients include cadmium in chip resistors and semiconductors, beryllium on motherboards and connectors, and brominated flame retardants in circuit boards and plastic casings. Plastics, including polyvinyl chloride (PVC), make up 6.3 kilograms of an average computer. The combination of various plastics makes recycling a challenge. PVC is especially difficult to recycle, and it contaminates other plastics during the process.[3]

The electronics industry is the world's largest and fastest-growing manufacturing industry, and due to high rates of product obsolescence, electronic waste (e-waste) is growing rapidly. By 2005, one computer will become obsolete for every new computer put on the U.S. market. Often, computers are discarded not because they are broken but because rapidly evolving technology makes them undesirable or incompatible with newer software. Americans replace their Pentium-class desktop PC after just two to three years of use. Large institutions often do regular upgrades; Microsoft's 50,000 employees worldwide receive a new computer every three years, on average.[4]

Government researchers estimate that three quarters of all computers ever sold in the United States are lying in basements and office closets, awaiting disposal. The ones that are junked often end up in landfills or incinerators. About 70 percent of the heavy metals found in U.S. landfills comes from e-waste. These toxins can leach into soil and groundwater and, if people are exposed to them, cause damage to the central nervous system, endocrine disruption, interference with brain development, and organ damage. Incineration is just as harmful. The burning of PVC and brominated flame retardants, for example, releases dioxins and furans—two of the most deadly persistent organic pollutants.[5]

Old computers from the industrial world make their way to foreign shores via the recycling industry, which estimates that 50–80 percent of U.S. e-waste collected for recycling is sent to Asia, mainly to China, India, and Pakistan. According to the U.S. Environmental Protection Agency, it is 10 times cheaper to ship CRT monitors to China than it is to recycle them domestically. This lower cost, along with the weak regulatory system in receiving nations, is driving the toxics trade despite an international ban on it under the main treaty on hazardous wastes, the Basel Convention. (The United States is the only industrial country that has not ratified the Basel Convention; the U.S. export of hazardous materials remains legal and recently became exempt from export regulations.)[6]

An investigation by the Basel Action Network and Greenpeace China in December 2001 found that most computers in Guiyu, an e-waste processing center in China, are from North America and, to a lesser degree, Japan, South Korea, and Europe. The study found that computers in these "recycling" facilities are dismantled using hammers, chisels, screwdrivers, and even bare hands. Workers crack CRT monitors to remove the copper yoke, while the rest of the CRT is dumped on open land or pushed into rivers. Local residents say the water now tastes foul from lead and other contaminants.[7]

Without any protective clothing or respiratory equipment, workers use paint brushes or their bare hands to open empty printer cartridges and brush any residual toner into buckets. According to both Xerox and Canon, carbon black and other toner ingredients cause lung and respiratory irritation. Workers are also exposed to toxic lead-tin solder fumes while heating circuit boards to recover gold-containing chips, and the acid baths used to dissolve and precipitate out the gold emit chlorine and sulfur dioxide gases. Heaps of PVC cables are burned in open areas to recover copper wiring. Ironically, China banned the import of solid wastes in 1996 and added a specific prohibition in 2000 against old computers, monitors, and CRTs, but these laws are poorly enforced.[8]

As industrial countries adopt stricter laws regulating the dumping of e-waste in domestic landfills and incinerators, the flow of computers to developing countries will likely increase unless other measures are introduced to deal with the waste at home. In 2002, the European Union parliament adopted two "extended producer responsibility" directives, requiring electronics manufacturers to phase out the use of hazardous materials and to be responsible for the recovery and recycling of e-waste.[9]

—*Radhika Sarin*

CHAPTER 3

Boosting Water Productivity

Sandra Postel and Amy Vickers

In this morning's cup of coffee and this evening's sip of tea reside molecules of water that have cycled through Earth's atmosphere thousands upon thousands of times. Liquid water has been on Earth for at least 3 billion years, circulating between the sea, air, and land. Powered by the sun, this cycling creates an illusion of plenty: fresh water seems limitless because it falls from the sky year after year.

Over the last two decades, however, that illusion has been shattered by the scale of human influences on Earth's freshwater ecosystems—the rivers, lakes, wetlands, and underground aquifers that store, move, and cleanse water as it cycles. Water tables are falling from the overpumping of groundwater in large portions of China, India, Iran, Mexico, the Middle East, North Africa, Saudi Arabia, and the United States. Many streams and rivers—including major ones such as the Amu Dar'ya, Colorado, Ganges, Indus, Rio Grande, and Yellow—now run dry for portions of the year. Large inland lakes, notably Central Asia's Aral Sea and northern Africa's Lake Chad, have shrunk to shadows of their former dimensions. Worldwide, freshwater wetlands—ecosystems that do a remarkable job of purifying water—have diminished in area by about half. At least 20 percent of Earth's 10,000 freshwater fish species are at risk of extinction or are already extinct.[1]

The scale and pace of human impacts on freshwater systems accelerated over the past half-century along with population and consumption growth. Worldwide, water demands roughly tripled. The number of large dams (those at least 15 meters high) climbed from 5,000 in 1950 to more than 45,000 today—an average construction rate of two large

Sandra Postel is co-author of *Rivers for Life: Managing Water for People and Nature* (Island Press, 2003) and director of the Global Water Policy Project in Amherst, MA. Amy Vickers, author of the award-winning *Handbook of Water Use and Conservation: Homes, Landscapes, Businesses, Industries, Farms* (WaterPlow Press, 2001), is an engineer and water conservation specialist based in Amherst, MA.

dams a day for 50 years. For a time, we registered only the benefits of these engineering projects and paid little attention to the social and ecological costs. We measured the additional hectares irrigated, kilowatt-hours generated, and populations served, but not the fisheries destroyed, aquatic species imperiled, people displaced from their homes, or sustainability of the water use patterns created by large-scale water development.[2]

A sustainable and secure society is one that meets its water needs without destroying the ecosystems upon which it depends or the prospects of generations yet to come. The good news is that it is possible to achieve this goal.

Currently agriculture accounts for about 70 percent of world water use, industry for about 22 percent, and towns and municipalities for 8 percent. Opportunities to increase the efficiency of water use on farms, in factories, and in cities and homes have barely been tapped. Efficiency improvements alone, however, will not suffice. In light of population growth and rising affluence, individuals have an important role to play by making responsible choices about their consumption patterns—from diets to material purchases.[3]

A New Mindset for Managing Water

Unlike copper, oil, and most other commodities, fresh water is not just a resource that acquires value only when it is extracted and put to human use. Most fundamentally, fresh water is a life support. When we pump or divert water to meet human demands, we tap into a living system that myriad other species depend on for their survival and that performs valuable services for the human economy. The work carried out by wetlands alone can be worth on the order of $20,000 per hectare per year.[4]

The fact that our economic balance sheets do not reflect these services means that the true cost of our water use is much higher than we realize. As more and more water is diverted to agriculture, industries, and cities, the amount left to do nature's work gets smaller and smaller. Eventually, ecosystems stop functioning. The tragic health and economic conditions surrounding the Aral Sea, which has lost more than 80 percent of its volume because of excessive river diversions, are a clear warning about the fateful endpoint of this trajectory.[5]

Scientists now know that healthy ecosystems require not just a minimum quantity and quality of water but a pattern of water flow that resembles their natural flow regime. This is because species have spent millennia becoming adapted to nature's flow variability—the natural cycle of highs and lows, floods and droughts—and their lives are keyed to it. They migrate, spawn, nest, and feed when nature cues them to do so. By disrupting natural flow patterns through the construction of dams, reservoirs, and diversion projects, humans have unwittingly destroyed many of the habitat and life-support conditions that our earthly companions—and the ecological services they provide for us—require.[6]

What does this imply for the consumption and management of fresh water? It means that the old goal of continuously striving to meet ever-rising demands is a losing proposition. Achieving an optimal balance between meeting human needs and protecting valuable ecosystem functions requires allocating sufficient water throughout the year to sustain those functions. Once that ecosystem allocation is established, the challenge is to use the remaining water to satisfy human demands efficiently, equitably, and productively.

Making this shift is easier said than done. But here and there, it is beginning to happen. In Australia, water extractions from the Murray-Darling river basin—the nation's largest and most economically important—have been capped in an effort to arrest the severe deterioration of that river's ecological health. These efforts come none too soon: the Murray's flow dropped so low in 2003 that its mouth became clogged with sand. South Africa's innovative 1998 water law calls for meeting the basic water requirements of both people and ecosystems before water is allocated to non-essential uses. This freshwater "reserve" gets top priority and, if implemented as intended, will ensure that water withdrawals remain within the ecological limits defined by scientists and communities. In the United States, in a case involving water allocation on the island of Oahu, the Hawaii Supreme Court ruled in August 2000 that underlying every private diversion of water is "a superior public interest in this natural bounty" and that public trust interests, which include ecosystem protection, are to take priority over private commercial uses in water allocation decisions.[7]

Setting limits on the use of rivers and other freshwater ecosystems is the key to sustainable economic progress because it protects the ecosystems underpinning the economy while spurring improvements in water productivity— the net benefit derived from each unit of water extracted from the natural environment. Just as improvements in labor productivity—the output per worker—help an economy, so do improvements in water productivity—the output per cubic meter of water. (One cubic meter equals 1,000 liters.)

Measured roughly as the value of economic goods and services per cubic meter of water used, water productivity tends to increase with national income for three main reasons. First, because crop production is so water-intensive and crop prices are so low relative to most other goods, a shift toward a more industrial economy increases economic output per cubic meter of water. Second, pollution control laws such as those adopted in Japan, the United States, and many European countries often make it more economical for factories to recycle and reuse their process water than to release it to the environment. Third, as economies shift from manufacturing to service industries, water productivity tends to rise even further. Germany's economy, for instance, now generates $40 of output per cubic meter of water, more than 10 times that of India's. (See Figure 3–1.)[8]

Water productivity in the United States (which has a much larger portion of water devoted to irrigated agriculture than Germany does) registers about $18 per cubic meter. Today, the U.S. economy generates 2.6 times more economic value per cubic meter

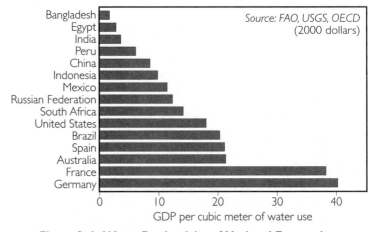

Figure 3–1. Water Productivity of National Economies, Selected Countries, 2000

withdrawn from its rivers, lakes, and aquifers than it did in 1960. (See Figure 3–2.) Still, despite this progress, the United States has all the telltale signs of unsustainable water use—including groundwater depletion, the loss of wetlands, decimated fisheries, and dried-up rivers. Why? Policymakers have yet to limit human water use to ecologically sustainable levels—a boundary that would actually foster much higher levels of water productivity.[9]

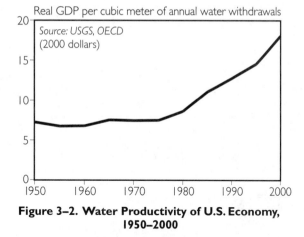

Figure 3–2. Water Productivity of U.S. Economy, 1950–2000

Water-Rich, Water-Poor

Earth's hydrologic cycle distributes water very unevenly across the planet. Just six countries—Brazil, Russia, Canada, Indonesia, China, and Colombia—account for half of Earth's total renewable freshwater supply of 40,700 cubic kilometers (counting only river and groundwater runoff, not evaporation and transpiration by plants). Whether a region is hydrologically rich or poor depends in part on how much of the global endowment it receives relative to its population size. Canada, for instance, ranks near the top of water wealth, with more than 92,000 cubic meters of water per inhabitant. At the water-poor end of the spectrum are Jordan with annual renewable supplies of 138 cubic meters per person, Israel with 124, and Kuwait with essentially none.[10]

National figures mask much of the world's water stress, however, because water is often distributed unevenly within countries as well. China, for instance, has 21 percent of the world's people but only 7 percent of Earth's renewable fresh water—and most of that supply is in the southern portion of the country. The North China Plain, which includes the Yellow River, is one of the most populous regions of water scarcity in the world. Home to some 450 million people, it has a per capita

renewable water supply of less than 500 cubic meters per year, roughly on a par with Algeria. Water use on the North China Plain already exceeds the sustainable supply. In most years, the lower Yellow River now runs out before reaching the sea. And across much of the plain, which produces one quarter of China's grain, underground water tables are falling by 1–1.5 meters annually. As water economist Jeremy Berkoff observes, water scarcity in the North China Plain will "tend to fall on those least able to afford it—on smaller farmers growing grains in more isolated locations."[11]

Water-poor places usually make heavier demands on rivers and aquifers than water-rich ones do (see Table 3–1) because in drier climates crop production—a very water-intensive activity—requires irrigation. Egypt's water use per capita is nearly twice that of Russia's not because Egyptians are water gluttons (although they use more than their fair share of the Nile), but because all of their cropland needs irrigation, whereas only 4 percent of Russia's does. The United States, however, does emerge as a water glutton: it has one of the highest per capita usage rates in the world even though only 11 percent of its cropland is irrigated.[12]

Table 3–1. Estimated Annual Water Withdrawals Per Capita, Selected Countries, 2000

Country	Water Withdrawals Per Capita
	(cubic meters per person per year)
Ethiopia	42
Nigeria	70
Brazil	348
South Africa	354
Indonesia	390
China	491
Russian Federation	527
Germany	574
Bangladesh	578
India	640
France	675
Peru	784
Mexico	791
Spain	893
Egypt	1,011
Australia	1,250
United States	1,932

SOURCE: See endnote 12.

So the picture is still incomplete without considering affluence and poverty. All it takes is a flight into Phoenix, Arizona, in the southwestern United States to see an oasis city that defies its natural water endowment. Although it gets just 19 centimeters of rainfall a year, Phoenix boasts a landscape lush with green lawns, golf courses, and backyard swimming pools. But this luxury comes at a high price—the depletion of aquifers and water imports from the distant Colorado River at U.S. taxpayer expense. An overflight of the East African nation of Ethiopia, on the other hand, where in 2003 more than 12 million people faced famine, reveals a land thirsting for water even though 84 percent of the Nile River's flow originates within its territory. Because of the influence of power, politics, and money, natural scarcity of water

does not imply deprivation; nor does natural abundance imply access.[13]

Easing both overconsumption and underconsumption are flip sides of the global water challenge. The most urgent task is to provide all people with at least the minimum amount of clean water and sanitation needed for good health. Today, one out of five people in the developing world—1.1 billion in all—daily face risks of disease and death because they lack "reasonable access" to safe drinking water, defined by the United Nations as the availability of at least 20 liters per person per day from a source within 1 kilometer of the user's home. The large gap in coverage has almost nothing to do with water scarcity: Indonesia, for instance, has a natural water endowment exceeding 13,000 cubic meters per person, yet one quarter of its people do not have safe drinking water. Globally, providing universal access to 50 liters per person per day by 2015 would require less than 1 percent of current global water withdrawals. There is more than enough water, but so far the political will and financial commitments to provide the poor with access to it have not been sufficient.[14]

In 2000, the United Nations General Assembly adopted as one of the Millennium Development Goals the target of halving, by 2015, the proportion of people lacking affordable access to safe water. Two years later, at the World Summit on Sustainable Development in Johannesburg, nations similarly committed to halving by 2015 the proportion of people lacking access to adequate sanitation. The spread of sanitation services has lagged well behind the provision of household water, leaving 2.4 billion people worldwide without basic sanitation. (See Table 3–2.) To meet the new commitments, water services will need to reach an additional 100 million people and adequate sanitation an additional 125 million people each year

Table 3–2. Populations Lacking Access to Safe Drinking Water and Sanitation, 2000

Region	Share of Population Without Access to	
	Safe Drinking Water	Adequate Sanitation
	(percent)	
Africa	36	40
Asia	19	53
Latin America & Caribbean	13	22

SOURCE: See endnote 15.

between 2000 and 2015.[15]

While ambitious, these achievable targets are essential milestones on the path to universal water and sanitation coverage. According to U.N. statistics, five countries—Bangladesh, Comoros, Guatemala, Iran, and Sri Lanka—succeeded in halving the proportion of their populations lacking safe drinking water between 1990 and 2000. (These statistics do not account, however, for the discovery of poisonous levels of arsenic in groundwater wells across large areas of Bangladesh.)[16]

South Africa has progressed in the provision of water services as well. When the African National Congress assumed power in 1994, some 14 million South Africans had no access to safe drinking water. The post-apartheid constitution ratified in 1996 made clean water a universal right, and the 1998 water law that established a two-part water reserve—to meet the basic water requirements of all people and ecosystems—made the extension of water supply services a high priority. Between 1994 and April 2003, the nation's Community Water Supply and Sanitation Programme provided access to 8 million people at an average cost of $80 per

person. Officials estimate that the remaining 6 million people will have access by 2008.[17]

To service the poorest residents in South Africa yet achieve reasonable cost-recovery, a low life-line price was set for the first 25 liters per day, with rates much higher above that level. Because even the minimum rates burdened poor families, officials have reportedly begun to make the life-line quantity free. In the handful of regions where the government has contracted with private corporations to manage water systems, however, cost-recovery appears to take priority over the constitutionally mandated right to water, causing residents to protest. In Johannesburg, for instance, where the water utility signed a management contract with the French corporation Suez, pre-pay water meters have been installed that dispense only as much water as families have paid for in advance. Private water corporations, concerned primarily with increasing shareholder profits, have little incentive to meet the basic needs of the poor unless required by public authorities to do so.[18]

Water, Crops, and Diets

Agriculture uses about 70 percent of all the water extracted from Earth's rivers, lakes, and underground aquifers, and as much as 90 percent in many developing countries. Recent projections indicate that by 2025 numerous river basins and countries will face a situation in which 30 percent or more of their irrigation demands cannot reliably be met because of water shortages. These include most river basins in India, the Hai and Yellow basins in China, the Indus in Pakistan, and many river basins in Central Asia, sub-Saharan Africa, North Africa, Bangladesh, and Mexico.[19]

Raising the productivity of agricultural water use is critical to meeting people's food needs as water stress deepens and spreads.

This challenge has three major parts: delivering and applying water to crops more efficiently, increasing yields per liter of water consumed both by irrigated and rain-fed crops, and shifting diets so as to satisfy nutritional needs with less water.

A large portion of water that is stored behind dams and diverted through canals for irrigation never benefits a crop. A 2000 review found that surface water irrigation efficiency ranges between 25 and 40 percent in India, Mexico, Pakistan, the Philippines, and Thailand; between 40 and 45 percent in Malaysia and Morocco; and between 50 and 60 percent in Israel, Japan, and Taiwan. The large share of water not reaching the roots of crops is not necessarily lost or wasted: it may, for instance, seep through a field or canal and recharge groundwater, becoming the supply for another farmer. Yet some of it is lost to evaporation from soil or canal surfaces. Either way, these inefficiencies carry high costs: water is not available when and where needed, aquatic habitats are destroyed unnecessarily, more land becomes salinized, and more fresh water is polluted by salts and pesticides.[20]

Most regions have made only modest gains in improving irrigation efficiency. With irrigation water often priced at less than one fifth of its true cost and with groundwater pumping largely unregulated, farmers and irrigation managers have little incentive to upgrade their practices. Improvements in the timing and reliability of water deliveries are a prerequisite to many of the efficiency measures farmers themselves can take. Growers in some districts of California, for instance, would like to shift to more-efficient irrigation systems but need greater certainty about the frequency, flow rate, and duration of their water deliveries to do so.[21]

There is a rich menu of options for improving the productivity of irrigation water, including a suite of technical, managerial, institutional, and agronomic measures. A growing number of farmers around the world are finding, for instance, that drip irrigation systems—which deliver water directly to the roots of plants at low volumes through perforated tubing installed on or below the soil surface—can save water and improve harvests at the same time. Compared with conventional flood or furrow irrigation, drip methods often reduce the volume of water applied to fields by 30–70 percent and increase crop yields by 20–90 percent. The combination can mean a doubling or tripling of water productivity.[22]

Worldwide, micro-irrigation methods (including drip and micro-sprinklers) are used on some 3.2 million hectares, only slightly more than 1 percent of irrigated land. A handful of water-short countries now rely on it heavily, however. (See Table 3–3.) Moreover, the area under drip and other micro-irrigation techniques has expanded markedly in a number of countries over the last decade, including more than a doubling in Mexico and South Africa, a 3.5-fold increase in Spain, and a nearly ninefold increase in Brazil. Although starting from a small base, China and India have also been expanding the use of drip irrigation to cope with growing water shortages.[23]

Changes in cropping patterns and growing methods also offer opportunities to get more crop per drop. This challenge is particularly salient for the production of rice, the preferred staple of about half the human population. More than 90 percent of the world's rice is grown in Asia, where many rivers and aquifers already are overtapped and the pressure to shift water from farms to cities is escalating. Over the last quarter-century, the widespread adoption of high-yielding and early-maturing rice varieties led to a 2.5- to 3.5-fold increase in the amount of rice harvested per unit of water consumed—an

Table 3–3. Use of Drip and Micro-irrigation, Selected Countries, 1991 and Circa 2000[1]

Country	Area Irrigated by Drip and Other Micro-irrigation Methods 1991	Circa 2000	Share of Total Irrigated Area Under Drip and Micro-irrigation, Circa 2000
	(thousand hectares)		(percent)
Cyprus	25.0	35.6	90
Israel	104.3	125.0	66
Jordan	12.0	38.3	55
South Africa	102.3	220.0	17
Spain	160.0	562.8	17
Brazil	20.2	176.1	6
United States	606.0	850.3	4
Chile	8.8	62.1	3
Egypt	68.5	104.0	3
Mexico	60.6	143.1	2
China	19.0	267.0	<1
India	17.0	260.0	<1

[1]Micro-irrigation typically includes drip (both surface and subsurface) methods and micro-sprinklers; year of reporting varies by country.
SOURCE: See endnote 23.

impressive achievement. Further gains will be harder to win. Many studies have shown, however, that the traditional practice of flooding rice fields throughout the growing season is not essential for high yields. Applying a thinner water layer or even letting rice fields dry out between irrigations can in some cases reduce water applications by 40–70 percent without significantly lowering yields.[24]

Similarly, researchers have found that grain yields can often be sustained with 25 percent less irrigation water than normally applied as long as the crops receive sufficient water during their critical growth stages. Called deficit irrigation, this practice is becoming a necessity in some water-short areas. On the North China Plain, for example, farmers now irrigate wheat three times a season rather than five.[25]

For many poor farmers, the question is not how to irrigate more efficiently, but how to irrigate at all. Most of the roughly 800 million people who are hungry or malnourished belong to farm families in sub-Saharan Africa and South Asia. For them, conventional irrigation equipment is too expensive, yet access to irrigation water is their key to more stable and productive harvests, greater food security, and higher incomes. Increasing poor-farmer access to irrigation through the spread of affordable technologies for small plots would vastly improve water productivity—yielding great health and social benefits per liter of water consumed.[26]

One model of success is found in Bangladesh, where poor farmers have bought more than 1.2 million human-powered devices called treadle pumps that give them access to shallow groundwater and let them grow crops for market during the dry season, boosting incomes an average of $100 per $35 pump in the first year. Colorado-based International Development Enterprises is now parlaying its experience in Bangladesh and numerous other countries into a global multidonor effort called the Smallholder Irrigation Market Initiative, which aims to provide poor farmers with access to affordable irrigation—including low-cost drip systems and treadle pumps—toward a goal of lifting 30 million rural farm families out of poverty by 2015.[27]

Across large parts of India, community groups are reviving the use of traditional tanks (ponds), check dams, and other structures to collect and store rainwater to irrigate their crops during the dry season and to recharge groundwater. In the Alwar district of Rajasthan, 2,500 ponds (called *johads*) have been built in 500 villages, increasing crop and milk production markedly. By replenishing groundwater, the *johads* also have lifted the water table from an average of 60 meters below the surface to 6 meters.[28]

These examples only scratch the surface of the many ways farmers and water managers can improve the efficiency of irrigation, make better use of natural rainfall, and increase crop yields per liter of water consumed. Through their dietary choices, individual consumers also have an important role to play—one that will prove critical to doubling agricultural water productivity.

The various foods we eat require vastly different amounts of water to produce. They also vary in the amount of nutrition they offer—including energy, protein, calcium, fat, vitamins, and iron. Combining these two features provides a measure of nutritional water productivity—how much nutritional value is derived from each unit of water consumed. Using crop water requirements and yields for California, researchers Daniel Renault and Wes Wallender estimated nutritional water productivity for principal crops and food products. The results were revealing: it takes five times more water to supply 10 grams of protein from beef than from rice, and nearly 20 times more water to supply 500 Calories from beef than from rice. (See Table 3–4.)[29]

With its high meat content, the average U.S. diet requires 5.4 cubic meters of water per person per day—twice as much as an equally (or more) nutritious vegetarian diet. Even a partial shift away from animal products would make a large difference. For example, cutting the intake of animal products in half and replacing them with highly nutritious vegetable products would reduce the water intensity of the U.S. diet by 37 percent. Mak-

Table 3–4. Water Consumed to Supply Protein and Calories, Selected Foods[1]

Food	Water Consumed to Supply 10 Grams of Protein	Water Consumed to Supply 500 Calories
	(liters)	
Potatoes	67	89
Groundnut	90	210
Onions	118	221
Maize (corn)	130	130
Pulses (beans)	132	421
Wheat	135	219
Rice	204	251
Eggs	244	963
Milk	250	758
Poultry	303	1,515
Pork	476	1,225
Beef	1,000	4,902

[1]Based on California crop yields and water productivity; takes into account only the crops' water requirements, not irrigation efficiencies or other factors.
SOURCE: See endnote 29.

ing this transition by 2025, when the U.S. population is projected to total more than 350 million people, would lower the nation's dietary water requirements at that time by 256 billion cubic meters per year—a savings equal to the annual flow of 14 Colorado Rivers. Many other benefits would result as well—including reduced heart disease, less cruelty to animals, and less pollution of streams and bays from industrial animal feedlots.[30]

Worldwide, ensuring a healthy diet for all people in the face of growing water scarcity will require adjustments at both the high and low ends of the diet spectrum. The nearly 1 billion people who are malnourished need to eat more in order to live healthy lives. Expanding access to minimum levels of irrigation water can help achieve this goal. More equitably sharing the water embodied in food, through trade and aid, will also be important.

And the sensible dietary shift just described for the U.S. population would free up enough water to provide healthy diets for nearly 400 million people, nearly one quarter of the number expected to be added to the developing world's population by 2025.[31]

Cities and Homes

The water demands—and shortages—of many cities throughout the world are expanding rapidly. With nearly half of the global population now living in urban areas, a figure that will increase to 60 percent by 2030, meeting the growing water desires of the rich and the water needs of the poor is now a significant challenge. (See Box 3–1.) While cities claim less than 10 percent of the world's freshwater withdrawals, their concentrated consumption requires complex, capital-intensive infrastructure that draws deeply from finite surface and subsurface water supplies.[32]

Excessive water demands have come at a cost. The majority of the world's 16 megacities—those with 10 million or more inhabitants—lie within regions experiencing mild to severe water stress, a condition where withdrawals are outstripping available supplies. As urban water demands increase, the pressure on agricultural and rural areas to sell or surrender their water rights will intensify.[33]

The headline story on urban water use and management can be summed up in one word: waste. "We need…to reduce leakage, especially in the many cities where water losses are an astonishing 40 percent or more of total water supply," declares U.N. Secretary-General Kofi Annan. Leakage and other losses are often overlooked and sometimes hidden sources of waste: most water system managers are unable to account for 15–40 percent of their supplies. In developing regions such as Africa, it is common for 50–70 percent of total water extracted to be dissi-

> **BOX 3–1. DESALINATION— SOLUTION OR SYMPTOM?**
>
> A growing number of cities are looking to de-salted seawater or brackish water to save them from future water shortages. There are currently about 9,500 desalination plants worldwide, with an estimated capacity of 11.8 billion cubic meters per year—0.3 percent of current global water use. An energy-intensive process, desalination is concentrated in the oil-rich Persian Gulf and Middle East, which account for about half of the global capacity. Both energy requirements and costs have been falling with improved technologies, however, and worldwide desalination capacity is expanding at an annual rate of some 11 percent. Israel's plan to generate as much as half of its urban water supplies from desalination by 2008 could conceivably free up other water sources for equitable sharing with Palestinians.
>
> But for most of the world, is desalination a sensible choice or another expensive supply-side solution? On a unit basis, most conservation and efficiency measures can meet new water needs for 10–25 percent of the cost of producing desalinated water. It makes little sense to de-salt the sea, and to add more greenhouse gases to the atmosphere in the process, when reducing waste and increasing efficiency can supply water more cost-effectively and with less ecological harm.
>
> SOURCE: See endnote 32.

pated through leakage, illegal hook-ups, and poor accounting. As much as a third of the water supplies in a typical Arabian Gulf coast city may be lost to leaky pipes and mains. Taiwan loses nearly 2 million cubic meters of water daily to leakage, about the same volume as 325 million toilet flushes. These losses are estimated to cost $200 million per year.[34]

"Water accountability" is a prime indicator of a utility's efficiency and management, yet water utilities commonly fail at this most basic maintenance task. (See Table 3–5.) Often the poorest countries, whose people lack adequate supplies, have the highest rates of water waste, although the privatized water industry's record in industrial countries is hardly stellar. (See Box 3–2.) Water system leakage and other losses—commonly referred to as "unaccounted-for water" (UFW) or "nonrevenue water"—is the volume that is withdrawn but that never reaches, or is never recorded as having served, an end-user. It is usually calculated as the difference between the water "produced" (as measured by a meter at the point of extraction or treatment plant) and the water sold (based on customer meter readings), although the water industry has long lacked consistent standards for defining, measuring, and reporting UFW. Most UFW is due to leakage from neglected mains and pipes, but theft and meter inaccuracies also play a role, particularly in poor and aging systems. Thus a good deal of UFW represents water that could beneficially serve other users, and another portion of it results in lost revenue because the water is used but not paid for. The economic value of lost water due to meter reading failures or theft is often up to 10 times the marginal operating cost associated with leakage.[35]

U.S. cities, considered to have some of the most modern water technologies and infrastructure, have UFWs ranging from 10 to 30 percent, and sometimes higher. In the absence of national codes to define and measure water losses, some states set their own standards. These range from 7.5 to 20 percent but are not well enforced. Only a few states report water loss figures to the public. For example, Kansas, whose western section overlies the declining Ogallala aquifer, has a

Table 3–5. Water System Leakage and Losses, Selected Countries

Country	Service Area	Estimated Average Losses of Total Water Supplied
		(percent)
Albania	nationwide	up to 75
Canada	Kingston, Ontario	38
Czech Republic	nationwide	20–30
Denmark	Copenhagen	3
France	Paris	30
	nationwide	up to 50
Japan	Fukuoka	5
Jordan	nationwide	48
Kenya	Nairobi	40
Singapore	nationwide	5
South Africa	Johannesburg	42
	Tshwane (Pretoria)	24
Spain	nationwide	24–34
Taiwan	nationwide	25
	Taipei	42
United States	nationwide	10–30
	Bethlehem, PA	27

SOURCE: See endnote 35.

15-percent UFW standard, yet the most recently reported figures on statewide water losses list 52 water suppliers with UFWs at 30 percent or higher. To its credit, the Kansas Water Plan has set UFW reduction as one of its primary objectives.[36]

Recovery of water "lost" due to leakage, faulty measurement, or corrupt accounting constitutes a great untapped water supply that could help cities and regions facing water scarcity meet their true water needs. Arguments that water lost to leakage is not significant because it recharges aquifers or supplies users elsewhere ignore the fact that

BOX 3–2. PRIVATIZATION AND LEAKAGE: ACCOUNTABILITY LACKING

Despite the promises of greater efficiencies and "smart management" systems that are supposed to come with privatized water systems, a number of investor-owned water companies fail to account for the massive volumes of water their systems lose to leaks and other unmeasured or unexplained uses.

The much-touted water loss reduction goals for the privatized British water system have yet to be realized, and the reality is that some "companies have still not achieved their economic level of leakage," according to a report by the House of Commons. Measuring leakage accurately is all the more difficult in the United Kingdom because only 20 percent of households are metered, which makes company leakage estimates "subject to manipulation," according to the report. Following the privatization of water systems in 1989, leakage levels across the U.K. water industry rose to an average of 30 percent by 1995. The Office of Water Services, which regulates the water and sewerage industry in England and Wales, intervened and set mandatory leakage reduction targets. Several companies with high loss levels, notably Thames Water Utilities LTD, serve areas facing supply shortfalls. In 2003, water leakage and losses by Thames Water accounted for over 25 percent of all water leakage in England and Wales, yet the company provides water services to only 15 percent of billed customers there.

SOURCE: See endnote 35.

water extractions come at a cost. Water that is displaced from its original "service area" in nature and then squandered in the "service" of leaky pipes causes rivers to run dry, habitats to wither, and wildlife to disappear. Like dental cavities, decaying pipes can be ignored for a time but ultimately cannot be avoided; the longer the problem is neglected, the more costly its repair. Unless existing infrastructure is water-tight, proposed capital projects designed to meet water "needs" are specious.

Copenhagen, Denmark, with only 3 percent UFW (roughly 1.6 cubic meters per person per year, or about a gallon a day), is a bright exception to the water industry's historically poor recordkeeping. The Copenhagen water department has also had a steady decline in daily per capita household use among its half-million residents since it set conservation targets and initiated a series of educational campaigns and water rate increases. Perhaps the strongest incentive for maintaining tight water systems in Denmark is that by law, utilities are taxed (0.7 euros, about 85¢, per cubic meter) if their leakage rate exceeds 10 percent. In 2000, only 8 out of 40 of the largest water suppliers in Denmark reported a loss over 10 percent. (See Box 3–3 for descriptions of other urban efficiency programs.)[37]

Reducing leakage and using water more efficiently also saves energy, since pumping, treating, and distributing water requires energy at each stage. California's water systems are one of the state's largest energy consumers because they move water long distances and over high elevations. On average, pumping one acre-foot (1,234 cubic meters) of water through the Colorado River Aqueduct to southern California takes approximately 2,000 kilowatt-hours (kWh) of electricity, while sending an acre-foot there through the State Water Project requires about 3,000 kWh. In a typical southern California home, the energy required to provide potable water can rank third behind that required to run the air conditioner and refrigerator. Since using water more efficiently

BOX 3–3. URBAN WATER CONSERVATION PROGRAMS THAT SAVE WATER AND MONEY

A number of cities and water systems have launched water efficiency programs in recent years, and several have achieved impressive water and cost savings:

• Singapore reduced unaccounted-for water from 10.6 to 6.2 percent from 1989 to 1995 and saved over $26 million in avoided capital facility expansions through aggressive leak detection and repair, pipe renewal, and 100-percent metering (including the fire department). By 2003, UFW had dropped to only 5 percent. Industrial and commercial meters are replaced every four years and residential meters every seven years to ensure accurate billing and to minimize unmetered water losses. Singapore water managers also promote public education, school programs, water audits, and reuse of nonpotable water by industries. Illegal connections bring fines of up to $50,000 or three years in prison. In 1995, Singapore's 3 million residents used an average of 1.2 million cubic meters per day; by 2003, total water demand had increased only 8 percent although the population had grown by 40 percent to 4.2 million.

• Fukuoka, Japan, known as the Water Conservation Conscious City, has one of the lowest system leakage rates (about 5 percent) in Japan, and its per capita water use is about 20 percent less than other comparably sized cities. Fukuoka has accomplished these water savings through active leak detection and repair, sophisticated metering techniques, rainwater harvesting, use of reclaimed water for toilet flushing, installation of efficient faucet devices in over 90 percent of households, and promotion of citizen awareness about water issues.

• Since the late 1980s, the Massachusetts Water Resources Authority (MWRA), which provides wholesale water supplies to over 40 cities and towns in the Boston area, has reduced systemwide water demands by about 25 percent by implementing a comprehensive demand-reduction program, including aggressive leak repairs and the installation of water-efficient plumbing fixtures and devices. This allowed the cancellation of a plan to dam the Connecticut River—a politically controversial proposal—and saved MWRA's 2.1 million customers more than a half-billion dollars in capital expenditures alone.

SOURCE: See endnote 37.

---

reduces energy use, it also reduces the output of climate-altering greenhouse gases that threaten to disrupt river flows and hydrological systems around the world.[38]

Conserving water clearly conserves energy, but conserving energy also conserves water. Thermoelectric power plants (coal, oil, natural gas, nuclear, or geothermal) use up water through evaporation, as excess heat is removed from condensers. Mining the fuels used to run these plants also consumes water. Hydroelectric power generation results in evaporation of water from storage reservoirs.

All together, the water required to service energy demands is substantial—in the United States, an estimated 8.3 liters per kWh of delivered electricity. Thus the average U.S. household, using 10,000 kWh of electricity per year, is indirectly also consuming an additional 83 cubic meters of water—a volume equivalent to nearly 14,000 flushes of an efficient toilet.[39]

Household water use varies greatly worldwide and says a great deal about variations in wealth and culture. (See Figure 3–3.) For example, people living in the United Kingdom use only about 70 percent as much water as the most water-thrifty Americans do. Indoor water use in U.S. homes is estimated to average 262 liters per capita a day (lpcd). Households that install water-efficient fixtures (toilets, showerheads, and faucets) and appliances (clothes washers and dishwashers) and that reduce leakage use only 151–170 lpcd.

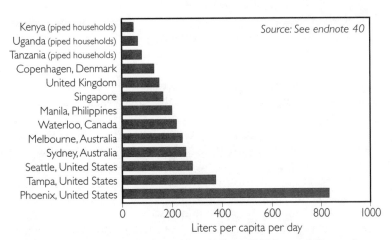

Figure 3–3. Household Water Use, Selected Cities and Countries

Since 1997, all toilets, urinals, faucets, and showerheads installed in the United States have been required to meet water efficiency standards established by the U.S. Energy Policy Act (EPAct) of 1992. By 2020, these efficiency standards are projected to save some 23–34 million cubic meters per day, enough water to supply four to six cities the size of New York City.[40]

Studies of 16 U.S. localities show that the water reductions from the EPAct standards will save water utilities between $166 million and $231 million over the next 15 years as a result of deferred or avoided investments in new or expanded drinking water treatment or storage capacity. The energy requirements of water and wastewater treatment facilities are projected to decline by 6 billion kWh annually. Some of these water, energy, and cost savings are now under threat, however: several major fixture manufacturers are actively promoting sales of tower-like shower stalls with multi-headed nozzles, some of which deliver over 300 liters per minute—more than most people in the world use in a day.[41]

When it comes to water use and costs paid by rich and poor, there is typically an inverse relationship: those who use the most pay the least per liter, and those who use the least pay the most. Low-income and poor city dwellers who are not connected to water systems often must turn to alternative and costly supplies, such as water vendors who may charge many times more than customers pay for piped water service. For example, the poor in Delhi pay informal vendors $4.50 per cubic meter of water, nearly 500 times the 1¢ per cubic meter paid by those with a house connection. In Manila, water vendors charge the poor 42 times more than the price paid by domestic users with piped water.[42]

The domestic water demands of the affluent take a dramatic upward trajectory with the presence of irrigated lawns. By volume, the biggest drinking problem in the United States is not alcohol but lawn watering. The irrigation of U.S. lawns and landscapes daily claims an estimated 30 billion liters of water—a volume that would fill 14 billion six-packs of beer. The average irrigated lawn uses about 38,000 liters per summer. Even worse, one resident of water-strapped Orange County, Florida, was billed for 15.9 million liters of water one year, most of it used to irrigate his 2.4-hectare property. That volume of water roughly equals what 900 Kenyans use in a year.[43]

Manicured lawns and turf-carpeted corporate, government, and roadside areas across the United States cover between 12 million and 20 million hectares, an area larger than the state of Louisiana—and more than is planted in any single agricultural crop. The United States also has about 60 percent of the world's golf courses; the 700,000 hectares they cover soak up some 15 billion liters of water per day. Lawns and golf courses not only guzzle vast amounts of water, they do so during the hottest months of summer, when flows in many rivers and streams are at their lowest levels.[44]

U.S. lawn and garden enthusiasts annually apply more than 45 million kilograms of fertilizers and chemicals to kill bugs, weeds, and fungi. In fact, homeowners use nearly 10 times more pesticide per hectare of turf than farmers use on crops. Fertilizers and chemicals not taken up directly by grasses and plants often run off into streams or seep through to aquifers, where they may contaminate drinking water and eutrophy lakes and ponds. (See Box 3–4.)[45]

While more-efficient sprinklers and irrigation systems can reduce lawn water consumption, a more fundamental reform of Americans' lawn-addiction is taking place through a burgeoning natural landscape and native plant movement. Homeowners and corporations are realizing lasting and substantial water savings by planting native and drought-adaptive grasses, groundcovers, wildflowers, and plants that thrive naturally in their local climates. The landscapes of Prairie Crossing, a subdivision outside of Chicago, and of the Sears, Roebuck & Company headquarters in Hoffman Estates, Illinois, for instance, are designed to embrace natural features instead of overriding them. Similarly, golf courses such as Prairie Dunes Country Club in Hutchinson, Kansas, and The

Landings in Savannah, Georgia, are reducing water use through measures such as weather-controlled irrigation, limited watering of tees and fairways, use of native plants and natural features in roughs, and organic soil and plant maintenance.[46]

Memberships in natural landscaping organizations such as the Wild Ones and Ecological Landscaper are growing rapidly, pointing to people's desire for a healthier relationship to the land. Others are mindful of the financial benefits. CIGNA Corporation in Bloomfield, Connecticut, spent about $63,000 over five years to convert much of its 120-hectare conventional corporate lawn to attractive walking meadows and wildflower patches, reaping the company several hundred thousand dollars annually in cost savings from reduced water, fertilizer, pesticide, and equipment and maintenance needs. As CIGNA's landscape manager explained, "What are you going to do, spend $5,000 on dandelion control?"[47]

Industrial Water Use and Material Goods Consumption

Industries account for about 22 percent of the world's total freshwater withdrawals, but they claim a far higher share in industrial countries (59 percent on average) than they do in developing ones (10 percent). Industrial demands in developing and emerging economies are growing rapidly and will compete for scarce water supplies with both cities and farms. Moreover, industries generate large volumes of wastewater, and in developing countries much of it is currently released untreated into nearby rivers and streams, polluting scarce supplies.[48]

The total volume of industrial water demand is not well understood because large industries often tap—and do not meter—water directly from their own wells or nearby

BOX 3–4. DRINKING YOUR NEIGHBOR'S LAWN AND MEDICINE CABINET

"It's early morning, do you know where your drugs are?" asks the U.S. Environmental Protection Agency's Christian Daughton in an article in *The Lancet.* "More than likely, some are on their way to local streams, rivers, and perhaps even farms, as sewage biosolids used as fertiliser." In a study of 139 streams sampled in 30 states, the U.S. Geological Survey found that 80 percent contained traces of at least one drug, endocrine-disrupting hormone, insecticide, or other chemical—some at levels that have been shown to harm fish and other aquatic life. This may not be surprising, given that the United States is the world's largest user of pesticides and that more than 3 billion prescriptions are written each year for the nearly half of all Americans who take at least one medication daily. Studies in Canada, the United Kingdom, and Germany also have found residues of pharmaceutical and personal care products (PPCPs) in fresh water, including sun-screen agents, antibiotics, and plasticizers.

Virtually no medical literature documents the extent, risks, or solutions to the problem of drugs as pollutants and what they are doing to human health and the environment. At present, PPCP contaminants in drinking water remain largely unregulated. At least with pesticides, some communities are not taking any chances. In Canada, both the Montreal suburb of Hudson and Halifax in Nova Scotia prohibit the cosmetic (purely aesthetic) use of pesticides, such as for lawns. "Better to err on the side of safety than suffer while awaiting some scientific proof," pointed out one community leader. Despite a challenge to the law by the lawn and chemical industries, the Canadian Supreme Court ruled that municipalities across Canada have the right to ban pesticide use on public and private property.

SOURCE: See endnote 45.

rivers and lakes. Worldwide, the major water-using industries include thermal electric power, iron and steel, pulp and paper, chemicals, petroleum, and machinery manufacture. Most use their largest volumes of water for cooling, washing, processing, and heating.[49]

An impressive number of industrial and commercial users have cut their water demands anywhere from 10 to 90 percent while boosting productivity and profits. (See Table 3–6.) Often these investments in water efficiency pay themselves back within two years and yield energy savings and pollution prevention benefits as well. For example, Unilever, a multinational producer of food, home, and personal care products, used an average of 4.3 cubic meters of water per ton of production in 2002, a one-third drop from the 6.5 cubic meters per ton used in 1998.[50]

While cost savings will be the primary motivation for efficiency investments for many industrial facilities, other incentives exist as well, including the need to comply with permit requirements, advances in onsite treatment technologies that allow process water to be recycled and reused, and the availability of low-cost reclaimed nonpotable water. For example, all of Singapore's sewage is treated at six water reclamation plants for reuse by industries, helping to conserve high-quality water for drinking and other purposes. Increased water and sewer rates also can act as an incentive for manufacturers to conserve; such pricing strategies sometimes backfire on water suppliers, however, by motivating customers to stop drawing municipal water and switch to an onsite well.[51]

With the expansion of manufacturing enterprises in developing countries, pollutant loads are rising along with industrial

Table 3–6. Examples of Industrial Water Savings from Conservation

Industrial Category or Product	Company	Savings	Water Efficiency Measures
Dairy (milk and other dairy products)	United Milk Plc, England	657,000 cubic meters per year; $405,000 per year	Reverse osmosis (RO) membrane system recovers and treats milk condensate for reuse throughout the plant, eliminating the need for an external supply. Excess recovered water will be offered for sale to other users in plant area.
Computer (plants and labs)	IBM, worldwide	690,000 cubic meters per year	Water savings in 2000 were 4.6 percent of total used; 375,000 cubic meters per year saved from multiple water efficiency projects and 315,000 cubic meters saved from recycling and reuse.
Steel	Columbia Steel Casting Co., Inc., North Portland, OR, U.S.A.	1.63 million cubic meters per year; $588,000 per year	Replaced once-through cooling system with recirculated cooling towers. Installed recycling system and storage tanks for rainwater capture and reuse of nonpotable wash water. Optimized manufacturing practices.
Pharmaceutical (life science research and biopharmaceuticals)	Millipore Corp., Jaffrey, NH, U.S.A.	31,000 cubic meters per year; $55,000 per year	Process wastewater recycled using RO technology; $61,000 investment was paid back in 1.2 years in reduced water, wastewater, and energy costs.
Chocolate	Ghirardelli Chocolate Co., San Leandro, CA, U.S.A.	78,840 cubic meters per year	Installed a recirculating cooling water loop, eliminating potable water use for cooling chocolate in large tanks.
Home Construction	Gusto Homes, England	50 percent water savings by households (50 cubic meters per year)	Millennium Green project involved installation of a rainwater harvesting system and underground storage in 24 homes and the company's offices. Dual flush toilets, aerated showerheads and toilets, and solar water heaters also installed.
Produce (pesticide-free fresh fruit, vegetables, and herbs)	Unigro, Plc, England	9,000–18,000 cubic meters per year; $7,400 per year	Sealed, climate-controlled facility uses the Greengro Farming system and includes precision irrigation and rainwater harvesting, requiring 30 percent less water per unit crop yield than conventional irrigation.
Beer	Anheuser-Busch Inc., nationwide, U.S.A.	90,850 cubic meters per year	Water meters installed throughout facilities to track use. Recalibrated bottle and can rinsing equipment.

SOURCE: See endnote 50.

water demands, posing risks to aquatic life and human health. The food and beverage, pulp and paper, and textiles industries account for more than three fourths of organic water-pollutant loads in developing countries. Textile rinse water contains dyes, for example, that deplete oxygen levels in rivers and lakes when discharged untreated. By capturing and recycling these dyes within the manufacturing process, factories can reduce pollution loads and save on input costs. In the West African nation of Ghana, a pilot program called the Waste Stock Exchange Management System aims to increase the reuse and recycling of industrial waste products in order to protect coastal and freshwater ecosystems. With the slogan "one person's waste, another person's raw material," the initiative reportedly has received an enthusiastic response from local manufacturers.[52]

Just as individual choices about diets and landscapes can make a big difference on the total human impact on water bodies, so can choices about the consumption of material goods. (See Box 3–5.) Virtually everything people buy—from clothes to computers to cars—takes water to make, and the manufacturing process may result in pollution of streams and lakes as well. People who drive gas-guzzling sport-utility vehicles rather than fuel-efficient cars, for instance, are not only consuming about three times more gasoline per kilometer driven, they are also indirectly using much more water since it takes 18 liters of water to produce just one liter of gasoline.[53]

In the environmentalist's credo of reduce, reuse, recycle, reducing material purchases always stands on top. When people do buy something, however, they can lower their water and energy impacts by choosing products made from recycled materials. Buying recycled paper rather than virgin paper products, for instance, saves not only trees and energy but also the water used in paper man-

---

BOX 3–5. ACTIONS INDIVIDUALS CAN TAKE TO REDUCE THEIR IMPACTS ON FRESH WATER

- Purchase fewer material goods.
- Eat a nutritious, less meat-intensive diet.
- Select native plants and grasses for lawns and landscapes and rely on natural rainfall only.
- Install water- and energy-efficient appliances and fixtures.
- Push for local land use ordinances that protect wetlands, aquifers, and watersheds.
- Serve on local water management boards to monitor and enforce water protection strategies.

---

ufacturing. And aluminum products made with scrap aluminum require just 17 percent as much water as the same product made with raw aluminum.[54]

Policy Priorities

There is no mystery about why so much of the water extracted for human use is wasted and mismanaged: the policies that drive water decisions in most cases foster inefficiency and misallocation rather than conservation and sustainable use. Instead of despairing about a new era of water scarcity, we need to confront old errors of waste.

First, it is essential for governments to fulfill their obligation to protect the public trust in water. Most freshwater ecosystems are not priced or valued in the marketplace, yet they support our economies and lives with services worth hundreds of billions of dollars a year. Laws and regulations that safeguard these functions are critical because market forces alone—including water pricing and trading—will never adequately protect non-market values. The European Union's 2000 water directive, South Africa's 1998 water law, and a handful of state laws in the United States are promising examples of governments attempting to assume their responsibilities to protect the public trust in water.[55]

Governments and communal authorities need to institute or strengthen groundwater regulations. A classic common-pool resource, groundwater is susceptible to overuse because the collective impact of each user acting out of self-interest is the depletion of the supply for all. Sustainable use of renewable aquifers requires that total withdrawals not exceed the level of replenishment. As researchers at the Sri Lanka-based International Water Management Institute point out, however, "nowhere in the world do we find such an ideal regime actually in operation…. Precious little is being done to reduce demand for groundwater or to economize on its use."[56]

In eastern Massachusetts, residents have drained the Ipswich River dry during several recent years because of their heavy pumping of groundwater for lawn irrigation.

Not only is groundwater insufficiently regulated, its use is often subsidized in various ways. In Texas, farmers who pump water from the diminishing Ogallala aquifer can claim a depletion allowance on their tax returns. India's farmers get subsidized energy worth $4.5–5 billion a year to pump 150 billion cubic meters of groundwater—a perverse incentive to deplete the nation's aquifers. While propping up production in the short run, these subsidies only hasten the rate of overexploitation and the ultimate day of reckoning. With groundwater contributing

$25–30 billion a year to Asia's agricultural economy, the adoption and enforcement of policies that lead to its sustainable use are urgently needed.[57]

Tiered water pricing is an economic tool that can promote more efficient and equitable use of water. With this method, the unit price of water to a user increases along with the volume used. This allows a basic-needs level of household water use to be priced very low, with greater use priced at a much higher rate in a stair-step fashion. A 2002 study of 300 Indian cities found that only 13 percent use such increasing block rate structures. Moreover, even when they are used, the lowest-priced blocks sometimes include far more water than needed to satisfy basic household requirements. In Bangalore, for example, the first two blocks together covered up to 50 cubic meters of water per month, a usage commensurate with average household use in the United States.[58]

Particularly in wealthy localities, pricing alone is unlikely to discourage profligate water use. For high-income households with large lawns, for instance, keeping grass green all year long is often more important to them than their water bill. In such areas, restricting water use is the next step. In eastern Massachusetts, residents have drained the Ipswich River dry during several recent years because their heavy pumping of groundwater for lawn irrigation depleted its summer base flows. In 2003, the conservation group American Rivers listed the Ipswich as one of the 10 most endangered rivers in the country. That May, the state Department of Environmental Protection set mandatory water withdrawal restrictions on each town permitted to use the Ipswich. When river flows drop to a specified level, these communities must institute mandatory water conservation measures. Because of a wet summer in 2003, the true test of the policy is yet to come. But it makes clear that the state's interest in protecting the river's flow takes priority over private homeowners' interests in having a green lawn.[59]

Along with strong regulations and more effective pricing, markets for water can help improve the efficiency of use and allocation. With a cap placed on extractions from the Murray-Darling river basin in Australia, for example, water trading among willing sellers and buyers is helping reallocate the available supply. The city of Adelaide may soon purchase water from farmers since it has reached the limit of what it can extract from the river. The ability to trade water encourages users to conserve, because they can sell their saved water and receive extra income. Where clear property rights or entitlements to water exist, "cap-conserve-and-trade" can be an effective strategy for protecting ecosystems and boosting water productivity.

Finally, individual consumers have important personal policy choices to make as well. By choosing a healthy and less water-intensive diet, an attractive and climate-appropriate landscape, and a lifestyle with fewer material goods, individuals can lessen their impact on Earth's freshwater systems without sacrificing personal satisfaction. Such choices can turn water consumers into water stewards.

Antibacterial Soap

"Getting rid of germs is now more fun than ever," reads the label on a bottle of fruit-scented Dial liquid soap, the leading antibacterial brand in the United States. In reality, however, the skyrocketing global production and use of such cleansers poses some not-such-fun health and environmental risks.

Liquid soaps, shower gels, and body washes with antibacterial properties have grown increasingly popular in recent years. In the United States, 75 percent of liquid soaps and nearly 30 percent of bar soaps now contain triclosan and other chemical compounds engineered to attack surface germs. Although labeled antibacterial, most of these are actually antimicrobial, attacking viruses as well as bacteria.[1]

The world market for soap is projected to grow steadily, from $5.5 billion in 2003 to $6.1 billion in 2008, reports the Icon Group, a global market research firm. The biggest growth is expected in Asia and the Pacific, where the soap industry anticipates that economic growth will spur consumer demand for enhanced soap products, including antimicrobials. In India and China, where liquid soap is viewed as an expensive luxury product, Procter & Gamble is now producing an antibacterial version of its Safeguard bar soap.[2]

All soap is produced through a chemical reaction known as saponification, in which an alkali, such as caustic soda (sodium hydroxide), potash (potassium hydroxide), or old-fashioned wood ash (lye), is heated with vegetable or animal fats (tallow) and water. In the process, the fats are broken down into liquid glycerol (glycerin—which is usually removed for other cosmetic and pharmaceutical uses) and fatty-acid salts, which make up the crude soap curds. These curds are boiled in water to remove impurities, then poured into molds and cut into bars.[3]

Some of the earliest traces of soap were found in Babylonian clay pots dating to 2800 BC. Before germ-killing versions came along in 1948, soap got rid of microorganisms by making surface dirt and oils slippery enough to be rubbed and rinsed off. Since World War II, human-made chemicals have altered the traditional recipe. They include surfactants that enhance sudsing and solubility, antimicrobial compounds such as triclosan, and plasticizers known as phthalates.[4]

Like any industry, soap manufacturing consumes raw materials and uses energy, such as fossil fuels, to heat the boilers, and these fuels create air pollution as they are burned. Other byproducts include solid fatty wastes and chemicals that can run off, polluting waterways. But it does not have to be this way: in Tunisia, one factory that makes soap from olive oil pressings has installed energy-efficient boilers and controls on releases of waste into air and water—and the plant has saved more money every year than the initial retrofit cost.[5]

In addition to industrial effluent, there's the problem of used soap rinsing down the drain after consumers have washed with it.

A 2002 study by the U.S. Geological Survey found that chemicals in pharmaceuticals and detergents—among them triclosan and phthalates—are entering water bodies across the United States in low concentrations through wastewater. These are a matter of concern, as safe drinking water levels have not been set for most of these chemicals.[6]

Triclosan and other antimicrobials raise troubling health and environmental issues. The production of triclosan can create highly toxic dioxins—hormone-disrupting, carcinogenic chlorine compounds that readily disperse in the environment and collect in the food chain. Triclosan can also cause nausea, vomiting, and diarrhea if swallowed—which the soap's "fruit flavors" may tempt children to try.[7]

Most urgently, though, the American Medical Association and the U.S. Centers for Disease Control and Prevention (CDC) are cautioning against home use of antibacterial soaps because these products are contributing to the rise of drug-resistant bacteria. The World Health Organization has launched a campaign against the misuse of antibiotics, noting that diseases such as tuberculosis, pneumonia, and malaria have become resistant to several antibiotics normally used to treat them. Triclosan acts by destroying enzymes in bacteria cell walls so that they cannot replicate; it targets the same enzyme as does the antibiotic isoniazid, used to treat tuberculosis.[8]

Moreover, studies have shown that antimicrobial soaps are not any more effective at getting rid of germs than regular soaps. "We found antimicrobial or antibacterial soaps provide no added value over plain soap," said Elaine Larson, associate professor at the Columbia University School of Nursing and lead author of a 1992 National Institutes of Health report on the trend. The authors recommend washing hands with ordinary soap and warm water after using the toilet and before preparing food as the best way to prevent colds and food-borne disease.[9]

Nor, despite modern obsessions with disinfection, is a home stripped of bacteria necessarily a good thing. Actually, it may have the opposite effect: a recent study found that adolescents who lived on farms and were regularly exposed to dust and germs were less likely to have asthma and allergy symptoms than teens raised in rural but non-farm settings. Researchers suggest that exposure to bacteria, fungi, and dust might actually help strengthen children's immune systems.[10]

The solution? Consumers should stop buying antimicrobial-spiked soaps and other home cleaning products, a move that could eventually push industry to cut back on their heavy promotion and production worldwide. "Antibacterial soaps and lotions should be reserved for the sick patients, not the healthy household," noted Dr. Stuart Levy of the Alliance for Prudent Use of Antibiotics at Tufts University. To stop the spread of germs in hospitals, the CDC advises health care professionals to use alcohol-based hand-rub gels, which do not pose the same risk of antibiotic resistance as antimicrobials. The gels can also be used in homes where a family member has AIDS or another immune system problem. But because these products cannot wash off plain old dirt, they are no substitute for plain old soap.[11]

—*Mindy Pennybacker, The Green Guide*

CHAPTER 4

Watching What We Eat

Brian Halweil and Danielle Nierenberg

In the mid-1980s, Mexican coffee farmers told the Dutch aid organization Solidaridad that they were barely making ends meet. As long as an international glut of coffee kept prices for the raw beans down, aid from industrial countries was not much help. In addition, Solidaridad learned that the growers and their families became sick when using fungicides and other toxic chemicals that were in vogue around the world. All of this was being endured for each cup of espresso and cappuccino enjoyed by Dutch citizens thousands of miles away—not by Mexicans.[1]

Solidaridad responded by joining up with other Dutch aid groups to create the Max Havelaar Foundation. (Max Havelaar was the conscientious Dutch protagonist in a nineteenth-century novel that depicts the harsh colonial treatment of the Dutch East Indies.) The Foundation developed a "fair trade" label that guaranteed coffee growers a set price above world market levels—a price that would cover their production costs and assure a decent living—and set a range of other social and environmental conditions,

from the right to organize into cooperatives to certain basic safety requirements. In contrast to the unfair relationship in which the coffee drinker seemed to benefit—albeit unwittingly—from the suffering of the coffee grower, buyers of Max Havelaar coffee paid a small premium to ensure a better life for farmers and communities at the other end of the transaction.[2]

The idea was not completely new. As early as the 1950s, groups like Oxfam in the United Kingdom, Ten Thousand Villages in the United States, and Stichting Ideele Import in the Netherlands offered products from developing countries. But the influence of these "Third World shops" was limited and the market was small.[3]

The real innovation of Max Havelaar was "to introduce fair-trade food into the mass market and work with commercial businesses," according to Rita Oppenhuizen, who handles public relations for the Foundation. Fifteen years after the first pack of Max Havelaar–labeled coffee arrived in Rotterdam harbor, the brand can be found in at least 90 percent of Dutch supermarkets and holds over 3 percent of the coffee market. The lower house of Parliament, most government ministries, and most provincial government halls serve Max Havelaar coffee. Max Havelaar–labeled chocolate was introduced in 1993 and is used by four major Dutch chocolate makers. Honey followed in 1995. The first Max Havelaar–certified bananas arrived in 1996 (and now account for 5 percent of the Dutch market), and tea was introduced in 1998.[4]

More recently, the notion that food could be fair has evolved even further. In the United States, the United Fruit Workers unrolled a fair apples campaign, which is a collaboration between farm workers, apple growers, and supermarkets to assure workers on apple farms—many of whom are recent immigrants—a living wage, the right to organize, and access to basic employee benefits. Farmers and the Soil Association in the United Kingdom are working to extend the "fair trade" distinction to locally grown produce, arguing that the caprices of free trade and agribusiness consolidation have crippled Britain's rural areas as badly as Africa's.[5]

A Revolution in Every Bite

People do not eat just to survive, of course, but also to socialize, to feel pleasure and satisfaction, and to define who they are. Increasingly, people eat to make a political statement and to help change the way farmers raise their crops. "Fair" food is just one of the growing number of distinctions that eaters now use to assure that their eating habits do not destroy the planet or farmers. "Certified organic" for fruits and vegetables, "pasture-raised" for beef, "sustainably caught" for seafood, and "bird friendly" for coffee, cocoa, and rainforest crops are a few other labels seen more often these days. Consumers seek-

ing these out are not simply looking for a bargain or flashy packaging: they are proactive and inquisitive. Still, these distinctions remain on the fringe—in spite of a rapidly growing but relatively small counter cuisine—and most people are not ready to view themselves as activist eaters or to know the intimate origins of their next meal. The rise in international food trade and the proliferation of heavily processed and packaged foods has further distanced most people from what they eat, both geographically and psychologically.[6]

But because humanity devotes such a large share of the planet's surface to food production—25 percent, more than the world's forested area—it is impossible to separate the way farmers raise food from the health of rivers, wetlands, forests, and our living environment. According to a report from the Union of Concerned Scientists, our food choices rival transportation as the human activity with the greatest impact on the environment. One European study found that food consumption accounts for between 10 and 20 percent of the environmental impact of the average household. When Annika Carlsson-Kanyama at the University of Stockholm compared the amount of greenhouse gas emissions generated by different food choices, she found that a meat-rich meal made with imported ingredients emits nine times as much carbon as a vegetarian meal made with domestically produced ingredients that do not have to be hauled long distances.[7]

Whether someone buys Chilean seabass from factory trawlers emptying the ocean of fish, pesticide-laden apples shipped halfway around the planet, or meat raised in giant factories swollen with manure, many food purchases currently support destructive forms of agriculture. (See Box 4–1.) For people living in wealthier nations, where hunger is not widespread, ubiquity and cheapness obscure many of these problems. The "glamour" associated with luxury foods has also encouraged the wealthy to turn a blind eye to how such items reach their tables. (See Box 4–2.)[8]

Heavy dependence on chemicals, from antibiotics and pesticides to fertilizers and food preservatives, represents the "conventional" way to raise food—that is, most agricultural ministries, colleges, and farm extension offices promote monocultural fields and a related cocktail of chemicals. Farmers accept exposure to toxic chemicals as an inevitable risk they would prefer not to take, and consumers accept residues of these toxins as an unfortunate truth they would pre-

fer to forget. Many of the risks endured by farmers, consumers, and food businesses are wrapped up in the same syndrome of conspicuous consumption that pervades other aspects of the economy. (See Chapter 1.) For instance, proponents of genetically modified crops, which are often created by combining the genetic material of wholly unrelated species that would not reproduce in nature, suggest that these crops are essential to help feed the growing global population and to drive the cost of food down. Any risks carried by this latest generation of agricultural technology, they argue, pale compared with the

---

BOX 4–1. SCOURING THE SEA

Industrial fleets have fished out at least 90 percent of all large ocean predators—tuna, marlin, swordfish, sharks, cod, halibut, skates, and flounder—in just the past 50 years, according to a study in *Nature* in 2003. "We're so good at killing things," says lead researcher Ransom Myers of Dalhousie University in Canada, "that we don't even know how much we've lost." According to oceanographers not connected with the study, this research gives the "best evidence yet that recent fish harvests have been sustained at high levels only because the fleets have sought and heavily exploited ever more distant fish populations."

The widespread use of massive trawlers (huge ships that literally scrape the bottom of the ocean) and long-liners (boats that drag lines with baited hooks up to several miles long) has been a recipe for disaster for most of the world's large predatory fish. When these predators disappeared from the oceans, smaller fish were able to rebound their populations, but only for a short time before they, too, became overfished. It is those larger fish, according to co-author Boris Worm of the Institute for Marine Sciences in Germany, that "we most value" for their economic and ecosystem services. And if the fish disappear, so

will the millions of communities and businesses that depend on them for food and income.

Reversing this problem, say experts, requires international cooperation. In 2002 at the World Summit on Sustainable Development, 192 nations signed on to a non-binding agreement to restore fisheries stocks to their maximum sustainable yield by 2015. This will mean lowering the percentage of fish killed each year by reducing quotas, cutting subsidies, reducing bycatch (fish that are thrown away because they are not economically profitable), and creating networks of marine preserves to protect fish stocks.

On the local level, marine conservation organizations are helping consumers find sustainable fish at their local grocery stores and restaurants by providing easy-to-carry information cards. Small enough to fit into a wallet, these cards provide consumers with a short list of eco-friendly fish species to choose from. The Seafood Choices Alliance, a coalition of chefs, hoteliers, wholesalers, retailers, and fishers, has gone one step farther. It is encouraging restaurants, hotels, and markets not to sell species known to be sharply in decline.

SOURCE: See endnote 8.

---

benefits of more food and cheaper food. (Currently, this technology is used not to feed hungry people, but primarily to raise corn and soybeans to feed livestock and the growing human appetite for meat.)[9]

Perhaps the most overt example of consumption gone awry in the food supply is the expanding waistlines and the crippling rise in

---

BOX 4–2. LUXURY FOODS

From foie gras to shark fin soup to caviar, consumers around the world have always craved rare and exotic foods as symbols of wealth and glamour. People will pay dearly for them, despite a sometimes marginal nutritional value: the $57-billion trade in coffee, cocoa, wine, and tobacco is worth more than international trade in grain. Growing consumer classes in nations like China and India mean that more people around the world can afford such foods. The cachet of such dishes draws partly from their high price and scarcity, although this invariably obscures the brutal and ecologically disastrous conditions behind their production.

Consider pâté de foie gras. Although the French consume 90 percent of all foie gras, it is considered a delicacy by wealthy consumers throughout the world. The name, which literally means fat liver, gives little indication of how it is produced. Foie gras is made by force-feeding ducks and geese large amounts of food through tubes or pipes. This causes the birds' livers to become abnormally large—they can weigh in at more than 10 times the size of a normal bird's liver—not to mention a range of other health problems, including liver hemorrhaging, throat bruising, and even suffocation.

The global trade in caviar has affected animal welfare in a different way. Caviar is the unfertilized eggs (the roe) of female sturgeons and, more recently, of salmon, paddlefish, and other species that have grown in popularity as sturgeon populations have shrunk. Overharvesting, habitat loss, pollution, and the slow reproduction rates of these large fish have all contributed to the declines, which are most notable in the Caspian Sea states of the former Soviet Union, the source of over 90 percent of the world's sturgeon roe. Fishery experts speculate that all sturgeon species are threatened to some degree; Beluga sturgeon, the most famous source of caviar, may no longer be reproducing in the wild. Americans alone import more than 40,000 kilograms of caviar per year—accounting for over 40 percent of the world's caviar sales—despite a price tag of $2,000 per kilogram.

Fishers also kill as many as 100 million sharks each year to feed the world's appetite for shark meat and shark fin soup—a delicacy in China since AD 960 and today revered in Asian cuisine around the world. Hunters typically catch and defin sharks while the animals are still alive, throwing them back into the ocean, where they either drown or bleed to death. In Asia, traders might offer 30–40 different species and can sell fins for up to $400 per kilogram. But like sturgeon, sharks reproduce slowly, and overfishing is pushing their stocks into rapid decline.

Animal welfare groups, ecologists, marine biologists, chefs, and other concerned groups are spearheading efforts to ban and stigmatize certain types of luxury cuisine, and more generally to get people to think before they eat. Animal welfare advocates are campaigning for U.S. and British restaurants and chefs to get foie gras off their menus. In the Netherlands, chefs voluntarily did this (although customers can still ask for it), and other countries have banned the force-feeding of ducks and geese. The Convention on International Trade in Endangered Species of Wild Fauna and Flora has called for stricter catch and export quotas, as well as a universal caviar labeling system. International bodies are also working to ban the wasteful practice of shark fin harvesting.

SOURCE: See endnote 8.

obesity that is becoming epidemic not only in the richest nations but in the urban centers of poor countries as well. Nutritionists, psychologists, and consumer advocates agree that at least one of the causes of the obesity epidemic has been the tendency of food companies looking for new customers to flood the media with advertising and to make food as ubiquitous as possible—a combination that makes overeating almost inevitable.[10]

Part of the evolving role of the consumer will include understanding that—as controversial as this may sound—cheap food might not always be desirable, particularly when the price stamped on an item does not reflect the subsidies that governments give to farmers or the cost of cleaning up environmental problems caused by agriculture. Recent surveys from Germany, the United States, and the United Kingdom estimate that people pay billions of dollars each year to clean up the pollution and cope with the other costs associated with modern farming: from removing pesticides from drinking water to repairing the damage from soil erosion to the loss of birds and other wildlife.[11]

Artificially low prices also obscure the fact that food grown nearby and eaten seasonally can often be cheaper, and more healthful, than foods grown and shipped from thousands of miles away. For instance, surveys from southwest England showed that food sold at farmers' markets and through a farm-to-home delivery scheme—including fruits, vegetables, meat, eggs, and certified organic produce—was on average 30–40 percent cheaper than similar products from the local supermarket.[12]

Although food industry officials and economists often point to consumer demand for cheap food as the ultimate driver of how we farm, consumers have had little direct impact on how food production has evolved. Yet this does not mean that consumers are powerless. Boycotts of food companies and food products, grassroots lobbying campaigns against certain pesticides, and selection of various ecolabeled foods all represent examples of the power that consumers can wield to influence farming.

At first blush, getting agribusiness behemoths to change might seem like a fantasy, yet McDonald's recently responded to concerns of animal rights activists and environmentalists by encouraging its meat suppliers to change certain industry-wide practices. And Kraft, the world's largest food company, announced plans to stop advertising directed at children, to shrink its portion sizes, and to eliminate some of its most unhealthy products. William Vorley of the International Institute for Environment and Development argues that the high level of concentration in agribusiness, in which just a handful of large companies control every step of the food chain, can actually facilitate this sort of activism because the targets are relatively few and obvious. This logic helped jumpstart the "Race to the Top" project to push the few supermarket chains that dominate the U.K. market "towards a greener and fairer food system." The concentration "makes retailers very sensitive to campaigns designed around ethics, safety or environment," Vorley notes, because no chain wants to appear to be the least ethical in the public eye.[13]

While these may look like isolated consumer actions, they are aimed at grabbing control of how food is produced and steering the global food system away from its current trajectory. The local manifestation of this "food democracy" will naturally vary around the world, and the motivations will not always be humanitarian but will include more selfish concerns like taste, food safety, personal health, and the preservation of open spaces. The choices in the average supermarket are, of course, endless. But some of the most profound changes "eaters" can make include rethinking their relationship with meat, selecting food produced without agrochemicals, and buying locally grown food. Meat represents the most resource-intensive segment of our diet; agrochemicals keep farmers tied to a monotonous agricultural landscape; and local food represents the best hope for returning power to the people who raise food as well as those who eat.

From Farm to Factory— and Back

Like most pigs in the midwestern United States, the more than 200 sows that live on Paul Willis's farm in Iowa love to eat corn. But Willis's animals have a diet and a lifestyle very different from the other 15 million hogs raised in that state. Along with the grain that Willis's hogs eat on a daily basis, they graze outside on pastures and are not confined in the concrete factories that dominate American pork production. Not only do Willis's animals get the chance to exhibit their natural and instinctive behaviors, like rooting for food, playing, and nest-making, but the meat they produce is healthier and tastes better than the pork produced on factory farms.[14]

Because pigs thrive under these more natural conditions, Willis can raise his meat without the use of antibiotics or growth promoters, which lowers his costs. And instead of selling his meat to one of the big companies that control most hog production in the United States—Smithfield or IBP—Willis markets his pork through the Niman Ranch, a California-based company started in 1982 to distribute humanely raised meat products to consumers and restaurants.[15]

Willis is part of a growing movement of farmers and consumers helping livestock go back to their roots. Although the shift might seem old-fashioned, farmers that raise animals

outside—and the consumers who buy this meat, which might be labeled as "pasture-fed" or "free-range"—are helping clean up what has become the most ecologically destructive and unhealthy sector of global farming—industrialized animal production. (See Figure 4–1.) Global meat production has increased more than fivefold since 1950, and factory farming is the fastest growing method of animal production worldwide. Industrial systems are responsible for 74 percent of the world's total poultry products, 50 percent of pork production, 43 percent of the beef, and 68 percent of the eggs. Industrial countries dominate production, but developing nations are rapidly expanding and intensifying their production systems. According to the U.N. Food and Agriculture Organization (FAO), Asia has the fastest-developing livestock sector, followed by Latin America and the Caribbean.[16]

Modern meat production, some might argue, is the only way to meet the growing appetite for meat around the world. By 2020, people in developing countries will consume more than 39 kilograms per person—twice as much as they did in the 1980s. People in industrial countries, however, will still consume the most meat—100 kilograms a year by 2020, the equivalent of a side of beef, 50 chickens, and 1 pig. Yet it is questionable whether the system that delivers all this meat to them can persist as its deficiencies mount and as alternatives such as vegetarianism and pasture-raised meat flourish.[17]

The cascading problems of animal factories begin with the cramped conditions and the feedlot diet. Cows are ruminants, meaning they digest grasses, legumes, and crop residues. But their feedlot ration consists of a mixture of corn and soybeans, since cows and other animals on this diet can quickly gain weight—and fatter livestock bring a higher market price. Although many con-

Feed
INPUTS
OUTPUTS
Manure
• A calorie of beef, pork, or poultry needs 11–17 calories of feed.
• 95 percent of soybean harvest is eaten by animals, not people.
• Feed containing meat and bone meal can cause mad cow disease, which has affected thousands of cattle in industrial countries.

• Manure from intensive pig operations stored in lagoons can leak into groundwater or pollute nearby surface water.

Methane
• Belching, flatulent livestock emit 16 percent of the world's annual production of methane, a powerful greenhouse gas.

Water
• Producing 8 ounces of beef requires 25,000 liters of water.

Additives
• Cows, pigs, and chickens get 70 percent of all antimicrobial drugs in the United States.

Disease
• Eating animal products high in saturated fat and cholesterol is linked to cancer, heart disease, other chronic illnesses.
• Factory farm conditions can spread *E. coli*, *Salmonella*, and other food-borne pathogens.
• Creutzfeldt-Jakob disease, the human variant of mad cow disease, has killed at least 100 people.

Fossil Fuels
• A calorie of beef takes 33 percent more fossil fuel energy to produce than a calorie of energy from potatoes would.

Source: See endnote 16.

Figure 4–1. Industrial Meat: Inputs and Outputs

sumers have come to expect the taste, texture, and appearance of industrial meat marbled with fat, this grain-fed beef has a number of hidden costs. First, cows tend to suffer from bloating, acidosis, liver abscesses, gas, and other symptoms of this rich diet. Second, the standard diet in factory farms has also been linked to the spread of food-borne pathogens, such as *Escherichia coli* 0157:H7, which can contaminate meat or even vegetables if the raw manure is used as fertilizer. It turns out that the grain diet also encourages the growth of the harmful microbe in the cow's stomach, whereas a grass diet eliminates the microbe.[18]

This is one reason that cattle and other livestock are fed low levels of antibiotics. In the United States, livestock consume eight times more antibiotics by volume than humans do. According to the World Health Organization and FAO, the widespread use of these drugs in the livestock industry is helping to breed antibiotic-resistant microbes, and making it harder to fight diseases among both animals and humans alike. But the crowded, unsanitary conditions weaken the animals further, and *Salmonella*, *E. coli*, and other lethal diseases can spread rapidly in the unhealthy herd or flock.[19]

Animals raised in crowded conditions, says Ian Langford of the University of East Anglia, encourage the growth and spread of microorganisms in meat because they often arrive at slaughterhouses covered in feces. "The problem," according to Langford, "isn't with the consumer looking after the food well enough, but…in the food production process."[20]

These kinds of modern factory farming innovations and technologies have the potential to create food safety disasters. For instance, bovine spongiform encephalopathy (BSE, known as mad cow disease) is a virus caused by feeding cattle the renderings of other ruminants, and it can be spread to humans who eat infected meat. Since it was first reported in the United Kingdom in 1986, BSE has been detected in 33 countries, and health officials estimate that 139 people worldwide have succumbed to variant Creutzfeldt-Jakob disease, a related illness in humans.[21]

Similarly, outbreaks of avian flu in densely populated chicken farms in Hong Kong during the past five years have led to massive culls of thousands of chickens. The disease jumped the species barrier for the first time in 1997, killing 6 of the 18 people infected. In 2003, avian flu spread to humans again, killing two. Dr. Gary Smith of the University of Pennsylvania School of Veterinary Medicine warns that these diseases and others will continue to spread because "the nature of farming nowadays is such that there is much more movement of animals between farms than there used to be…The problem is that the livestock industry is operating on a global, national, and county level." The recent foot-and-mouth disease epidemic in the United Kingdom is a perfect example of how just a few cows can spread a disease across an entire nation.[22]

Food safety outbreaks aside, nutritionists have found that grain-fed livestock are not as healthy as the grass-fed alternative. Animals raised in feedlots accumulate Omega 6 fatty acids (the bad fats), which have been linked with cancer, diabetes, obesity, and immune disorders. In contrast, grass-fed meat contains Omega 3 fatty acids, like those found in fatty fish, which help lower cholesterol. In addition, grass-fed products have higher levels of conjugated linoleic acid, which can block tumor growth and lower the risk of obesity and other diseases.[23]

These health claims have prompted many people to seek out meat raised on pasture without antibiotics, hormones, or any of the other inputs that are tied to factory farming. But people who have curbed their meat consumption might also be interested in the ecological implications of moving animals back outside. According to Canadian scientist Vaclav Smil, feeding animals grain is "highly inefficient and an absurd use of resources." Producing just one calorie of flesh—beef, pork, or poultry—requires 11–17 calories of feed, according to Smil, whereas animals raised on pasture require little if any grain. As a result, a diet high in grain-fed meat can require two to four times more land than a vegetarian diet. When people eat less meat, it is unlikely that the forgone feedgrain will reach hungry mouths, but it does mean considerably less pressure is put on farmland to raise vast monocultures of corn and soybeans.[24]

Reversing the human health and environmental problems caused by our appetite for modern meat will mean eating fewer animal products. Animals raised on pasture do not mature as quickly as feedlot animals do, and rangelands will support fewer total animals than can be squeezed into feedlots. But demand for meat is growing, especially in the developing world, where rising incomes and urbanization are changing diets. According to David Brubaker, former Executive Vice President and CEO of PennAg Industries, people in developing nations do not have the luxury of choosing expensive pasture-raised or organic meat. Instead, they are climbing up the protein ladder and following the bad example of producing and eating low-quality animal products set by the United States and other "fast-food nations." Curbing this appetite will mean encouraging developing nations to preserve traditional methods of raising livestock that both support local economies and enrich the environment.[25]

The inefficient inputs to factory farms are mirrored by inefficient outputs in terms of waste. When livestock waste can be used to fertilize crops, it enriches the soil and is a key part of a healthy farm—one of the main reasons that farmers around the world keep livestock. Yet the waste produced by thousands of animals in confinement facilities usually exceeds the amount of land available to handle it. As a result, manure goes from being a valuable agricultural resource to being a toxic waste. Manure contains nitrates, which at high concentrations can cause methemoglobinemia (blue baby syndrome), cancer, and algal blooms and eutrophication of surface waters. The lagoons used to store liquid waste are also vulnerable to natural disasters, as discovered in North Carolina when they burst during Hurricane Floyd in 1999, flooding miles of waterway with excrement and causing massive fish kills.[26]

Farmers who begin to see the role of animals differently often enjoy a range of unexpected benefits. In the Philippines, Bobby Inocencio has transformed the way many Filipinos produce and eat chicken. Once a factory farmer, Inocencio raised white chickens for Pure Foods, one of the largest companies in the Philippines, and followed the standard model of squeezing tens of thousands of birds into cage-lined buildings. But in 1997, he decided to revive village-level poultry enterprises that support family-size farms. He began raising free-range chickens and teaching other farmers how to do the same. His birds roam freely in large tree-covered areas of his farm that he encloses using recycled fishing nets. And Inocencio's farm is profitable—in part because his costs

per bird are considerably lower: no antibiotics, growth promotants, pricey feed, or huge sheds to maintain. But he has also found a niche in the Filipino market by giving consumers a taste of how things used to be. His chickens are part native and part Sasso (a French breed) and are better adapted to the climate of the Philippines, unlike white chickens that are vulnerable to heat. Not only are Inocencio's chickens raised humanely, they are nutritious and taste good. They are just 5 percent fat, compared with 35 percent in white chicken, and they do not contain any antibiotics.[27]

Creating the space and market for these sorts of farms will sometimes require more than just actions by farmers. In Poland, where almost every farm raises a few pigs on pasture or hay, large meat corporations have begun stepping in. Animex, the Polish subsidiary of Smithfield, the world's largest pork producer, has plans to turn some of the country's richest and most productive land into concentrated animal feeding operations (CAFOs) like those that now dot the landscapes of North Carolina and Iowa in the United States. However, activists from the U.S.-based Animal Welfare Institute have teamed up with Andrzej Lepper, head of the Polish Farmers' Union, to oppose Smithfield's attempt to take over the Polish hog industry. By showing Polish farmers how CAFOs have destroyed many small-scale U.S. livestock farms, they hope they can convince them and the Polish government to resist corporate agriculture.[28]

Such coalitions are prompting some corporations to change their minds about how meat is made. In 2002, bowing to pressure from animal rights and public health groups, McDonald's announced that it would stop buying eggs from chickens confined in battery cages and forced to lay additional eggs through starvation—practices already banned in Europe but still permitted in the United States. By 2004, McDonald's will require chicken suppliers to stop giving their birds antibiotics to promote growth and will choose indirect suppliers who do not use antibiotics over those who do.[29]

Since McDonald's is one of the largest buyers of chicken in the United States, the company's decision to change its standards will have a domino effect on the entire meat industry. And Wendy's, Burger King, and Kentucky Fried Chicken have recently hired animal welfare specialists to research and devise new standards to ensure better animal welfare. The World Bank, too, has changed its mind about funding large-scale livestock projects in developing nations. In 2001, the Bank said that as the livestock sector grows "there is a significant danger that the poor are being crowded out, the environment eroded, and global food safety and security threatened." It promised to use a "people-centered approach" to livestock development projects that will reduce poverty, protect environmental sustainability, ensure food security, and promote animal welfare.[30]

Food Without Pollution

Just a few years ago, drinking the water in Lithuania was a public health risk. Concentrations of nitrate, a fertilizer byproduct that is poisonous in high doses, were well above safe limits—six times the acceptable level in some regions. Since the 1950s the Lithuanian Ministry of Agriculture and Ministry of Environmental Protection had struggled to reduce the high rates of fertilizer and pesticide applications in the northern Karst region, the nation's agricultural epicenter, where the groundwater had become severely contaminated. But in 1993 they began to encourage farmers to forgo chemicals.[31]

They offered classes in organic produc-

tion, provided expert support in the field, and paid farmers in their first years of conversion. The program grew from 9 certified organic farms in 1993 to 106 in 1998, and then to 290 farms in 2001 covering 6,469 hectares of land, together with 8 certified organic processing companies and 11 other certified organic companies. This sector still represents a small fraction of the country's total area and food market. Nonetheless, groundwater contamination rates in the communities surrounding the converted farms have dropped substantially, and locals are enjoying a new source of chemical-free food.[32]

Other regions around the world have also used organic farming to prevent groundwater pollution. Since 1992, local authorities in Munich and Leipzig have offered financial incentives to farmers who switch to organic methods and have watched the nitrate levels in untreated groundwater fall from well over 40 milligrams per liter in the 1980s to less than 26 milligrams in 1996. The utilities in these German cities do not just provide payments and consultation: the Munich water company helps market the organic produce grown in its district and uses only local, organic produce in its canteen. (The authorities estimate that the total spent so far is just one seventh of what they would have spent on new water purification and processing technology.)[33]

As these government agencies have found, although organic produce generally costs more at the grocery store—a result, industry analysts say, of limited distribution and marketing—organic farming can actually be much cheaper in a range of other ways. Researchers at the University in Essex found that the cost to the public of removing pesticides from drinking water supplies in England is equivalent to one quarter of what farmers pay for the chemicals. They also estimated that organic farming cost society one third as much in terms of pesticide pollution, erosion, and other fallout as nonorganic farming did. A study from the Philippines found that the health costs to farmers from spraying pesticides—sick days, visits to clinics, and medication—exceeded the value of crops saved from pests, not to mention the cost of the sprays, powders, and other pesticides in the first place.[34]

Since agricultural chemicals running off of farm fields disrupt or simply kill the beneficial organisms living in the soil, streams, lakes, and coastal waterways, it is no surprise that studies from around the world have found that organic farms harbor a greater number and diversity of birds, insects, wild plants, and earthworms and other soil life than nearby nonorganic farms. In other words, organic agriculture is not just a reaction to industrial farming. It is a healthier way for humans to raise food. As the toll of chemical-intensive agriculture mounts, a more organic approach to farming might be the only option.[35]

Public interest in organic foods has already pushed global sales to an estimated $23 billion in 2002, more than a 10-percent increase from the previous year, according to Organic Monitor, a consulting firm that tracks the industry. Farmers from Australia to Argentina raise certified organic crops on nearly 23 million hectares, and many more raise crops without agrochemicals, either by choice or necessity, but are not certified as organic. North America and Europe still account for most of the sales, though markets are growing rapidly in all regions. (See Figures 4–2 and 4–3.)[36]

Some of the biggest obstacles to the continued spread of organic farming tend to be conceptual. Many farmers, agricultural researchers, and people who make farm policy simply believe that farming with fewer or no synthetic chemicals is not feasible on a

large scale. It is true that farmers converting to organic production often encounter lower yields in the first few years, as the soil quality, soil life, and insect populations recover from years of assault with chemicals. It may take several seasons to refine the new approach. And because of the emphasis on crop diversity as a means to reduce pest problems, organic farms will not raise the same crop every year, making it difficult to compete with other farms on total production of a single crop. But studies have shown that organic farming can be just as productive and generally more profitable.[37]

A recent survey comparing organic and nonorganic yields at agricultural research stations in the United States found that organic corn yields were 94 percent of conventional yields, organic wheat yields were 97 percent, organic soybean yields were 94 percent, and organic tomatoes showed no yield difference.[38]

A seven-year study from Maikaal District in central India involving 1,000 farmers cultivating 3,200 hectares found that average yields for cotton, wheat, chili, and soy were equal or up to 20 percent higher on the organic farms than on nearby conventionally managed ones. Farmers and agricultural scientists attributed the higher yields in this

Figure 4–2. Global Sales of Organic Foods, Circa 2002

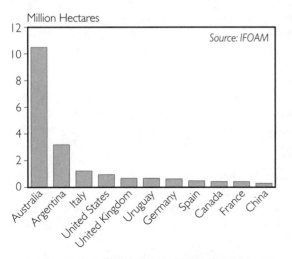

Million Hectares

Source: IFOAM

Figure 4–3. Top National Plantings of Certified Organic Land, Circa 2002

dry region to the emphasis on cover crops, compost, manure, and other practices that increased organic matter (which helps retain water) in the soils.[39]

A study from Kenya found that while organic farmers in "high potential areas" (those with above-average rainfall and high soil quality) had lower maize yields than nonorganic farmers, organic farmers in areas with poorer resource endowments consistently outyielded conventional growers. In both regions, organic farmers had higher net profits, return on capital, and return on labor.[40]

A 2002 FAO report noted that "organic systems can double or triple the productivity of traditional systems" in developing nations but suggested that yield comparisons offer a "limited, narrow, and often misleading picture" since "the multiple environmental benefits of organic farming, difficult to quantify in monetary terms, are essential ingredients in any comparison." Nick Parrott of Cardiff University, who recently assessed the potential of organic farming in the developing world, found numerous examples from Asia,

Africa, and Latin America of the adoption of organic agriculture significantly increasing yields in comparison with "unimproved traditional practices." He notes, "Many cases show that organic farming increases food security and farm incomes. This holds true for certified systems aimed mainly at northern markets and informal ones aimed at local markets." Parrot describes several mechanisms at work, including the use of manure and compost to help conserve water and buffer farmers against drought conditions and the elimination of expensive artificial inputs, which can reduce indebtedness.[41]

Perhaps a more important question than whether organic farming is feasible is how much longer farmers can continue to depend on heavy doses of agrochemicals. Pests have shown an uncanny ability to dodge, resist, and evolve around anything we throw at them, so that today farmers actually lose a greater share of their crop to pests than they did 50 years ago. Even genetically engineered crops, which were touted as helping to eliminate pesticide use, are vulnerable to the resistance treadmill. Researchers at Iowa State University have discovered at least four species of common weeds that have developed resistance to Roundup herbicide, the product used with herbicide-tolerant plants that have been grown in the Midwest for less than a decade. This will necessitate further pesticide use. In this futile effort, farmers have spent billions of dollars attacking increasingly resistant pests with increasingly potent chemicals, and most pesticides have ended up in our water, air, soil, and bodies.[42]

Although the public benefits of organic farming—reduced water pollution or increased wildlife—are fueling some of the growth in sales, the greatest interest has come from consumers with more personal concerns. Parents, for instance, may choose to feed their newborns organic baby food—knowing that small, developing bodies are more sensitive to endocrine-disrupting pesticides, residues of antibiotics and growth hormones, and other synthetic ingredients used routinely in food production. Then they may decide to make a shift for the entire household. Organic farming is the only system of food production in which consumers have a clear sense of what practices are allowed and forbidden, and farmers not only must demonstrate that they are not spraying known pollutants on the land but also must follow any number of practices that actually restore the landscape, from crop rotation to cover-cropping to composting. This level of transparency does not exist with most food production, where farmers are allowed to use a cocktail of chemicals, feed indiscriminate amounts of antibiotics and hormones to animals, and even apply sewage sludge to their fields.

Consumers eating organic fruits and vegetables are exposed to one third as many pesticide residues as they would get in conventional produce.

There is abundant evidence that farmers regularly exposed to pesticides run a higher risk of certain cancers, immune system malfunctions, mental illness, and a range of other conditions. Tests on animals have demonstrated that large doses of common chemicals are highly toxic. But most experts agree that it is harder to pin down the health effects of chronic exposure to lower levels of pesticides in the food supply or groundwater. Government regulators generally consider the safe level for humans to be 100 times, or even 1,000 times, less than levels that have no

adverse effects in animal studies. But human dietary exposure can exceed these conservative definitions of acceptable risk.

"The possibility that you are getting to an unsafe level increases because of the multiple residues in the diet," on top of exposure from drinking water, air, and other sources, says Edward Groth, a senior scientist with the Consumers Union. Groth notes that 40 different organophosphate pesticides are approved for use on crops in the United States alone, and that since all organophosphates share the same mechanism of toxic effects, "it's reasonable to assume that the impacts are additive or synergistic." Researchers recently found that men with higher levels of three common pesticides in their urine showed dramatically lower sperm counts and a higher incidence of irregular sperm. And toxicologists are now finding that a mixture of chemical fertilizers (nitrates) and pesticides, the two major inputs in industrial agriculture that often end up in groundwater together, can actually exacerbate the adverse health effects of exposure to each. Groth adds that exposure for young children, because of their smaller bodies and greater sensitivity, is more likely to be in the dangerous range. "But the kind of harm we're talking about"—damage to the nervous system that shows up later in life, for instance, as a learning disability—"is subtle and unlikely to be detected without careful studies of large populations," he says.[43]

Exposure is clearly exacerbated by eating food grown with pesticides. Researchers analyzing thousands of U.S. Department of Agriculture food samples found that consumers eating organic fruits and vegetables are exposed to one third as many pesticide residues as they would get in conventional produce, which was also six times as likely to carry multiple pesticide residues. And a recent study found that children fed predominantly organic produce and juice had only one sixth as many pesticide byproducts in their urine as children who ate conventionally farmed foods.[44]

Most of the studies on the health and ecological impacts of pesticide use have been conducted in the industrial world. But some of these concerns are most acute in the developing world, not simply because farmers there continue to use some of the most toxic pesticides—the ones banned in wealthier nations—but also because these same farmers are finding heavy chemical use less affordable and appropriate for their conditions. In India, according to the Ministry of Agriculture, 32 of the 180 pesticides registered for use have been banned in other countries due to health concerns. Between 1998 and 2001, India produced 40,000 tons of these compounds each year. Monocrotophos, an insecticide highly toxic to the neurological system whose registration was canceled in the United States in 1988, is India's top-selling pesticide.[45]

Although most people select organic food for what it does not have, there is recent evidence that such produce contains substantially higher concentrations of antioxidants and other health-promoting compounds than crops produced with pesticides. A University of California study confirmed a long-held suspicion among some nutritionists and agricultural scientists that heavy use of pesticides and chemical fertilizers can disrupt the ability of crops to synthesize certain phytochemicals—compounds that have antioxidant properties and are associated with reduced risk for cancer, stroke, heart disease, and other illnesses. Some observers noted the irony of conventional produce carrying both traces of chemicals that are known or suspected carcinogens and fewer compounds that help our bodies ward off cancer.[46]

Eat Here

One of the hottest concepts in the food industry is "traceability." The term describes the ability of a restaurant or grocery store or hungry shopper to know where a food item came from, who produced it, what chemicals were sprayed on it, and any number of other characteristics that reach beyond traditional concerns of taste, price, and packaging. Getting this information depends, to a large extent, on shortening the distance between the farmer and the eater.

The motivation to eat local food is as varied as the foods themselves. Homemakers responding to recent food scares and craving fresh food. Urbanites rebelling against an anonymous, long-distance food chain. Environmentalists trying to halt urban sprawl and the loss of green spaces. Nutritionists pushing less processed foods. Farmers trying to salvage their livelihood. Politicians in developing countries hoping homegrown food can help them retain precious foreign exchange. Chefs, restaurateurs, and food connoisseurs awakening to the pleasures of regional cuisines and artisinal dishes.

Preserving distinct flavors and "the right to taste," is only part of the mission of a new international movement called Slow Food. This 17-year-old group, which has 75,000 members in 80 nations, views the social interactions between eaters and bakers, butchers, and farmers, as well as meals shared with friends and family, as inseparable from the joy of eating. Carlo Petrini, founder and president, notes that the price societies have paid for having access to every possible food at any time of year is "the deliberate development of species with characteristics functional only to the food industry and not to the pleasure of food, and the consequent sacrifice of many varieties and breeds on the altar of mass-production."[47]

In the Peruvian Andes, the Association for Nature Conservation and Sustainable Development (ANDES) is not only trying to preserve traditional cropping patterns as a way to improve farm incomes, it is also reviving an east-to-west food trading corridor started by the Incas thousands of years ago. In Choquecancha, the central town along this corridor, people exchange highland foods (potatoes, guinea pigs, lama, lima beans, amaranth, and local tubers such as *ulloco, oca,* and *mashua*) for lowland foods (cocoa, coca, mangos, papaya, and coconut). For Peruvians who have moved to lowland urban areas, where fast food and processed cuisines are nudging out the local fare, this market allows them to share in the diverse mountain foods enjoyed in the area for thousands of years. In this case, people are paying not just for the nutritional value, according to Alejandro Argumedo of ANDES, but also "to preserve the spiritual stewardship of the mountain cultures, the highland cultures, and the indigenous varieties that end up providing the best nutrition." ANDES plans to open a restaurant in Cuzco that would feature local foods.[48]

The movement to preserve farms, farmland, and cuisines is evolving at a time when food travels farther and is controlled by a smaller number of global entities than ever before. The value of international food trade has tripled since 1960 and the volume has quadrupled. In the United States, the average food item travels 2,500–4,000 kilometers, about 25 percent farther than in 1980. In the United Kingdom, food travels 50 percent farther than two decades ago. A "traditional" Sunday meal in Great Britain made from imported ingredients generates nearly 650 times the transport-related carbon emissions as the same meal made from locally grown ingredients. (See Figure 4–4.) As a result, people who eat local produce can help save

lots of energy, reduce greenhouse gas emissions, keep money in their community, and gain a certain peace of mind that comes from knowing their farmers.[49]

A good rule of thumb is that the farther food travels, the less money is retained by the farmer and the rural community. The hauling, packaging, processing, and brokering of the food gobbles more and more of the final price. This leakage of money out of communities—and the ability of local food to help staunch it—can be particularly relevant where people are still engaged in agriculture. A study by the New Economics Foundation in London found that every £10 spent at a local food business is worth £25 for the local area, compared with just £14 when the same amount is spent in a supermarket—that is, a pound or a dollar, a peso, or a rupee spent locally generates nearly twice as much income for the local economy.[50]

This sort of multiplier is part of the motivation behind the Navdanya (Nine Seeds) movement in India founded in 1987 by the Research Foundation for Science, Technology and Ecology to protect local varieties of wheat, rice, and other crops from patents by cataloguing them and declaring them common property. "We started the movement to anticipate genetic engineering and patent monopolies in agriculture," explained Indian activist and scientist Vandana Shiva, who heads Navdanya, "but also to bolster the village economies." Navdanya started setting up locally owned seed banks, farm supply shops, and storage facilities, and it encouraged a shift to organic agriculture to reduce dependence on imported chemicals. "Right now we have over 3,000 villages in which farmers have basically created what we call 'Freedom Zones,' those are agricultures that are free of chemicals, free of corporate inputs, free of hybrid seeds, free in the future of patents and genetically engineered crops." Working with farmers' organizations, women's groups, and church groups, Navdanya has set up over 20 seed banks in seven states. Movement organizers estimate that they serve more than 10,000 farmers and have rescued more than 1,500 hundred varieties of rice, hundreds of millets, pulses, oilseeds, and vegetable varieties.[51]

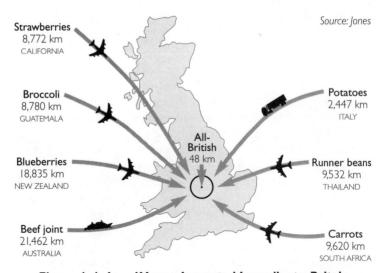

Figure 4–4. Local Versus Imported Ingredients: Britain

Local crop diversity might provide all that is available to eat in many of the poorest nations, where people cannot readily afford imported food. In Zimbabwe, urban farmers have found a market for indigenous vegetables in city-dwellers who crave a gastronomic link to their country's cultural identity. Households depend on roughly 25 indigenous vegetables, including Lady's finger (also known as okra), Spider flower, Horned cucumber, and bottle gourd, which provide a source of highly nutritious leafy vegetables from August to December—the typical season of scarcity—giving the poor both a source of income and nutrition. The government's Crop Breeding Institute is helping farmers raise these crops by distributing seed and developing processing and preservation technologies. Governments can also encourage domestic farm economies with procurement programs that get local crops into government offices, hospitals, and schools.[52]

"School lunch programs, for example, can provide a significant stimulus to the expansion of commercial food markets, if the produce involved is locally grown," Nobel laureate and agricultural scientist Norman

Borlaug recently wrote in the *New York Times* when he made a plea for greater food self-sufficiency in Africa. In 2000, several school districts in northern Italy passed new laws mandating that the region's schools favor local and organic produce when buying for their cafeterias. There are now over 300 organic school meal services in Italy, and hundreds more local meal services. Officials and citizens pushed for this shift partly to preserve the rural landscape and farm livelihoods, but they also found that the fresher meals with fewer processed ingredients could save money, were healthier, and tasted better.[53]

Greater self-sufficiency, in turn, means that nations, regions, and communities command greater control over how food is produced. "In the present food marketplace, there are great inequalities with respect to voting power, and more fundamentally, with respect to control," according to sociologist JoAnn Jaffe of the University of Regina in Canada, a situation due in part to how the food system has sprawled. Jaffe suggests a retaliatory strategy of "eating lower on the marketing chain" by buying food as locally as possible in order to regain sovereignty and to create a direct route for feedback between farmer and eater. (Eating lower on the marketing chain will often be healthier, because buying more food directly generally means eating more fresh fruits and vegetables, and because many of the extra steps between the farmer and the consumer remove nutrients and fiber and add fat, sugar, salt, and other fillers.) In contrast to the backroom decisionmaking behind fewer corporate doors, farmers' markets, community-supported agriculture, and locally owned food businesses all tend to return decisionmaking power to the local community.[54]

Eating is not a choice, but a necessity. But we do have the right—and the responsibility—to choose how our food is produced.

In many communities, these food businesses no longer exist. The local butcher and baker, the dairy and the cannery, have folded under waves of consolidation. As restaurants, school cafeterias, supermarkets, and other food businesses begin buying more food locally—and as consumers demand it—the forgotten infrastructure can gradually re-emerge. Navdanya recently opened a local cafe in Delhi, similar to the planned ANDES cafe in Cuzco, linking the countryside and city by promoting Indian food traditions and celebrations that revolve around the seasonal harvests. In *Hope's Edge*, Navdanya's Maya Jani explains, "Navdanya wants to retrieve indigenous food and drink from extinction through pleasure—and fast, before our taste buds are completely stolen by Pepsi and Coke." During the scorching summer months, Navdanya's *panna* festival

celebrates traditional cooling beverages, including brews made from coconut, mango, litchi, barley, and rhododendron. "Our festivals are a way to help people regain confidence in their traditions."[55]

The Rise of Food Democracy

Food democracy is a term that might best describe the growing number of farmers, consumers, chefs, and food businesses resisting the temptation to eat blindly and instead eating deliberately. Yet rethinking our relationship with food is not simply about giving up meat or the so-called convenience of shopping at a chain grocery store.

Changing our diets is about adding something back to our lives that has been lost—our connection to food and the people who produce it. Whether someone is a farmer, a restaurateur, a politician, a banker, an entrepreneur, a student looking for a career, or a concerned parent, we all need to know more about the food that we buy and eat. And there are an infinite number of entry points to eat more deliberately and reinforce food democracy. Eating is not a choice, but a necessity. But we do have the right—and the responsibility—to choose how our food is produced. From shopping at a local farmers' market to preparing meatless meals to buying fair-trade coffee and cocoa, small but growing groups of consumers all over the world are voting with their forks and their wallets for a healthier food system.

The typical consumer will not necessarily take these steps alone. And although few people would tolerate their governments mandating what they eat, governments command considerable power to change the way we grow food—through everything from regulations on what chemicals farmers use to the sort of research promoted at agricultural universities. (See Box 4–3.) As noted

---

BOX 4–3. POLICY PRIORITIES FOR RETHINKING OUR RELATIONSHIP WITH FOOD

- Governments should shift the money spent on agricultural subsidies each year—more than $300 billion—into support for ecological farming.
- Governments should consider taxing pesticides, synthetic fertilizers, factory farms, and other polluting inputs or farm practices.
- Governments should work with farming organizations to increase the share of their land under organic production to 10 percent over the next 10 years by improving organic certification programs, by boosting organic know-how at agricultural universities, research centers, and extension agencies, and by providing subsidies or tax credits to farmers in the first few years of conversion.
- Governments should reform international trade agreements to eliminate export subsidies, food dumping, and other unfair trade practices that restrict the ability of nations to protect and build domestic farm economies by forbidding domestic price support and tariffs on imported goods.
- Governments, from the national to the local level, should use food procurement for schools, hospitals, government offices, and other institutions to support ecologically raised crops from local farmers.

SOURCE: See endnote 56.

---

ple—and can use these purchases to encourage certain agricultural markets. (See also Chapter 6.) The Swedish Environmental Protection Agency recently joined with the Swedish Food Administration and the Swedish Consumer Agency on a campaign to link food habits not only with nutrition but also with the environment. This collaborative effort resulted in a cookbook called *Mat Med Känsla för Miljö—Food with a Sense for Environment*—which argues that consumers can substantially reduce energy use in the food chain by making the right food choices.[56]

In the realm of food, governments and corporations quite often lag behind consumers and are slow to make change without some widespread and persistent public outcry. Historically, the biggest food-related victories in the consumer movement, including mandatory nutritional and ingredients labels, grew out of consumer efforts despite reluctance from governments and the food industry. In hindsight, the changes always seemed logical and well overdue. The required grassroots energy, in turn, often originates from a shift in mindset. Changing our collective menu, Stuart Laidlaw writes in *Secret Ingredients: The Brave New World of Industrial Food*, means producing food that "doesn't kill fish or send children running inside at recess to escape the pesticides.... We should do these things...not because the food on our plate would be better for us, but because it would be better for the planet." The potential for recreating the collective menu is vast—and so is the need. But the work will always depend on motivated individuals searching for a more secure livelihood, a stronger community, a healthier environment, or simply a delicious meal.[57]

earlier, governments buy considerable amounts of food for schoolchildren, government offices, and armies—the U.S. government provides more than 26 million meals each day to schoolchildren, for exam-

Bottled Water

The next time you visit your local food store, you may be astonished by the proliferation of bottled water choices, from high-end names like Perrier and Evian to their low-rent store-brand cousins. Worldwide, bottled water consumption is growing at an annual rate of 12 percent, though in newer markets like India, it is increasing by as much as 50 percent annually. Consumers across the globe now spend an estimated $35 billion a year on this water.[1]

Although its contents might appear the same everywhere, bottled water essentially comes in three different forms: natural mineral water, spring water, and purified water. Under the European Union's definition, natural mineral water is "microbiologically wholesome water, originating in an underground water table or deposit and emerging from a spring tapped at one or more natural or bore exits." In Europe, mineral water's reputation for health benefits dates back to the Roman Empire. The actual benefits of these minerals, however, are regarded today as minimal. While the sources of these waters are protected from pollution, since the water is not disinfected it can contain naturally occurring bacteria. And though bottlers guard against it, contamination is always possible, as seen in the 1990 worldwide recall of Perrier due to high benzene levels.[2]

In the United States, the Food and Drug Administration defines natural mineral water as having 250 parts per million total dissolved solids and deriving from a protected underground water source. Spring water, in contrast, need not have a constant mineral composition and is usually cheaper. Purified water, also called drinking water, is taken from lakes, rivers, or underground springs and has been treated—making it almost identical to tap water.[3]

Bottled water's skyrocketing popularity has a number of causes. In Asia and the Pacific, population growth and problems with local water quality and supply are the biggest factors. (Currently 1.5 billion people worldwide have no access to safe drinking water, and 12 million people die each year from diseases brought on by unsanitary water.) Bulk packaging made bottled water more affordable in India, the United States, and many other countries in the early 1990s. And, prompted by advertising, many consumers buy bottled water as an alternative to soft drinks and alcohol because it is perceived to be safer than tap water and, particularly in France, because it tastes better than tap water.[4]

Yet many people are concerned about the environmental costs of producing bottled water. A leading concern is that growing demand for the water could put a strain on existing water resources. In recent years, several international beverage companies have been exploring water-rich Canada as a source for bottled water. To prevent this, several Canadian provinces have banned, or are considering banning, the bulk export of fresh water.[5]

The Container Recycling Institute reports that sales of virgin resin PET (polyethylene terephthalate), the plastic most commonly used in water bottles, shot up to 738 million kilograms in 1999, more than double the amount in 1990. Producing 1 kilogram of PET plastic requires 17.5 kilograms of water and results in air emissions of 40 grams of hydrocarbons, 25 grams of sulfur oxides, 18 grams of carbon monoxide, 20 grams of nitrogen oxides, and 2.3 kilograms of carbon dioxide. In terms of water use alone, much more is consumed in making the bottles than will ever go into them.[6]

As for distribution, one large difference between bottled water and tap water comes from the fossil fuels burned to transport it by truck, train, or boat instead of by pipe. The World Wildlife Fund, while noting that 75 percent of bottled water is produced for local consumption, argues that international companies should invest in bottlers aiming at local markets and ship bottled water in bulk containers. Yet even this would be more inefficient than public drinking water systems.[7]

Among the largest issues besetting bottled water is plastic waste. According to the Container Recycling Institute, in 2002 some 14 billion water bottles were sold in the United States, 90 percent of which were thrown in the trash—even though most of them were made of recyclable PET plastic. In June 2003, the Pollution Control Board of West Bengal, India, determined that bottle producers were responsible for collecting used bottles and recycling them. Effective bottle bills promoting recycling also exist in Austria, Bel-

gium, Canada, Denmark, Finland, Germany, the Netherlands, Norway, Sweden, Switzerland, and 11 states in the United States.[8]

Americans say one main reason they drink bottled water is because it is safer than tap water. Yet a four-year Natural Resources Defense Council study tested a thousand bottles sold in the United States and found that about one fifth contained chemicals such as toluene, xylene, or styrene—known or possible carcinogens and neurotoxins. In India, tests by the Centre for Science and Environment in February 2003 found high pesticides levels in sampled waters, resulting in governmental quality certificates being taken away from a number of brands and warnings issued to Coca-Cola and PepsiCo.[9]

The United Nations declared 2003 to be the International Year of Freshwater, and it is working to improve the quality of fresh water worldwide. One of the targets under the U.N. Millennium Development Goal of ensuring environmental sustainability is to halve the proportion of people without safe drinking water by 2015. Yet given the environmental impacts of the use and disposal of bottled water, it is worth asking if there is not a better way to distribute this vital resource. For those of us fortunate enough to have the option, tap water (filtered, if necessary) is the cheaper, less polluting choice.[10]

—*Paul McRandle, The Green Guide*

Chicken

Most chickens follow one of two paths: they are raised to lay eggs (layers) or only for their meat (broilers).

They begin their journey along the industrial food chain in breeding farms owned by Tyson Foods, Perdue Chicken, or some other agribusiness company. There, eggs are kept warm by carefully controlled incubators. Breeders make sure that chicks all hatch at close to the same time by artificially inseminating the mother hens. After hatching, chicks destined to be layers come into contact with humans for the first, and often the only, time. Workers sex the chicks when they are one day old, throwing the males into large bins. These unlucky chicks are ground up (sometimes while they are still alive) for use as fertilizer or animal feed.[1]

Females are put on an assembly line and painfully debeaked with hot blades. After 18–20 weeks, the chicks (along with feed, antibiotics, and other inputs) are shipped to contract growers. The layers are housed in 60-by-360–foot barns (as are the broilers)—about half the size of a football field. Each barn can hold more than 90,000 chickens; since the business of raising chickens has gone high-tech, one farmer can usually manage an entire barn with little help. Although these farmers own their own land and incur most of the financial risk, they never own the chicks they raise. From start to finish, the chicks are branded as company property. The barn costs about $250,000, plus another $200,000 for the equipment to keep it going; once you throw in the chickens, feed, and other miscellaneous expenses, the start-up costs in industrial countries reach at least $1 million.[2]

Once on the farm, each laying hen is put into a wire battery cage with as many as nine other birds. These layers will each produce about 300 eggs per year—more than three times as many as chickens did a century ago, thanks to genetic manipulation and growth-promoting drugs mixed into their feed. Hens are also tricked into laying more eggs by round-the-clock artificial lighting. Their cages, stacked one on top of the other and covered in dripping feces, allow for little movement. The hens are easily startled because they rarely have any human contact. Usually the only birds that producers have to touch are those that have somehow escaped from a cage or have died from stress.[3]

Not surprisingly, chickens kept in these conditions are more susceptible to disease and tend to die much earlier than traditionally raised chickens. In fact, after a year or so most hens are so worn out that their egg production declines. Producers used to send hens off to be processed into cat and dog food, chicken nuggets, and even baby food. But in some places they are killed on-farm or transported to specialty and live-animal markets, where meat from birds at the end of their laying life is still valued for its taste.[4]

Broiler chickens have an even shorter lifespan. Although they are not kept in individual cages, broilers are packed tightly into sheds without much leg or wing room—each bird is given about 9 by 9 inches of floor space. They are not exposed to outside light or fresh air, and they have unnaturally long days because the windowless sheds are lit for up to 23 hours a day.[5]

Each day these chickens eat about 0.86 kilograms (1.9 pounds) of specially designed feed, which may contain antibiotics or growth promoters. Although chickens are efficient at converting grain into protein, their living conditions make them very susceptible to respiratory diseases. So producers have a long history of adding to the feed antibiotics much like those used to treat human disease. (A study in 2002 found that 37 percent of the broilers found in major grocery stores are contaminated with antibiotic-resistant pathogens.) Often these chickens gain so much weight so quickly that they cannot stand up. Chickens raised in factory farms often suffer from lameness, and many die of heart attacks because their hearts are not strong enough to support their disproportioned bodies.[6]

When they weigh about 2 kilograms, broilers are rounded up by workers known as catchers, stuffed into cages, and taken to processing plants. Workers sort, cut, and weigh the chickens for distribution to grocery stores and restaurants. Wrapped in plastic, the packages of thighs, wings, and legs bear little resemblance to a live animal. Some packages carry warnings to consumers to cook chicken completely to prevent the meat, often contaminated with feces, from spreading foodborne illnesses, such as *Escherichia coli* and

Salmonella, which are common in industrial growing environments.[7]

But not all farmers are raising chickens in a factory. According to the U.N. Food and Agriculture Organization (FAO), backyard and free-range chickens account for as much as 70 percent of both egg and meat production in some of the poorest countries. These chickens not only provide food, they are also a source of economic security. Farmers can use them, says Robyn Alders of FAO, as a "kind of credit card, instantly available for sale or barter in societies where cash is not abundant." They are also an important source of pest control and fertilizer. Projects in Bangladesh and South Africa are improving poultry health, providing income for members of poor communities, and giving native chicken breeds—which are already adapted to heat and low-input conditions—a chance to survive.[8]

Some chicken and egg producers in rich countries are also responding to consumer demand for organic, humanely raised chickens. At West Wind Farms, the only farm in the state of Tennessee providing certified organic meats and poultry, Ralph and Kimberlie Cole raise 600 chickens every year on organic pasture and grains. The chickens range on small patches of grass near mobile chicken houses that can be moved from field to field. The Coles refer to the chickens as part of their "soil improvement crew" because they fertilize the soil and control pests. Raising chickens this way—instead of in factory farms—can help the the environment, and it is certainly kinder to the chickens.[9]

—*Danielle Nierenberg*

Chocolate

The next time you bite into a bar of chocolate, think of that taste as a link to some of the world's most endangered forests—and to the millions of farmers who live near them. Chocolate comes from the seeds of a small rainforest tree, the cacao (*Theobroma cacao*). The cacao is native to northern South America and perhaps also to southern Central America. Its fruit is about the size of a small melon and is packed with those seeds—the cocoa beans. These are processed in various ways to make cocoa, cocoa butter, and chocolate.[1]

Cocoa is grown commercially in nearly 60 countries, but production is concentrated in just a few of these. Côte d'Ivoire, the world's leading producer, accounted for 35 percent of the 2002 cocoa bean harvest, down from its peak share of 41 percent in 1999 and 2001. The top five producers in 2002—Côte d'Ivoire, Ghana, Indonesia, Nigeria, and Brazil—accounted for 79 percent of global production. At present, over 70,000 square kilometers are in cocoa production worldwide, an area a little larger than the country of Ireland. Production area has grown substantially in recent decades, expanding by nearly a quarter since just 1990.[2]

Because the cacao tree requires a plentiful and constant water supply, it can only be cultivated commercially in rainforest biomes. This limitation is a kind of economic blessing: whatever value cocoa adds to rainforest areas cannot be undermined by growing it else-where. It is also of great significance to conservation because all the major cocoa areas—in the Caribbean, Central and South America, the Indonesian-Malaysian archipelago, and West Africa—are "biodiversity hotspots." These are regions that have been identified as global conservation priorities because they are unusually rich in biodiversity and highly threatened. Cocoa is a hotspot crop.[3]

The cacao tree is shade-tolerant, so it can be grown under forest canopy. In rainforest areas, agriculture generally displaces forest, but cocoa allows farmers to earn a living under the trees—or at least, under some trees. (Reasonable yields can probably be achieved while maintaining 50–60 percent of the original canopy cover.) Unfortunately, most of the world's cocoa is grown on lands that have largely lost their original cover—either to the cocoa itself or to some activity that preceded its introduction. In Indonesia, for example, cocoa farming has often followed the logging of primary forest. So despite its shade tolerance, cocoa has often been an agent of deforestation, although usually the result is a kind of "deforestation lite." That is because cocoa is often grown together with other tree crops, both native and introduced. Some of these "agroforests" are quite complex and support a considerable share of local wildlife. On the other hand, cocoa is sometimes grown as a monoculture in full sun, an arrangement that supports far less diversity.[4]

On a global scale, cocoa's contribution to tropical deforestation is minute—perhaps one third of 1 percent of the world's original tropical forest area is now in cocoa. But on a regional scale, cocoa farming has sometimes been a major force in the landscape. For example, cocoa accounts for more than 13 percent of the original forestlands of Côte d'Ivoire, and it is still chewing up forest in parts of West Africa and Indonesia. But it does not have to work this way. In some places, cocoa farming has already become a de facto conservation system. In Bahia, Brazil, for instance, and south central Cameroon, cocoa is cultivated under thinned native forest in areas where little other forest remains. By default, the farms have in effect become the forests.[5]

Cocoa has important social potential too. Outside of Brazil and Malaysia, where it is mostly grown on large farms, cocoa is generally a smallholder crop. Thousands of West African cocoa farms consist of less than 1 hectare (about 2.5 acres), and the average farm size in Côte d'Ivoire is under 3 hectares. Cocoa works well on a small scale because of its relatively high value and because the cacao tree responds to extra care. Skilled smallholders, whose cocoa orchards are small enough to "garden," can achieve rates of productivity beyond the reach of big farms, which have too many trees to look after individually. Potentially, at least, cocoa rewards labor.[6]

But from the farmers' point of view, those rewards are poorly distributed. On the retail level, the chocolate business is worth $42–60 billion annually, depending on how "chocolate product" is defined. It is difficult to determine how much of this money actually makes its way back to the farms, but a very generous estimate would be 6–8 percent, and it is likely considerably less. Even this small share of cocoa's wealth, however, has meant a better life for millions of farmers and their families. Yet the economy of cocoa, like its ecology, has a bleak side as well. Labor abuse is apparently rife in the Ivoirian cocoa region. Persistent reports that some farmers are enslaving thousands of child migrant workers have sparked widespread criticism of the industry. In 2002, Côte d'Ivoire responded by ratifying a treaty against labor abuse of children, and the big chocolate companies launched an initiative that aims to certify Ivoirian chocolate as "slavery-free" by 2005. (It is not clear what effect the country's civil war might have on that goal.)[7]

What should consumers make of all these issues? The next time you decide to indulge your taste for chocolate, you might explore the shelves for a label that promises three things. First, look for a high cocoa content. Generally, more cocoa means higher quality and—at least potentially—more farm income. Next, look for a "fair trade" brand or the mark of a similar socially responsible producer. And even though cocoa does not generally have a "shade-grown" certification, it is well worth looking for an organic brand. One of the most common cocoa pesticides in West Africa, for example, is lindane, an organochlorine cousin of DDT. Eliminating such chemicals would be a boon to both farmworkers and forests.[8]

—*Chris Bright*

Shrimp

Shrimp has long been on the menu of coast-dwelling humans. Tomb paintings from ancient Egypt depict scenes of fishermen pulling shrimp from the Nile. And for centuries Southeast Asian farmers have kept a few wild shrimp cordoned off in coastal ponds for easy picking.[1]

Today's multibillion-dollar shrimp industry bears little resemblance to shrimp harvests of old. For one thing, the small, leggy crustacean is no longer a delicacy enjoyed primarily by those living close to the source. Today, huge quantities of shrimp are produced in the developing world for consumption in Japan, the United States, and Western Europe. Shrimp production is no small matter: in 2001, over 4.2 million tons of shrimp swam into the global marketplace.[2]

China produces more shrimp than any other country, hauling in over 1.2 million tons in 2000, more than double its total from a decade before and over three times as much as each of its nearest competitors—India, Thailand, and Indonesia. But the bulk of China's harvest stays in that country. The distinction of being the top shrimp exporter goes to Thailand.[3]

In the late 1990s, the United States passed Japan as the primary customer in the shrimp trade, with yearly imports reaching 300,000 tons. In fact, by 2001 shrimp had displaced canned tuna as the top seafood choice on American dinner plates. The Japanese are still first in per capita consumption, however, despite a recent economic downturn that helped drive annual shrimp consumption below 3 kilograms per person.[4]

From modest beginnings a few decades ago, the shrimp industry has become one of the most lucrative fisheries in the world. The United States and Japan alone imported $7 billion worth of shrimp in 2000. Yet this industry is also among the most destructive. Roughly three quarters of the shrimp on the market is "wild captured"—mostly by fishing boats dragging huge conical nets (trawls) over estuaries, bays, and continental shelves. Trawlers scour the seabed in a manner likened to clearcutting—destroying habitat and scooping up whatever lies in the paths of the trawls. This method wreaks havoc on some of the most biologically productive spots within the marine ecosystem.[5]

Shrimp fishing as practiced today is not only destructive, it is incredibly wasteful as well. Turtles, fish, and other marine species swept up in the nets are considered unprofitable "bycatch" and are generally deposited—dead—back into the ocean. In temperate areas, the bycatch-to-shrimp ratio might be 5:1. In the tropics, the bycatch ratio reaches 10:1, and it runs even higher in some fisheries. All told, shrimping accounts for one third of the world's discarded catch, while producing less than 2 percent of global seafood.[6]

In the 1980s, technological innovations triggered a boom in shrimp aquaculture to supplement ocean captures. By 1989, shrimp farms had blossomed along tropical coastlines around the world and were churning out one quarter of the world's shrimp harvest. Since then, the market share for farmed shrimp has leveled off, its growth hampered in part by widespread disease outbreaks in densely stocked shrimp pens.[7]

Shrimp aquaculture has been no more ecologically benign than wild capture. A typical shrimp farm produces copious amounts of waste, some of it highly toxic. Chemicals and fertilizers used in the farms seep into local water sources and estuaries, while farmers dump much of the waste directly into the ocean.

Where shrimp farms are planted, native mangrove forests often are rooted out. Mangroves have many functions, serving as a breeding ground and home to many species (including providing nurseries for 85 percent of the tropics' commercial fish species), acting as water filters, and offering critical protection against shoreline erosion and violent tropical storms. Nearly one quarter of the world's remaining tropical mangrove forests were destroyed over the past two decades, in major part to make way for shrimp farms.[8]

A host of human rights abuses have accompanied the serious environmental degradations of shrimp farming, as powerful interests running the farms clash with local people harmed by the operations. Typically, domestic and foreign investors with few or no ties to local communities come in to establish the farms, in the process destroying vital resources, draining away livelihoods, and leaving locals destitute. Land seizures, violent intimidation of local fishers, even

murders are all too common.[9]

Indian physicist and environmental advocate Vandana Shiva once estimated that the average shrimp farm provided perhaps 15 jobs on the farm and 50 security jobs around the farm, while displacing 50,000 people through loss of land and loss of traditional fish and agriculture. One Filipino fisherman lamented: "The shrimp live better than we do. They have electricity, but we don't. The shrimp have clean water, but we don't. The shrimp have lots of food, but we are hungry."[10]

The shrimp industry has a long way to go before it could be considered remotely sustainable, and many advocacy groups suggest consumers simply avoid shrimp to ease the burden on both ecosystems and people. On a positive note, a consortium involving the World Bank, the U.N. Food and Agriculture Organization, and the World Wide Fund for Nature is exploring environmental certification standards for aquaculture. And the Sea Turtle Restoration Project and others are working with industry to develop and promote devices that drastically reduce bycatch. Meanwhile, grassroots environmental groups in farmed areas are teaming up with international nongovernmental groups to promote more ecologically sound shrimp farming. In one case, the Mangrove Action Project and the Small Fishers Federation of Sri Lanka bring fishing communities and other stakeholders together to promote conservation and work with shrimp farmers to curb mangrove destruction and protect fish habitat.[11]

—*Dave Tilford,*
Center for a New American Dream

Soda

Soda, with its decadent sweetness, has shown itself to have universal appeal. In 2002, people drank 185 billion liters of soda, making it the third most popular commercial beverage in the world, after tea and milk. Yet unlike these two, soda is a complex blend of ingredients, including water, sweeteners, carbon dioxide, dozens of natural and artificial flavorings, and frequently caffeine. Once these are mixed, bottlers neatly package the drink in attractive bottles or cans, market it, and distribute it for the enjoyment of people in just about every country in the world.[1]

Water is soda's main ingredient and is also central to processing its other ingredients and packaging materials. The average bottling plant churns out more than 300,000 liters of soft drinks each day—a process that demands up to 1.5 million liters of water, enough to meet the minimum requirements of at least 20,000 people. In some water-stressed areas, bottlers have actually come into conflict with local communities. In Plachimada, India, for example, the local authorities revoked a Coca-Cola bottling plant's license in April 2003 after residents complained about wells drying up, the diminished quality of the remaining water, and the release of toxic effluents. But after pressure from the Coca-Cola Company, one of the largest foreign investors in the Indian economy, the national government is considering repealing this decision. More recently, new troubles have sprouted for soda companies in India when scientists from the environmental group the Centre for Science and the Environment found pesticides in major soda brands throughout the country—a finding later confirmed by the government.[2]

Soda gets much of its taste, its texture, and all of its calories from a generous dose of sweeteners. The average can of non-diet soda, 355 milliliters, has 38 grams (or 150 calories) of added sweeteners. Along with promoting tooth decay, sugars displace healthier foods or, when consumed along with a regular diet, increase total caloric intake. Thus, added sugars can lead to nutrient deficiencies or obesity. This is of particular concern with children and adolescents, who are more susceptible to dietary deficiencies and whose dietary habits are especially malleable. In the United States, as annual soda consumption doubled to 185 liters per person between 1970 and 2001, milk consumption fell 30 percent. At the same time, total calcium intake by adolescents fell significantly, while overweight and obesity rates nearly tripled, to 14 percent (while reaching 61 percent in adults). A recent study showed that children who drink sugar-sweetened drinks are more often obese and that this risk increases another 60 percent with each additional beverage consumed.[3]

Caffeine is one of soda's other main ingredients—found in 80 percent of the global volume of the top 10 major carbonated soft drinks. While the soda industry claims it uses caffeine to enhance flavor, studies have shown that people cannot find significant differences between caffeinated and non-caffeinated samples. More likely, caffeine is added for its stimulant properties, which gives soda its extra kick as well as helping to produce loyal customers. Caffeine is physiologically addictive at just 100 milligrams per day—and at less for children. One can of Pepsi contains 41 milligrams of caffeine.[4]

While bottlers in some countries still rely partially on reusable glass bottles, more commonly they use plastic bottles or aluminum cans. In 2001, bottlers around the world filled some 159 billion plastic bottles, 112 billion cans, and 72 billion glass bottles. Once packaged, the sodas are shipped regionally by truck to stores, restaurants, schools, and vending machines. To make sure people buy these beverages, soda companies spend billions on advertising—on television, billboards, and the Internet, among other media. The Coca-Cola Company and PepsiCo., the two largest soft drink firms, are the thirteenth and twentieth largest advertisers in the world. Together they spent $2.4 billion on ads in 2001. Companies also work behind the scenes to ensure a ready supply of soda for all who suddenly have an impulse for something sweet. For example, bottlers in the United States often sign exclusive contracts with school boards, offering a share of the profits for selling a certain amount—a strategy that is being replicated around the world.[5]

The average can of soda, once opened, lasts perhaps 20 minutes before being tossed. In the United States, as often as not, it ends up in the garbage. If Americans had recycled the 32 billion soda cans they discarded in 2002, they would have saved 435 thousand tons of aluminum—enough to rebuild the world's entire commercial air fleet more than one and a half times. A coalition of U.S. environmental groups is currently working to create a new law that would set a national goal of 80-percent recovery of beverage containers and would allow the industry to design its own system to achieve this. This strategy has been a success in Sweden, where a national goal has maintained a recovery rate of 86 percent, driven primarily by an industry-imposed bottle deposit of 10¢. Michigan, the one state in the United States with a 10¢ bottle deposit, has a 95-percent recovery rate.[6]

While environmentalists are hoping to reduce container waste, nutritionists and government officials seek to moderate overall soda consumption, largely to combat the growing obesity epidemic in children. California, for instance, passed a law that will phase out the sale of all junk food (including soda) in its public elementary schools by the start of 2004. California also currently taxes junk food, which helps to reduce overall consumption while also serving as a potential source of revenue for more health education. Some countries, such as Sweden and Poland, have gone as far as banning commercials on children's television, recognizing that audience's susceptibility to marketing messages. Nevertheless, in 2002 soda sales grew 2.1 percent globally. Industry experts predict that soda will pass milk as the world's second most consumed beverage within the next five years.[7]

—Erik Assadourian

CHAPTER 5

Moving Toward a Less Consumptive Economy

Michael Renner

In 1895, traveling salesman King Camp Gillette came up with the idea of disposable razor blades—a product consumers would have to keep coming back for again and again. Sales soon soared, reaching more than 70 million by 1915, and today Gillette is a company with $10 billion annual turnover. What started out as one businessman's high-profit vehicle for ensuring an endless stream of sales became a widely embraced concept of great endurance—planned obsolescence.[1]

Fast-forward to the present: in mid-2003, the Walt Disney company announced that it would soon test-market a new DVD that is intended to replace rental video discs and cassettes and that stops working after a short, pre-set time. Opening the DVD's airtight package kicks off a chemical countdown that renders the disc unusable after a mere 48 hours. The sophisticated technologies involved may be strictly from the twenty-first century, but the underlying philosophy hews to that time-honored concept pioneered by Gillette and his contemporaries.[2]

Consumption as a Way of Life

The technological advances of the past century or so have made it possible "to produce more than was demanded and to offer more than was needed," as journalist Edward Rothstein pointed out recently in the *New York Times*. Endless economic growth driven by unbridled consumption has been elevated to the status of a modern religion. This is as much an objective of corporate executives wanting to keep shareholders happy as it is a goal of political leaders with an eye on winning the next election.[3]

Brushing aside doubts about whether material possessions and human happiness travel along a common trajectory (see Chapter 8), some observers argue that the mass-production, mass-consumption, mass-disposal system is no less than sheer economic necessity. In 1950, for instance, U.S. marketing analyst Victor Lebow wrote that "Our enormously productive economy…demands that we make consumption a way of life…. We need things consumed, burned up, worn out, replaced, and discarded at an ever-increasing rate."[4]

But the compulsive worship at the altar of consumption has brought humanity right to the edge of an environmental abyss—depleting resources, spreading dangerous pollutants, undermining ecosystems, and threatening to unhinge the planet's climate balance. Stepping back from this precipice will require a major reduction in human claims on Earth's resources.

Humanity's deep divides complicate this task. Even as evidence grows that the global consumer class of about 1.7 billion people (see Chapter 1) will need to rein in its voracious material appetite, an equally large number of people in an emerging global middle class are striving to emulate the perceived "good life." And close to 3 billion people—the global poor—struggle to survive on just a few dollars a day.[5]

It is often said these days that the planet cannot bear the burden of everyone in the developing world owning as many cars, refrigerators, and other consumer goods as Americans, Europeans, or Japanese on average do. From the standpoint of global justice and equality, however, the solution cannot be a system of consumer apartheid that upholds western binge habits but denies the poor a decent standard of living. Instead, the rich need to curb their outsized material appetites. Rough calculations suggest that in order to accommodate the twin imperatives of environmental protection and social equity, the rich nations may need to cut their use of materials by as much as 90 percent over the next few decades.[6]

At the moment, the world is hurtling in the opposite direction. Modern economies are capable of producing huge quantities of goods at very low cost. This leads both producers and consumers to regard more and more products as little more than commodities that can be discarded relatively quickly rather than items that embody valuable energy and materials and that should be well maintained and designed for long life spans.

Cheap raw materials, many of which originate in developing countries, sustain the consumer cornucopia. The overall quantities of raw materials traded internationally are expanding sharply, but commodity prices have been on a downward trajectory since the mid-1970s, part of an ongoing slide that reaches back to the beginning of the twentieth century. The massive extraction of fuels, minerals, and timber ravages ecosystems in developing countries, triggers social upheaval, and in some cases has led to devastating resource wars, even as most people in the areas affected derive little benefit.[7]

> **Cheap raw materials, many of which originate in developing countries, sustain the consumer cornucopia.**

Although the old industrial nations remain major producers, growing volumes of merchandise are made in poor countries. Particularly in labor-intensive industries like textiles and apparel, multinational corporations are continually searching for cheaper labor, and many developing countries seek to compete against one another by keeping workers' wages low. China has emerged as a major producer of cheap, primarily "low-end" consumer goods, exported mainly to the North American market. Its trade surplus with the United States skyrocketed from a little more than $10 billion in 1990 to $103 billion in 2002. Even Mexico, long a center of low-cost factories, finds itself increasingly unable to compete because Chinese wages on average are just a quarter of those prevalent in Mexico. Just since 2001, one seventh of Mexico's *maquiladora* export plants have shut down.[8]

Opinions diverge widely on whether such export strategies and the pursuit of liberalized trade more generally can bring about meaningful development. But whether low wages are seen as an inevitable ingredient of a successful export strategy or as a symbol of exploitation and an impediment to vibrant domestic markets, it seems clear that they sustain consumerism.

The global consumer class is obviously key to reshaping the relationship between consumption and sustainability—not only because it claims the bulk of the world's resources, but also because its actions echo around the world. Yet solutions need to take into account the ways in which developing countries are tied into the global economy and their desire to emulate the materials-intensive model that is still widely perceived as embodying "the good life." It is critical to devise ways to reduce the environmental burden associated with consumption, particularly so that an increase in consumption levels in poorer countries is fully compatible with the goal of sustainability.

To support the move toward a less consumptive economy, consumers and producers will need to pay close attention to the full lifecycle of products. This means they need to concern themselves not just with the characteristics of the product itself, such as how much energy its use may require, but also with the materials and production methods used to manufacture the product and the kinds and types of wastes generated in the process. In addition, both consumers and producers need to consider how effectively goods actually deliver wanted services and comforts, how long products last, and what happens to them once they reach the end of their useful life.

A range of tools is potentially at the disposal of governments, companies, and individual consumers to make progress toward the overall goal of a less consumptive econ-

omy. Many are not only under discussion, but are beginning to be implemented. To make a difference, however, these efforts will need to be scaled up considerably, and political and structural barriers to change must be struck down.

Government's Toolbox

Governments can take a number of steps to facilitate the transition to a less consumptive economy. Prominent among the needed measures are recalibrating tax and subsidy policies, pursuing pro-environment procurement rules, and establishing appropriate product standards and labeling programs.

Numerous subsidies allow the prices of fuels, timber, metals, and minerals (and products incorporating these commodities) to be far lower than they otherwise would be, encouraging greater consumption. Limits in data availability prevent a complete accounting of subsidies for environmentally harmful activities, and underlying methodologies and definitions may differ from study to study. But a recent report by the Organisation for Economic Co-operation and Development (OECD) estimates that global subsidies amount to something like $1 trillion a year, with OECD member states accounting for three quarters of the total.[9]

A study by researchers Norman Myers and Jennifer Kent puts perverse subsidies in six sectors—agriculture, energy, road transportation, water, fisheries, and forestry—at about $850 billion or more annually. In addition, there are about $1.1 trillion worth of quantifiable environmental "externalities." (See Table 5–1.) Although these are not subsidies in a formal sense, they do represent uncompensated costs that have to be borne by society at large and that, like subsidies, have distorting and detrimental impacts. For instance, the environmental and health costs

Table 5–1. Estimates of Global Environmentally Harmful Subsidies and Externalities

Sector	Subsidies	Quantifiable Externalities	Total
		(billion dollars)	
Agriculture	260	250	510
Fossil Fuels, Nuclear Energy	100	200	300
Road Transportation	400	380	780
Water	50	180	230
Fisheries	25	n.a.	25
Forestry	14	78	92
Total	849	1,088	1,937

SOURCE: See endnote 10.

associated with car use are not charged to motorists, which makes individual automobile travel cheap in comparison with rail and other modes of transportation. Subsidies and uncompensated externalities combined are equivalent to 5–6 percent of the global economy—about the same size as the German economy.[10]

Phasing out destructive subsidies and shifting even a portion of those funds to renewable energy, efficiency technologies, clean production methods, and public transit would give the transition toward sustainability a powerful boost.

Ecological tax reform is another key measure. The intent is to make market prices reflect the full environmental costs of eco-nomic activities more adequately. Carbon taxes, levies on the use of nonrenewable energy and virgin materials, landfill fees, and other waste and pollution charges provide an incentive for manufacturers to move away from heavy fossil fuel use, to boost energy and materials productivity, and to curtail the generation of wastes and emissions. Rather than merely imposing a new tax, though, the concept most discussed is tax shifting. Current tax systems make natural resource use far too cheap and render labor too expensive. Eco-tax revenues would be used to lighten the tax burden now falling on labor, which would encourage job creation.[11]

Discussed theoretically since the late 1970s, ecological tax shifting started to become a reality in the 1990s in a growing number of European countries, with most of the momentum occurring between 1990 and 1994. Denmark, Germany, Italy, the Netherlands, Norway, Sweden, and the United Kingdom introduced reforms linking a variety of green taxes to reductions in social security contributions. Before adjustment for inflation, environmental tax revenues in the European Union (EU) more than quadrupled between 1980 and 2001, to 238 billion euros. (See Table 5–2.) The bulk of these revenues are derived from taxes on gasoline, diesel, and motor vehicles.[12]

Notwithstanding these totals, eco-tax shifts to date remain relatively limited. Environmental taxes in OECD countries account on

Table 5–2. Environmental Tax Revenue, European Union, Selected Years

Environmental Taxes	1980	1990	2001
		(billion euros)	
Revenues	54.6	130.4	237.7
		(percent)	
Revenues as a Share of All Taxes and Social Contributions	5.8	6.2	6.5
Revenues as a Share of Gross Domestic Product	2.2	2.5	2.7

SOURCE: See endnote 12.

average for only 6–7 percent of all tax revenues. Payroll taxes and mandatory social security contributions, in contrast, weigh in at 25 percent. (In the European Union, with its extensive social programs, labor's tax burden is far higher—between 45 and 47 percent in the late 1990s.)[13]

This is not to say that nothing has been accomplished. In Germany, for instance, an eco-tax levied on different forms of energy consumption was first introduced in 1999, with four subsequent annual increases in the tax. By 2002, it had already helped avoid emissions of more than 7 million tons of carbon dioxide (CO_2). Annual revenues have grown from about $4 billion in 1999 to an estimated $19 billion in 2003. Reductions in social security contributions made possible by these funds helped create 60,000 additional jobs by 2002, a number expected to rise to 250,000 by 2010.[14]

All modern economies can be leaner without being doomed.

Unfortunately, eco-taxes are frequently weakened by a variety of loopholes—granting exemptions to certain industries or energy sources, applying reduced tax rates to energy-intensive firms, or making companies eligible for partial reimbursements. Often this is done in the name of preserving the competitiveness of domestic industries on the world market. Such arguments will lose traction if national policies can be harmonized. (See Chapter 7.) That is what the European Union is attempting to do with an energy taxation directive that is to take effect in 2004. So far, however, the deliberations that were begun in 1997 have yielded a disappointing compromise text that waters down the initial draft.[15]

In Germany, coal and jet fuels are not sub-

ject at all to the eco-tax. Companies in the mining and manufacturing sectors, utilities, construction firms, and agricultural enterprises were assessed at only 20 percent of the nominal tax rate levied on natural gas, heating oil, and electricity. At the beginning of 2003, however, that preferential rate rose to 60 percent, and the government said that industry may soon have to pay the full rate if it does not meet a voluntary 2010 goal of reducing CO_2 emissions by 35 percent.[16]

To grow into a major tool for sustainability, the scope of eco-tax reform needs to become far broader and loopholes need to be closed. This will require winning difficult political battles against those interested in maintaining the status quo. The challenge is illustrated by the German experience, where opposition politicians and parts of the media launched an intense campaign to discredit eco-taxes. Having risen rapidly to broad acceptance by all political parties and the general public in the 1990s, eco-tax reform suffered an equally fast decline in popularity once it was implemented.[17]

Another important tool that governments can wield is procurement, as described at length in Chapter 6. From the federal to the local level, government authorities in industrial countries spend trillions of dollars on public purchases every year. By buying environmentally preferable products, they can exert a powerful influence on how products are designed, how efficiently they function, how long they last, and whether they are handled responsibly at the end of their useful lives. Well-designed purchasing rules can drive technological innovation and help establish green markets.[18]

Governments further influence product development through regulatory tools. National standards have been adopted in a growing number of countries to save energy and water, for instance. By 2000, 43 countries had household appliance efficiency programs in place—seven times as many as in 1980. Most of these were in Europe and Asia.[19]

Standards, which "push" the market by requiring manufacturers to meet minimum standards, are well complemented by ecolabeling programs, which "pull" the market by providing consumers with the requisite information to make responsible purchasing decisions, and hence encourage manufacturers to design and market more eco-friendly products.[20]

Labeling schemes have been developed for a wide range of products, including appliances, electricity, wood products, and agricultural goods such as coffee and bananas. Some focus on a single product or product class, whereas others evaluate a broad range of items. The first, and most comprehensive, labeling program—Germany's Blue Angel—just celebrated its twenty-fifth anniversary. The number of products covered grew from about 100 in 1981 to 3,800 today. Both governmental and private labeling programs have mushroomed in recent years.[21]

In fact, there are competing labels in some product areas, which can confuse consumers and may even thwart eco-friendly consumption choices. Some programs, particularly industry-sponsored ones, may make vague or unsubstantiated claims concerning recycled content of a product, organic food-growing methods, biodegradability, and other issues. Others may be based on relatively low performance standards. Concerned about these problems, a recent OECD report argued: "To avoid a general discredit of labeling schemes, some kind of regulatory instruments may be needed to signal to consumers that certain schemes are more appropriate for certain issues than others." Such regulation may take the form of certification schemes. A range of qualified certification bodies (government agencies or accredited private groups) can evaluate whether a product conforms to existing standards or verify the accuracy of environmental claims made by manufacturers.[22]

All the tools discussed here—subsidy phaseouts, tax shifts, green procurement, product standards, and labeling programs—will need to be expanded dramatically to put consumption on a sustainable footing. But doing so is often an uphill battle. The failure of the international community to agree on reductions in agricultural subsidies during trade talks in September 2003 in Mexico demonstrated all too vividly the entrenched nature of vested interests.

Lean and Clean

Industrial economies mobilize enormous quantities of fuels, metals, minerals, construction materials, and forestry and agricultural raw materials. A study for the European Union found that in 1997, the per capita material throughput amounted to about 80 tons per American, 51 tons per citizen of the EU, and 45 tons per Japanese. Using different methodologies, a study by the World Resources Institute arrived at similar figures, though it pegged Japanese material flows at just 21 tons per person.[23]

None of today's industrial economies are truly sustainable. All modern economies can be leaner without being doomed. Given broadly comparable living standards, if Europeans can live on roughly half the material throughput mobilized on behalf of Americans (and the Japanese survive on even less), there is significant room for improvement in the United States—the paragon of consumption that much of the rest of the world strives to emulate.

In fact, most material flows in industrial economies serve no useful purpose whatsoever and never actually pass through the hands of any consumer. So-called hidden flows account for slightly more than 60 percent of total materials flow in the EU—a portion that has remained more or less unchanged in the last two decades. In the United States, hidden flows account for more than 70 percent, and in Japan, for just under half. (See Figure 5–1.) These hidden flows include waste materials from mining and other industries, overburden (earth removed by mining firms to reach desired ores), dredging materials, carbon dioxide and other emissions and pollutants, and soil loss from farmland erosion. The term "hidden flows" is apt as they are for the most part invisible to consumers. This is particularly the case with the rising quantities of wastes associated with the extraction from developing countries of natural resources that are then imported by industrial nations.[24]

Dealing with the hidden flows requires that some of the most destructive activities—mining, smelting, and logging, in particular—be downsized. This can be accomplished by improving energy and materials efficiency, boosting recycling and reuse, and lengthening the lifetime of products, so that there is far less need to extract virgin raw materials. But there is also ample space for reducing the environmental impact of the goods and services actually delivered to consumers. Dematerialization, clean production, and "zero-waste" closed-loop systems are some of the key concepts behind a new approach.

A range of studies and assessments have affirmed the potential of a "dematerialization" strategy—a concept pioneered by Rocky Mountain Institute co-founders Amory and Hunter Lovins, eco-entrepreneur Paul Hawken, and German researcher and politician Ernst Ulrich von Weizsäcker. It aims to reduce the amount of raw materials needed to create a product by, for example, making

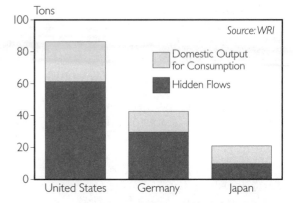

Figure 5–1. Material Requirements Per Person in the United States, Germany, and Japan, 1996

paper thinner and vehicles lighter and to cut the amount of energy needed to operate products—from light bulbs to washing machines and automobiles. Specifically, the advocates of dematerialization have pushed for Factor 10—policies that aim at providing a given volume of goods and services with one tenth as much material input.[25]

While the technological potential for dematerialization is far from exhausted, there has already been some de-coupling between economic growth and material throughput. In the European Union, for instance, resource productivity (measured as gross domestic product per total material requirements) improved 39 percent between 1980 and 1997. But this achievement has not translated into a lower overall claim on resources: in Western Europe, North America, and Japan, total resource consumption has remained fairly constant and at unsustainably high levels.[26]

Why is that? While per-unit materials consumption has declined, consumer tastes and wants keep spiraling upward: cars and houses are getting larger and fancier, vacation travel has increasingly shifted toward long-distance air travel, diets have become more meat-based, and there is a steady flow of ever-new gadgets and related "accessories." And industrial economies experienced an unexpected rebound effect: lower per-unit energy or materials requirements translate into lower consumer costs, which in turn encourage increased usage. Efficiency gains have repeatedly been canceled out or even overwhelmed. For instance, greater automobile fuel efficiency means that motorists can drive longer distances for the same cost. And ever-expanding numbers of automobiles mean that the auto industry's claim on fuel and materials like aluminum, copper, steel, and plastic keeps rising. As important as dematerialization is, therefore, it alone appears insufficient in the face of the consumption juggernaut.[27]

The toxic products of our material society are another pressing concern. Advocates of "clean production" say that there are plenty of opportunities to reduce and perhaps eliminate the reliance on toxic materials in manufacturing, to prevent air and water pollution, and to avoid hazardous waste generation.[28]

A pulp and paper mill on the Androscoggin River in Jay, Maine, presents an inspiring example. At the beginning of the 1990s, the mill, owned by industry giant International Paper, was a major polluter and had an antagonistic relationship with its work force and the local community. A management shakeup led to active cooperation with local stakeholders, and state and community environmental regulations proved major drivers of change. An initial focus on "end-of-pipe" pollution control quickly gave way to pollution prevention measures. The mill dramatically reduced releases of organic pollutants and mercury; eliminated dioxin, furan, and chloroform emissions; and cut particulate emissions in half. It reduced hazardous waste generation from 6 million pounds in 1990 to 300,000 pounds in 1998, and slashed the amount of solid waste going to landfills by 91 percent.[29]

More ambitiously, the mill managers made an effort to move away from industrial orthodoxy in which, in "cradle-to-grave" fashion, raw materials are extracted and processed, and the substances not directly useful to a factory become unwanted waste. An alternative "cradle-to-cradle" system seeks to build integrated, closed-loop systems, in which the byproducts of one factory become the feedstock of another, instead of becoming environmental time bombs. (See Box 5–1.) Some mill byproducts, including ash from sludge and bark incineration and carbon dioxide from a lime kiln, are being used by other local industries. In fact, several companies decided to locate near the mill to take advantage of its byproducts.[30]

In Western Europe, North America, and Japan, total resource consumption has remained at unsustainably high levels.

Environmentalists widely regard the community of Kalundborg in Denmark as a trailblazer of industrial ecology. An increasingly dense web of symbiotic relationships among a number of local companies there has slowly been woven over the past three decades, yielding both economic and environmental gains. For instance, natural gas previously flared off by Denmark's largest refinery is being used as feedstock in a plasterboard factory, desulfurized fly-ash from a coal-fired power plant (also the country's largest) goes to a cement manufacturer, and sludge containing nitrogen and phosphorus from a pharmaceutical plant is used as fertilizer by nearby farms.[31]

Instead of a master plan, the present network in Kalundborg actually evolved both slowly and spontaneously from a series of bilateral agreements, all of which were concluded in the first place because they were economically attractive. This experience presents a real-life alternative to industrial ortho-

BOX 5–1. THE CRADLE-TO-CRADLE ALTERNATIVE

Imagine a world in which all the things we make, use, and consume provide nutrition for nature and industry—a world in which growth is good and human activity generates a delightful, restorative ecological footprint.

While this may seem like heresy to many in the world of sustainable development, the destructive qualities of today's cradle-to-grave industrial system can be seen as the result of a fundamental design problem, not the inevitable outcome of consumption and economic activity. Indeed, good design—principled design based on the laws of nature—can transform the making and consumption of things into a regenerative force.

This new conception of design—known as cradle-to-cradle design—goes beyond retro-fitting industrial systems to reduce their harm. Conventional approaches to sustainability often make the efficient use of energy and materials their ultimate goal. While this can be a useful transitional strategy, it tends to reduce negative impacts without transforming harmful activity. Recycling carpet, for example, might reduce consumption, but if the attached carpet backing contains PVC, which most carpet backing does, the recycled product is still on a one-way trip to the landfill, where it becomes hazardous waste.

Cradle-to-cradle design, on the other hand, offers a framework in which the effective, regenerative cycles of nature provide models for wholly positive human designs. Within this framework we can create economies that purify air, land, and water, that rely on current solar income and generate no toxic waste, that use safe, healthful materials that replenish the earth or can be perpetually recycled, and that yield benefits that enhance all life.

Over the past decade, the cradle-to-cradle framework has evolved steadily from theory to practice. In the world of industry it is creating a new conception of materials and material flows. Just as in the natural world, in which one organism's "waste" cycles through an ecosystem to provide nourishment for other living things, cradle-to-cradle materials circulate in closed-loop cycles, providing nutrients for nature or industry. This model recognizes two metabolisms within which materials flow as healthy nutrients.

First, nature's nutrient cycles constitute the biological metabolism. Materials designed to flow optimally in the biological metabolism are biological nutrients. Products conceived as these nutrients, such as biodegradable packaging, are designed to be used and safely returned to the environment to nourish living systems. Second, the technical metabolism, designed to mirror the earth's cradle-to-cradle cycles, is a closed-loop system in which valuable, high-tech synthetics and mineral resources—technical nutrients—circulate in a perpetual cycle of production, recovery, and remanufacture. Ideally, all the human systems that make up the technical metabolism are powered by the renewable energy of the sun.

Biological and technical nutrients have already entered the marketplace. The upholstery fabric Climatex Lifecycle is a blend of pesticide-residue-free wool and organically grown ramie, dyed and processed entirely with nontoxic chemicals. All of its product and process inputs were defined and selected for their human and ecological safety within the biological metabolism. The result: the fabric trimmings are made into felt and used by garden clubs as mulch for growing fruits and vegetables, returning the textile's biological nutrients to the soil.

Honeywell, meanwhile, is marketing a textile for the technical metabolism, a high-quality carpet yarn called Zeftron Savant, which is made of perpetually recyclable nylon 6 fiber. Zeftron Savant is designed to be reclaimed and repolymerized—taken back to its constituent

BOX 5–1. (continued)

resins—to become new material for new carpets. In fact, Honeywell can retrieve old, conventional nylon 6 and transform it into Zeftron Savant, which is in effect "upcycling" rather than downcycling an industrial material. The nylon is rematerialized, not dematerialized—a true cradle-to-cradle product.

In the commercial carpet industry, material recovery systems are providing a model for the development of technical metabolisms. Shaw Industries, for example, has developed a technical nutrient carpet tile for its commercial customers. The company guarantees that all of its nylon 6 carpet fiber will be taken back and returned to nylon 6 carpet fiber, and its safe polyolefin backing returned to safe polyolefin backing. Raw material to raw material. A cradle-to-cradle cycle.

Shaw's technical nutrient carpet tile is conceived to be a product of service, a key element of the cradle-to-cradle strategy. Products of service are durable goods, such as carpets and washing machines, designed by their manufacturer to be taken back and used again. The product provides a service to the customer while the manufacturer maintains ownership of the product's material assets. At the end of a defined period of use, the manufacturer takes back the product and reuses its materials in another high-quality product. Widely practiced, the product-of-service concept can change the nature of consumption as human systems powered by renewable energy reuse valuable materials through many product lifecycles.

On a large scale, cradle-to-cradle thinking can transform the nature of economies. In Chicago, for example, these principles are serving as a reference point as Mayor Richard Daley strives to make the city the greenest in America, a hub of energy effectiveness and beneficial material flows.

In a cradle-to-cradle economy, cities are the principal home and source of technical nutrition—the place where metals are forged, polymers synthesized, and tractors, computers, and windmills designed and manufactured. Cities send these materials forth into the world and receive them back as they move through closed-loop cycles. The countryside, meanwhile, can be seen as the home of the biological metabolism. Materials generated there—food, wood, fibers—are created through interactions of solar energy, soil, and water and are the source of biological nutrition for rural communities and nearby cities. One of the city's fundamental roles in this metabolism is to return biological nutrition in a safe, healthy form, say as clean fertilizer, back to the rural soil. These flows of nutrients and energy are the twin metabolisms of the living city, the engines of the vibrant economies of the future.

Even nations as vast and influential as China have adopted cradle-to-cradle strategies. Building on a 4,000-year-old tradition of sustainable agriculture, Vice Minister of Science and Technology Deng Nan announced in September 2002 that China will begin to develop industries and products based on cradle-to-cradle principles through the China–U.S. Center for Sustainable Development. China is already developing a cradle-to-cradle village as well as solar and wind power enterprises.

The cradle-to-cradle strategy allows us to see our designs as delightful expressions of creativity, as life-support systems in harmony with energy flows, human souls, and other living things. When that becomes the hallmark of productive economies, consumption itself will have been transformed.

—William McDonough and Michael Braungart,
McDonough Braungart Design Chemistry

SOURCE: See endnote 30.

doxy. But replicating this model may not be all that easy. Setting up a zero-waste industrial symbiosis takes considerable time. And it may be more workable to construct such reciprocal webs piece by piece rather than drawing up overly ambitious plans from the outset. Still, the notion of clean production is attracting growing interest around the world. Among other places, companies are working toward this goal in China, Fiji, India, Japan, Namibia, the Philippines, Puerto Rico, and Thailand.[32]

Take It Back!

It is much more likely that resource consumption will be minimized and the generation of wastes and emissions avoided if manufacturers factor environmental considerations in from the very beginning when they design products, develop production technologies, and select materials. To encourage companies to move in this direction, a growing number of governments are adopting "extended producer responsibility" (EPR) laws that require companies to take back products at the end of their useful life. These typically ban the landfilling and incineration of most products, establish minimum reuse and recycling requirements, specify whether producers are to be individually or collectively responsible for returned products, and stipulate whether producers may charge a fee when they take back products.[33]

The goal of EPR is to induce manufacturers to assess the full lifecycle impacts of their products. Ideally, they will then eliminate unnecessary parts, forgo unneeded packaging, and design products that can easily be disassembled, recycled, remanufactured, or reused.[34]

Part of the challenge is to develop materials that can easily be reused or otherwise will not linger in a landfill for centuries. For instance, German chemical giant BASF invented a new material made from an infinitely recyclable nylon-6 fiber; it can be taken back to its constituent resins and made into new products. The Swiss textile firm Rohner and the textile design company DesignTex jointly developed an upholstery fabric that, once it has been removed from a chair at the end of its useful life, will naturally decompose.[35]

The EPR philosophy had its beginnings in Germany's Packaging Ordinance of 1991. Holding producers responsible for taking back and managing packaging waste, the law triggered steady reductions in packaging materials. More important, it is widely credited with motivating many other governments in Europe, Asia, and Latin America to embrace this concept. Since then, the EPR approach has spread far beyond packaging to encompass a growing range of products and industries, including consumer electronics and electric appliances, office machinery, cars, tires, furniture, paper goods, batteries, and construction materials. (See Table 5–3.)[36]

Europe remains at the center of the EPR movement. Many European governments have passed EPR legislation, and the European Union has promulgated directives for packaging, electronics, batteries, and automobiles in an effort to harmonize the sometimes divergent national efforts.[37]

Driven by concern over rapidly accumulating electrical and electronics waste from computers, cell phones, and similar equipment, the EU adopted an Electronic and Electrical Equipment Directive in February 2003. Member states are to enact implementing national legislation by August 2004 (they are free to impose more-restrictive policies), and equipment producers need to have systems in place to take back and manage electrical and electronic waste free of charge to consumers by August 2005. For products

Table 5–3. Extended Producer Responsibility Laws, Selected Industries[1]

Product Area or Industry	Countries with EPR Laws
Packaging	More than 30 countries, including Brazil, China, Czech Republic, Germany, Hungary, Japan, Netherlands, Peru, Poland, South Korea, Sweden, Taiwan, and Uruguay (beverage containers only)
Electric and Electronic Equipment	Currently more than a dozen countries, including Belgium, Brazil, China, Denmark, Germany (voluntary only), Italy, Japan, Netherlands, Norway, Portugal, South Korea, Sweden, Switzerland, and Taiwan
Vehicles	Brazil, Denmark, France, Germany, Japan, Netherlands, Sweden, and Taiwan
Tires	Brazil, Finland, South Korea, Sweden, and Taiwan; Uruguay is considering voluntary measures
Batteries	At least 15 countries, including Austria, Brazil, Germany, Japan, Netherlands, Norway, and Taiwan; Uruguay is considering voluntary measures

[1] EU Directives have been promulgated in all the sectors covered in the table except tires. In addition to national rules adopted by a number of EU members independently, these directives are binding on all current 15 members of the EU (and will be binding for 10 East European countries that are to become EU members).
SOURCE: See endnote 36.

marketed prior to August 2005, the costs are to be shared by all producers according to their market share; for items sold later, producers have individual responsibility. (While individual responsibility provides an incentive to undertake environment-friendly design changes that reduce company compliance costs, there is also a danger that separate take-back systems may bring a duplication of efforts and possibly higher costs.)[38]

A companion directive on Restrictions on Hazardous Substances requires that manufacturers of electronic and electrical equipment no longer use lead, mercury, cadmium, hexavalent chromium, and the brominated flame retardants PBDE and PBB in products sold after 1 July 2006. There is growing concern worldwide about these hazardous materials; Japan is the leader in eliminating such substances from electrical and electronic products, and China has announced that it will model its policy on the EU Directive.[39]

The United States is lagging behind on producer responsibility. Industry opposition

has prevented federal take-back legislation. It is tempting to assume that U.S. companies operating worldwide will eventually decide that, having to meet EPR requirements in Europe and elsewhere, they might as well embrace such policies in the United States as well. IBM began offering product take-back programs as early as 1989 in Europe, and then initiated a more-restricted program in the United States in 1997. But IBM may be an exception; the evidence to date provides limited hope for such a development.[40]

A number of state and local governments (including Florida, Maine, Massachusetts, Minnesota, South Carolina, and Wisconsin) have expressed interest in European-style take-back laws. If a patchwork of local regulations were to arise, it is conceivable that companies may decide that national (though perhaps still voluntary) rules are preferable. That is exactly what happened with regard to nickel-cadmium batteries—and the industry launched a nationwide take-back and recycling initiative in 1995.[41]

Some companies are launching voluntary take-back efforts to avert mandatory programs. Under increasing pressure from regulators and grassroots groups to deal with electronics waste, major computer manufacturers like Dell, Hewlett Packard, and IBM have set up voluntary programs. Return rates tend to be low, however, because consumers are typically charged $20–30 when they bring a product back.[42]

Other companies and industries see EPR as an opportunity to lower production costs or to marshal goodwill with environmentally conscious consumers. Carpet manufacturers and some of their major suppliers, for instance, see take-back as a vehicle to gain competitive advantage and accordingly have initiated a variety of programs for reusing and recycling used carpet materials. Kodak started a take-back program for disposable cameras in 1990 (but a quarter of the cameras are believed to still end up in landfills). Nike set up a "Reuse-a-Shoe" program to "downcycle" old sneakers. The outsole rubber and midsole foam from used shoes are converted into surface material for running tracks and other athletic facilities and playgrounds. The fabric from the shoe uppers becomes padding under carpets.[43]

Still, progress is limited in the United States. And despite substantial strides in Europe, a number of technical and political challenges remain. Plastics recycling has proved resistant to easy solutions, as have certain packaging materials that consist of a complex amalgam of layers of different materials. Industry opposition is far from defeated: In Germany, the retail industry is undermining an ambitious attempt to require the return of all beverage bottles and cans and to discourage the use of throwaways. Finally, the rapid pace at which many electronic devices, such as cell phones, palm pilots, and computers, become obsolete is a tremendous challenge:

it is difficult to set up workable collection systems when the turnover pace is so fast and the volume of materials mounts rapidly.[44]

Rethinking Products and Services

Today's industrial economies are able to churn out large quantities of goods with considerable ease and at such low cost that there is a great incentive to regard most merchandise as throwaways, intended to fall apart easily, rather than designing and manufacturing for durability. Many consumer products are made in such a way as to discourage repair and replacement of parts, and sometimes even render this plain impossible. And even when something can be repaired, the cost is often too high relative to a new item. That is because the cost of discarding the embedded valuable materials and labor is not accounted for, and the value of materials embedded in new products is not fully reflected in the purchase price.

Durability, repairability, and "upgradability" are essential to lessen the environmental impact of consumption. For easy refurbishing and upgrading (so that durability does not translate into a technological dead-end that prevents the introduction of more efficient designs), a "modular" approach permits access to individual parts and components, which allows them to be replaced easily. Companies like Xerox (in its copiers and printers) and Nortel (in telecommunications) have adopted this philosophy. By working to extend and deepen useful product life, companies can squeeze vastly better performance out of the resources embodied in products rather than selling the largest possible quantity. Although EPR laws do not as such address the issue of product longevity, they can be an incentive for companies to move in this direction.[45]

When goods do not wear out rapidly, they

need not be replaced as frequently. An obvious implication is that fewer goods will be produced, and that might mean that companies have less business. But there will be greater opportunity and incentive to maintain, repair, upgrade, recycle, reuse, and remanufacture products, and thus greater business and job potential throughout the life of a product.

Already, recycling and remanufacturing have become substantial industries. The Bureau of International Recycling in Brussels estimates that in at least 50 countries the recycling industry processes more than 600 million tons annually. With an annual turnover of $160 billion, the industry employs more than 1.5 million people. Not only does recycling keep materials out of landfills and incinerators, it provides substantial energy savings by replacing new raw materials extraction and processing with secondary materials. (See Table 5–4.)[46]

Remanufacturing is also becoming a serious business, particularly in areas like motor vehicle components. Remanufacturing operations worldwide save at least 11 million barrels of oil each year—an amount of electricity equal to that generated by five nuclear power plants—and a volume of raw materials that would fill 155,000 railroad cars annually. In the United States, remanufacturing is a $50-billion-plus annual business and employs close to half a million people directly in 73,000 different firms; this is roughly equal to employment in the entire U.S. consumer "durables" industry. According to Walter Stahel of the Product-Life Institute in Geneva, the remanufacturing sector in EU member countries accounts for about 4 percent of the region's gross domestic product.[47]

Xerox is one of the pioneers of this concept, having embarked on an Asset Recycle Management initiative in 1990. While the company had previously done some reman-

Table 5–4. Energy Savings Gained by Switching from Primary Production to Secondary Materials

Material	Savings
	(percent)
Aluminum	95
Copper	85
Plastics	80
Steel	74
Lead	65
Paper	64

SOURCE: See endnote 46.

ufacturing, this program led Xerox to design its products from the very beginning with remanufacturing in mind and to make every part reusable or recyclable. As a result, 70–90 percent of the equipment (measured by weight) returned to Xerox at the end of its life can be rebuilt. Like some of its competitors, Xerox also remanufactures spent cartridges for copy machines and printers; in 2001, it rebuilt or recycled about 90 percent of the 7 million cartridges and toner containers returned to it by consumers. All in all, the company estimates that environment-friendly design has kept at least half a million tons of electronic waste out of landfills between 1991 and 2001.[48]

Extended producer responsibility, remanufacturing, and related concepts logically lead to a whole new way of thinking about products, the way an economy functions, and what it is supposed to accomplish. Instead of merely selling goods—as many as possible, with little thought to what happens after a sale is made—manufacturers move on to provide a desired service. In the future, consumers may lease or rent products rather than buying them outright. By retaining ownership, manufacturers also remain responsible for proper upkeep and repair, take the necessary steps to extend product life, and ultimately recover the item's components and materials for recycling, reuse, or remanufacturing. Manufacturers may work directly with their customers or may rely on retailers. But the emphasis would be on "quality retail," advising consumers on the best lease options available and on quality and upkeep of products; counseling them on how to extend usefulness with the least amount of energy and materials use; and diagnosing whether upgrades or other changes would maximize the usefulness of a product. Such arrangements would amount to constructing an entirely new kind of service economy, quite unlike the service economy of today.[49]

Consuming better does not obviate the need to consider moderation in overall consumption levels.

Several companies have begun to translate this concept into reality. Xerox already leases three quarters of its equipment. Carrier Corp., instead of selling air-conditioning equipment, is creating a program to sell cooling services and advising customers on energy-efficiency measures that will help reduce air-conditioning needs. Dow Chemical and Safety-Kleen have begun to lease organic solvents to industrial and commercial customers, advising them on their proper use, and recovering these chemicals instead of making the customer responsible for disposing of them. This gives them a strong incentive to use fewer solvents.[50]

Some companies specialize in "performance contracting," helping institutional customers—private firms, government agencies, hospitals, and others—identify ways to cut their use of energy, raw materials, and water. In marked contrast to traditional business interests, it is avoided resource consumption and prevented waste and pollution that makes

such companies thrive.[51]

One often-cited example of a company seeking to reinvent itself in this way is Interface, one of the world's largest commercial carpet manufacturers. After its founder and CEO, Ray Anderson, had an environmental epiphany in the mid-1990s, the firm launched an effort to slash its environmental impact and to move from selling to leasing office carpets. It succeeded in substantially reducing its energy and water consumption and cutting its reliance on petroleum-based raw materials. In 1999, it introduced "solenium," a material that lasts four times as long as traditional carpets, uses up to 40 percent less raw material and embodied energy, and can be entirely remanufactured into new carpets instead of being thrown away or "down-cycled" into less valuable products.[52]

In perhaps its most audacious move, Interface launched an "Evergreen lease" in 1995. Under this, the company retains ownership of the carpet and remains responsible for keeping it clean in return for a monthly fee. Regular inspections would permit the company to focus on replacing just the carpet tiles that show most of the wear and tear instead of the entire carpet, as in the past. This more-targeted replacement helps reduce the amount of material required by some 80 percent.[53]

But only about a half-dozen or so leases were ever actually signed, as most customers opted for a traditional purchase instead. The program did not succeed for a variety of reasons, some specific to the carpet business. Some customers felt the lease agreement was too complex or too inflexible, locking them into a long-term arrangement that limited their future options. But perhaps the biggest problem was cost—a reflection of Interface's emphasis on high-quality material and high-quality maintenance services. In the end, the company felt compelled to drop the Evergreen lease.[54]

The Interface story is at once encouraging and cautionary. It is clear that the new business model the company was proposing is still facing enormous hurdles. As with all radical challenges to established practice, broad acceptance will not come quickly.

Public Consumption and Sustainable Credit

More-efficient and cleaner technologies are essential instruments in the sustainability toolbox. And the emergence of a new type of service economy will provide additional maneuvering space in the quest for a more sustainable economy. Sooner rather than later, however, we need to confront the specter of insatiable consumerism itself. There is a danger that the consumer juggernaut will overwhelm even the most sophisticated methods and technologies that can be devised to make consumption lean and super-efficient. Consuming better does not obviate the need to consider moderation in overall consumption levels. It is worth recalling ecological economist Herman Daly's warning that "to do more efficiently that which should not be done in the first place is no cause for rejoicing."[55]

How should societies go about the task of discouraging "excessive" consumption? While a well-designed luxury tax can play a useful role, there will always be controversy as to what constitutes an unnecessary luxury. Hewing to "consumer sovereignty," capitalist societies leave it almost entirely up to individuals to make purchasing decisions, considering government regulation as an unwelcome intrusion (while conveniently overlooking the incessant attempts to manipulate "sovereign consumers" by advertising campaigns). Clearly, a command-and-control approach is neither workable nor desirable. But while specific purchasing decisions are best left to

individuals and households, there is a broader, more structural, aspect that governments need to address.

The predominance of highly individualized consumption patterns inevitably leads to the multiplication of many goods and services on a grand scale. This virtually ensures redundancy and far greater material requirements than necessary. Governments and communities can take action to strike a better balance between private and public forms of consumption. Even in the most market-oriented societies there are public libraries, swimming pools, and parks. Such organized sharing of facilities and amenities can be expanded. For instance, car-sharing is rapidly gaining adherents in cities across Europe and elsewhere, providing a much-needed, if partial, alternative to private ownership of vehicles and the strictly commercial rentals of cars. Governments can facilitate such initiatives by granting favorable tax terms. Local communities can also set up tool-sharing arrangements, so that not everyone has to own a separate drill, circular saw, or lawnmower.[56]

Government action is also indispensable in overcoming the immense structural impediments to lower consumption levels and to more public forms of consumption. Nowhere is this more pronounced than in transportation: low-density, sprawling settlement patterns translate into large distances separating homes, workplaces, schools, and stores—rendering public transit, biking, and walking difficult or impossible. While the decision on what kind of automobile to buy is up to consumers, the need to buy one at all is frequently out of their control. Likewise in housing, homeowners have a range of choices for heating and air-conditioning. But it is in developers' and builders' hands whether a house incorporates adequate insulation and energy-efficient windows; these fundamental decisions dictate heating and cooling needs over the life of the house.

In recognition of these realities, the OECD has referred to an "infrastructure of consumption" that compels people to engage in involuntary patterns of consumption. As important as it is for consumers to choose more-efficient products, this alone cannot overcome these structural constraints. Forward-looking government policies—improved land use planning, environment-oriented norms and standards, and the creation of a reinvigorated public infrastructure that allows for greater social provision of certain goods and services—will help ensure that consumers are not overly compelled to make consumption-intensive "choices."[57]

Another key area where government action is needed is consumer credit. The unrelenting drumbeat of advertising, which insinuates that corporate brands symbolize desirable lifestyles and that individuals' happiness is innately related to the merchandise they own, has helped propel consumer tastes steadily upward. But advertisers' ability to project new "needs" easily outstrips the reach of consumers' wallets; "wants" always seem to be greater than available means.

Particularly since the 1990s, the savings rate in most OECD countries has been falling, while household debt is on the rise. Young adults, vulnerable to the aggressive marketing directed at them by banks and other credit card issuers, are increasingly sinking into a debt morass. The number of 20–24 year olds facing personal bankruptcy in Germany, for instance, rose one third just between 1999 and 2002.[58]

U.S. consumers' indebtedness is now growing twice as fast as their incomes. Outstanding U.S. consumer credit soared in the last two decades, reaching $1.8 trillion in July 2003. (See Figure 5–2.) The share of U.S. credit card holders with an outstanding balance each month has grown to 61 percent, and average credit card debt topped $12,000 in 2002. In the United Kingdom, consumer debt almost tripled (in current terms) between 1991 and 2001. In Germany, consumer credit doubled in 1989–99 to 216 billion euros, and in 2001 almost a quarter of all households had outstanding consumer debts. And the number of Dutch households seeking bankruptcy help doubled in 1992–99.[59]

Until the mid-1990s, consumers outside North America and Western Europe rarely ran up large amounts of personal debt. Now credit card spending is expanding rapidly among emerging middle-class consumers in Asia, Latin America, Eastern Europe, and even parts of Africa. The amount of money charged to credit cards by South Koreans more than doubled in 2001, but 2.5 million South Koreans have fallen into arrears on their payments. Personal bankruptcies are rising not only in South Korea, but also in Argentina, Brazil, Chile, China, Mexico, and Thailand.[60]

Whereas consumer credit is now geared to maintaining the hyper-throughput economy, which encourages people to carry high personal debts, finance in a sustainable consumption economy will need to devise ways to allow—and to reward—the purchase of efficient, high-quality, durable, and environment-friendly products. These undoubtedly have a higher up-front cost of purchase, but over time such items will be economically more advantageous to consumers than cheaper, flimsier items that must be replaced frequently.

Governments could help consumers by offering advantageous credit terms for certain purchases. The Japanese and German governments do this to support the installation of solar roofs on private homes, but many other eco-friendly purchases could be encouraged in the same way. Or governments can offer targeted rebates. The Canadian government, for instance, announced in

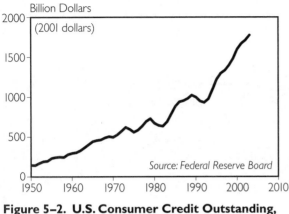

Billion Dollars

(2001 dollars)

Source: Federal Reserve Board

Figure 5–2. U.S. Consumer Credit Outstanding, 1950–2003

August 2003 that it would earmark C$131.4 million (US$95 million) for a program that offers an average rebate of C$1,000 per home to entice owners into making energy efficiency upgrades.[61]

To further encourage the manufacture and purchase of environmentally benign products, governments could design policies that offer tax rebates for the best-performing products while taxing those that fall short of standards. A graduated system could be constructed in which rates of both rebates and fees are scaled according to how efficient, long-lasting, or otherwise environment-friendly an item is. Such a blend, known as a "feebate," has been used to some extent vis-à-vis energy producers, but the concept has not yet been implemented in a consumer setting. A feebate system might even be more effective if hitched up with other policies, such as ecolabeling and EPR laws.[62]

Escaping the Work-and-Spend Trap

Industrial economies are extraordinarily productive—meaning that the same quantity of output can be produced with less and less

human work. In the United States, for example, only about 12 hours of work per week were needed in 2000 to produce as much as 40 hours did in 1950. In principle, this can translate into either of two objectives: raising wages (in line with productivity) while holding working hours constant, or providing greater leisure time while holding income from wages constant. In practice, it has mostly been the former. Most people have been locked into a "work-and-spend" pattern. Greater disposable income translates into greater consumer purchases. And the lure of advertising, "keeping up with the Joneses," and other factors make it seem as though every penny earned is needed to stay on the material treadmill.[63]

Since the rise of mass industrialization in the late nineteenth century, there has been an ongoing tug-of-war between employers and unions over working hours. Employees have struggled for less work time—in the form of shortened workdays or weeks, extended vacation time, earlier retirement, or paid leave. These efforts were primarily motivated by a desire to improve the quality of life and to create more jobs. While environmental issues have not played a central role, channeling productivity gains toward more leisure time instead of higher wages that can translate into ever-rising consumption also increasingly makes sense from an ecological perspective (assuming that greater leisure does not primarily translate into activities that are environmentally questionable, such as long-distance air travel for vacationing in "exotic" locales).[64]

It took close to a century to arrive at the 40-hour workweek in most industrial countries. Where once there was a common trend toward shorter hours throughout the industrial world, there is now an increasing divergence between the United States and Europe. In a reversal of the situation prior to the 1970s, Americans now work longer than most Europeans. (Japanese workers, meanwhile, still have by far the longest hours.) (See Figure 5–3.)[65]

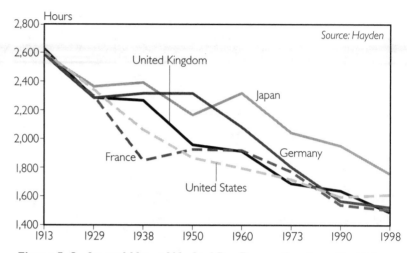

Figure 5–3. Annual Hours Worked Per Person Employed in Major Industrial Countries, Selected Years, 1913–98

Most employers have been very reluctant to agree to more reductions, and a shift in the employer-union balance of power, with waning union strength and rising pressure from globalization, has made further change difficult. By and large, a full-time job at something like 40 hours per week is still considered the norm for anyone wanting to be considered eligible for employment with career advancement opportunities. But the discussion has shifted from fixed weekly hours to introducing greater flexibility, with employers and employees promoting competing notions and interests. Employers are seeking the ability to turn the spigot of labor supply on and off according to fluctuations in the demand for their products. Employee demands center on more individual options to accommodate personal and family needs and to achieve greater "time sovereignty." Several promising models of this have emerged in Europe. (See Table 5–5.)[66]

More than Americans, Europeans prefer work time reductions over additional income growth. Even so, surveys show that almost two thirds of all employees in the United States were working longer-than-desired hours in the late 1990s, up from under half in 1992. At the same time, however, Boston College economist Juliet Schor reports that "during the first half of the 1990s, a fifth of all Americans went through some type of voluntary downshift, with just over half intending it to be a permanent change." "Downshifting" is shorthand for a withdrawal, or partial withdrawal, from the labor force, which is sometimes triggered by a quiet shift in values and behaviors away from consumerism. (Although some people reduce their hours voluntarily, however, others are compelled to accept part-time work against their wishes.)[67]

On the other hand, trading income for time is not a realistic option for many people in the face of adverse wage trends. In the United States, average hourly wages and salaries were largely stagnant between the early 1970s and the mid to late 1990s. And average figures mask a sharply increased degree of inequality in income distribution during the past quarter-century. For the bottom 10 percent of workers, wages in 2001

Table 5–5. New Approaches to Work Time in Europe

Country	Status
Belgium	Established a "time credit" system that allows individuals to work a four-day week for up to five years and to take a one-year leave of absence during a career while receiving a paid allowance from the state.
Denmark	Pioneered a system of paid educational, child care, and sabbatical leaves that allows job rotation between the employed and unemployed. (Variations on this scheme were later put in place by Belgium, Finland, and Sweden.)
Netherlands	In 1982 government, business, and labor agreed on work time reductions in return for wage moderation. Length of workweek was cut from 40 to 38 hours in the mid-1980s and to 36 hours in early 1990s. Voluntary part-time work expanded dramatically, with part-time workers legally entitled to the same hourly pay, benefits, and promotional opportunities as full-timers. Legislation in 2000 extended the right to reduce hours to all workers, while part-timers can request longer hours.

SOURCE: See endnote 66.

were no higher than in 1979; in fact, as many as 70 percent of the work force did not fare much better, seeing virtually no real wage gain until 1998.[68]

A significant fraction of the population has felt compelled to work additional hours, often taking on second jobs just to make ends meet. Overall, there are a number of contradictory trends. In Europe, wage trends have been more favorable, but wage growth has nonetheless lagged behind the expansion of labor productivity. To be workable, then, work time reduction policies need to be accompanied by wage increases to narrow the income differential between rich and poor.[69]

New Dynamics and Values

As described throughout this chapter, with the help of a wide range of policy tools modern economies can be made far less consumption-intensive. Still, what does consuming less mean for a capitalist economy geared toward perpetual economic expansion? After all, the consumerist culture does play an important role: it ensures that the goods produced by a hyper-productive economy will indeed be purchased. This means that capital accumulation can go forward, which drives technical innovation, which in turn results in growing labor productivity (and which provides, at least in principle, the rising incomes and purchasing power necessary for consumerism). This dynamic may fall apart if consumers do not spend enough.[70]

And there is a further complication: although sustainability requires that material appetites be curbed, the enormous overcapacities that have arisen in many industries seem to demand that consumption be stimulated. The global automobile industry, for instance, is operating at only 70 percent of its productive capacity. In the semiconductors industry, capacity utilization runs to just 65 percent, in telecommunications equipment, a mere 50 percent. If anything, the world economy faces growing contradictions. Export-driven economies in a number of developing countries are rapidly expanding their output. China's steel, chemical, constructions materials, and mobile phone industries, for example, are likely to double their production capacities in the next three years, bringing additional pressure to bear.[71]

A huge share of the world's exports is being absorbed by the land of consumerism-par-excellence, the United States. During the 1990s, the U.S. economy acted more and more like a giant vacuum sucking in much of the world's surplus production. (See Box 5–2.) Since 1995, U.S. domestic demand has grown at twice the rate of the other industrial nations. The U.S. current account balance (measuring trade flows and financial transfers) went from plus $3.7 billion in 1991 to minus $503 billion in 2002—a record level. The dollars that flow out of the United States to pay for swelling imports are circulating back to the United States, as foreign investors buy U.S. treasury notes, bonds, stocks, and real estate. These dollar flows have created a vast reservoir of global liquidity. This explosion of credit has been a major engine of world economic activity. But it has encouraged large overinvestment in virtually every major industry. And the well-being of the global economy has grown steadily more dependent on ever-expanding consumption in the United States. At least some observers, including analysts at the Economic Policy Institute in Washington and Stephen Roach, chief economist at Morgan Stanley, think that this system is inherently unstable and cannot go on expanding indefinitely.[72]

A sustainable economy needs a different way of measuring human activity and of providing signals to investors, producers, and consumers.

Arguably then, from both an environmental and an economic perspective, a course correction is in order. But is a recalibration feasible? Certainly, a large and sudden decline in consumer spending would likely send the world economy into a tailspin. But it is far more likely that a less consumptive economy will come about quite gradually. That would allow time to reorient how the economy functions, giving companies an opportunity to adjust.[73]

Smoothing a transition will be a series of investments and technological innovations to accomplish the shift toward sustainability. Promoting renewable energy sources; expanding public transit systems; replacing inefficient machinery, equipment, buildings, and vehicles with far more efficient models; redesigning products for durability—all these activities amount in effect to an ecological stimulus program for the economy.

It is crucial to retool not only the economy, but also economic thought. Right now, economic actors are primed to respond to quantitative growth signals. The concept of the gross domestic product, in which all economic activities are lumped together whether they contribute to or detract from well-being, reigns supreme. (See Chapters 1 and 8.) A sustainable economy needs a different way of measuring human activity and of providing signals to investors, producers, and consumers. It needs a different theory, abandoning the outdated assumption that quantitative growth is unconditionally desirable and embracing instead the notion of qualitative growth.

Most fundamental, though, is a shift in human perceptions of economic value. In *Natural Capitalism*, Amory Lovins and co-authors Hunter Lovins and Paul Hawken make the case for "a new perception of value, a shift from the acquisition of goods as a measure of affluence to an economy where the continuous receipt of quality, utility, and performance promotes well-being." In such an economy, corporate revenues and profits would no longer be associated with maximizing the quantity of stuff produced and

BOX 5–2. U.S. CONSUMERS, CHEAP MANUFACTURES, AND THE GLOBAL SWEATSHOP

During the last decade, American consumers increased their annual spending on goods and services at roughly 3 percent a year, about as fast as in earlier decades. If consumption is measured by quantities of items, however, rather than flows of money, rates of acquisition for a wide range of manufactured products are far higher. Standard economic thinking views this trend as a pure gain in consumers' welfare. But from an environmental point of view, it is a negative. Newer economic reasoning and evidence also recognize that when spending satisfies status and social goals rather than purely functional needs, there is much less true welfare to be gained from additional consumption. The main reason for high rates of product acquisition is that the prices of manufactured goods have fallen substantially in the past decade. This is not due to superior efficiency or technologies, however. The structure and rules governing the global economy have depressed labor costs and plundered natural resources.

Consider the case of clothing, a historically valuable commodity. In 1920, the average U.S. household spent 17 percent of its total expenditures on clothing. In 2001, the figure was a mere 4.4 percent, despite the fact that consumers were buying far more garments. Indeed, clothing has become so cheap that it is hard to give away. The surfeit of clothes is largely attributable to the exploitation of female labor in apparel factories throughout Asia and Central America. Labor's share of production is at historically low levels, and wages have fallen below subsistence. First-hand accounts from workers and western observers in factories producing for Disney, Nike, Liz Claiborne, and many other U.S. firms have found that people frequently work more than 100 hours a week. Workers are subjected to arbitrary supervisory authority; cases of physical, sexual, and verbal abuse are common and well documented; and unions are not permitted.

Developments in the new global economy have exacerbated these problems. The Asian financial crisis of the late 1990s was a direct result of neo-liberal reforms imposed by the U.S. Treasury through the International Monetary Fund, which led a number of Asian economies to collapse under the weight of privatization and decontrol of capital. Wages throughout the region plummeted after the crisis. Indonesian garment wages and benefits fell to 15¢ an hour. In Bangladesh, which has become the fourth largest apparel-exporter to the United States, wages fell to a range of 7–18¢ an hour. Wal-Mart, which controls 15 percent of the U.S. apparel market and is the world's largest clothing retailer, continuously squeezes labor costs in Chinese factories— they can be as low as 13¢ per hour, and the norm is below 25¢. (Higher-priced clothing lines, such as Ralph Lauren and Ellen Tracy also rely on cheap Chinese labor.) Workers have had little success resisting these conditions, because the transnational companies go elsewhere if workers make demands and because factory owners enjoy political protections from their governments. Furthermore, the displacement of rural workers from land and livelihood as a result of transnationals' activities ensures a steady stream of new recruits to urban factories.

Meanwhile, these factors have led to declining prices in the United States, where apparel prices fell by 10 percent over the last decade, with an especially sharp drop after the Asian downturn. The number of garments purchased has skyrocketed, increasing a stunning 73 percent between 1996 and 2001. Consumers reduced how frequently they wear new items and discarded their purchases at record rates. By 2001, the average U.S. consumer purchased 48 new pieces of apparel a year. Goodwill officials report that rates of consumer discard rose by 10 percent a year throughout the 1990s. Clothing became a disposable, and hence freely available, good.

BOX 5–2. (continued)

Similar, albeit less dramatic dynamics have occurred with other consumer goods. U.S. consumers spend about $30 billion a year on toys, 60 percent of which are made in Chinese sweatshops, where wages and working conditions are similar to those in apparel. Since 1994, toy prices have fallen by 33 percent, and children on average get 69 toys a year. The prices of personal computers and peripheral equipment have fallen by 81 percent since 1997 as a result of more powerful chips, low wages, and the offloading of environmental costs. In 2001, nearly 23 million new computers were purchased, with similar numbers being discarded. It is estimated that in 2005, 63 million personal computers will be retired in the United States alone. Falling prices and rising quantities also characterize appliances, sporting goods equipment, tools, hardware, and outdoor equipment and supplies. Department store prices declined by nearly one third since 1993, and durable goods prices fell by 57 percent. In part this is due to the steady downward pressure on prices from Wal-Mart as the company exploits labor abroad and domestically and as it takes advantage of taxpayer subsidies for trucking costs and unaccounted environmental degradation.

Sustainable consumption requires higher prices for consumer goods and a shift to higher quality, longer lived products made by well-remunerated workers under environmentally safe conditions. This will help satisfy elementary criteria of justice, such as the right of foreign and domestic laborers to a decent livelihood and the right of all people to share in Earth's bounty. But this is a political as well as a consumer issue. Current policies are moving us away from these sustainability conditions. Considerable research on the effects of the World Bank and the International Monetary Fund show that they mainly represent the interests of the U.S. government, and U.S. corporations, at the expense of workers and domestic industries in poor countries. As opposition to the global economy has mounted, the U.S. administration has responded with large increases in military expenditures. A just and healthy global economy will ultimately be rooted in a worldwide structure of wages that are high enough to support strong domestic demand, a closer balance of power between capital and labor, and more equal distributions of income and wealth. The need for environmentalists to make common cause with others concerned about global justice and peace has thus become a most urgent task.

—*Juliet Schor, Professor of Sociology,*
Boston College

SOURCE: See endnote 72.

---

sold but rather with deriving the most service and best performance out of a product, and therefore minimizing energy and materials consumption and maximizing quality.[74]

As much as the dot-com boom of the 1990s was an illusion, it did show the potential for perceiving value in new ways—ones less connected to the mobilization of physical resources per se. The future may not lie in "counting eyeballs"—that is, determining how many pairs of eyes can be attracted to a particular Web site—but consumers, manufacturers, financial institutions, and governments do need to develop a new understanding of what is truly valuable and what is therefore worth doing.

Without doubt, serious political obstacles need to be overcome. Vested interests, particularly in the energy and mining industries, are adept at defending lucrative subsidies and opposing meaningful environmental tax reform. Throughout the economy, many manufacturing companies are wedded to the business model they are familiar with and are inclined to stick with the assumptions of yesterday rather than venture into the still highly unfamiliar territory of product take-back and related concepts. And retailers, particularly in the United States, are strongly oriented toward maximizing sales of ostensibly cheap products rather than pursuing quality retail. The staying power of these interests should not be underestimated. A less consumption-oriented economy is possible, but it will take government action, consumer education, and growing numbers of corporate trailblazers to make it happen.

Cell Phones

Cell phones have rocketed into ubiquity. In 1992, less than 1 percent of people in the world had a mobile phone and only one third of all nations had a cellular network. Just 10 years later 18 percent of people had a cell phone—1.14 billion, more than the 1.10 billion with a conventional phone line—and more than 90 percent of countries had a network.[1]

More Europeans now send and receive short text messages with their mobile phones than use the Internet from personal computers. The Philippines leads the world in text messaging via cell phone; indeed, "txting" by protesters to organize rallies against former President Joseph Estrada was a factor in his ouster. In Africa, mobile phones outnumber fixed lines at a higher ratio than on any other continent; entrepreneurs there who sell the use of their cell phones are bringing service to villagers who previously had to walk miles to place a call.[2]

As people mainly chat on cell phones, the handset draws radio waves closer to their heads than most other electronic gadgets do. A 10-country study backed by the World Health Organization to determine whether cell phone use could be implicated in cancers of the head and neck is due to be finished by 2004. With long-term data not yet available, researchers advise people to plug an earpiece into their phone, so they can hold the handset further away. And a study group assembled by the British government discouraged excessive cell phone use by children.[3]

Like computers, cell phones are short-lived products that present the clearest threat to human and environmental health when they are being created or destroyed, as they contain toxics-rich semiconductor chips. Lifecycle analyses identify the phone's chip-containing circuit board, liquid crystal display, and batteries as the biggest hazards, followed by the hard-to-recycle plastic casing.

In the United States, the world's second largest market for cell phones after China, handsets are cast off on average after 18 months, and the research group INFORM estimates that by 2005, consumers will have stockpiled some 500 million used mobiles that are likely to end up in landfills, where they could leach some 312,000 pounds of lead.[4]

Their small size makes phones easier to discard than computers, but they are also more easily reused. Extending the life of phones in this way lessens their environmental toll. Charitable groups have partnered with companies to refurbish used handsets—some are programmed to dial emergency services and given to victims of domestic violence or the elderly, while others are resold in developing countries—and companies such as ReCellular buy and sell used phones in bulk.[5]

Competing standards for cellular networks are one reason mobile devices are discarded so quickly in the United States; in contrast, Europe has had a single standard since the early 1980s. As companies stuck with outmoded equipment would lose out, the industry has thwarted attempts by the

International Telecommunication Union to forge consensus on a single standard, but some industry watchers believe such a move is inevitable as the number of users grows.[6]

Ultimately, incentives for companies to design and recycle less-toxic cell phones hold the most promise for lightening the environmental burden. Since 1998, Japan has made producers take back major electrical appliances; this mandate now extends to computers, and regulations for other electronics are on the horizon. Companies must pay for recycling their products, prompting firms such as Sony to invest in technologies that are easily recyclable.[7]

The Netherlands, Norway, Sweden, and Switzerland all have "extended producer responsibility" programs that include cell phones, in which consumers pay advance disposal fees to fund recycling. Certification programs in a few countries let consumers know which mobiles are most environmentally friendly: in Germany, the Blue Angel label goes to phones that meet standards for toxic content, and in Sweden, TCO Development certifies handsets according to emissions, ergonomic, and environmental criteria, including whether they are easily recyclable.[8]

Two directives from the European Commission entered into force in 2003, sending the strongest environmental signal yet to electronics companies. The Waste Electrical and Electronic Equipment directive will make each company responsible for collecting and recycling its own new goods after August 13, 2005, while all firms will be collectively responsible for electronics put on the market before that date. A companion

rule prohibits the use of certain toxins in electronics, including lead, mercury, cadmium, hexavalent chromium, and certain brominated flame retardants.[9]

The new European laws are spurring study of eco-friendly technologies. For example, Nokia has been working with university scientists to develop biodegradable plastics and phones that disassemble for easy recycling when triggered by high temperature.[10]

The California-based Silicon Valley Toxics Coalition is campaigning for U.S. take-back legislation. In the absence of national laws, Minnesota has pioneered legislation making producers responsible for certain toxic materials, Massachusetts banned electronic waste from landfills and created a fund to recycle electronics, California introduced a limited ban and expects local governments to cover recycling costs, New York recently required vendors to accept and recycle any cell phones they sell, and other states are drafting bills to reduce the amount of electronic trash they must dispose of.[11]

At the international level, in late 2002 the secretariat of the Basel Convention on hazardous waste trade convened major manufacturers to launch a new mobile phone working group. In recent years, the health hazards of toxic electronic trash have received more media attention as environmental activists have documented the export of U.S. electronic scrap to Asia. The convention secretariat intends the new mobile phone group to be the first of several efforts to work with industry on the waste problems associated with particular products.[12]

—*Molly O'Meara Sheehan*

CHAPTER 6

Purchasing for People and the Planet

Lisa Mastny

In the spring of 2001, two students at Connecticut College in the northeastern United States circulated an ambitious petition. Concerned about the school's emissions of harmful air pollutants, they urged their classmates to support a voluntary hike in student activity fees to fund the university's membership in a local renewable energy cooperative. More than three quarters of the students backed the proposal, and it won unanimous support from both the student government and the Board of Trustees. Although the coop closed down a year later, the seeds for the transition had been planted. By January 2003, Connecticut College was meeting 22 percent of its electricity needs from renewable wind energy—the largest share of energy obtained this way by a U.S. college or university.[1]

Around the world, growing numbers of universities, corporations, government agencies, and other institutions are reviewing their purchasing habits and incorporating environmental concerns into all stages of their procurements. In doing so, they are helping to spur markets for a wide range of environmentally preferable products. Global sales of energy-efficient compact fluorescent lamps (CFLs) have increased nearly 13-fold since 1990, for example, to some 606 million units in 2001. The use of solar energy and wind power has surged by more than 30 percent annually over the past five years in countries like Japan, Germany, and Spain. Retail sales of organic produce in the United States have grown by at least 20 percent annually since 1990, to $11 billion per year, while U.S. sales of hybrid-electric cars doubled in 2001.[2]

Even so, green markets are tiny relative to conventional ones. The U.S.-based Natural Marketing Institute estimates that the global demand for "health and sustainability" products—from alternative transport to organic foods—reached a record $546 billion in 2000. But this still represents only about 1

An expanded version of this chapter appeared as Worldwatch Paper 166, *Purchasing Power: Harnessing Institutional Procurement for People and the Planet* (Washington, DC: Worldwatch Institute, July 2003).

percent of the total world economy.[3]

Green markets are faring better in some regions than others. In Europe, for instance, recycled paper now costs the same as or less than virgin paper, mainly because consumers have increasingly sought it out. Municipal purchasers in Dunkerque, France, save roughly 50¢ per ream (about 16 percent) by buying recycled. In the United States, however, paper buyers generally still pay a premium of 4–8 percent for recycled content. Despite extensive efforts to boost the market share of recycled paper, an estimated 95 percent of U.S. printing and writing grade paper (which accounts for more than a quarter of the U.S. paper market) is still made from virgin wood fiber. Domestic use of recycled paper has actually declined in recent years, and unless demand for it picks up, the manufacturing infrastructure could soon disappear.[4]

These markets are even smaller in the developing world, though interest in renewable energy and other product areas is growing in many countries. Overall resource use in the developing world is still low relative to the industrial world, but rising consumer demand for everything from cars to computers will make strengthening local markets for environmentally sound technologies increasingly important.[5]

Greening Institutional Procurement

Through the things they buy, institutions wield great influence over the future of our planet. Nearly every purchase an organization makes, whether it is a ream of copier paper or a new office building, has hidden costs for the natural environment and the world's people. Many products require huge inputs of water, wood, energy, metals, and other resources that are not always renewable. And they often contain toxic chemicals that, when released

into the environment, endanger the health of humans and the ecological systems we depend on. These impacts can occur at any stage of a product's lifecycle: obtaining the raw materials, manufacturing, packaging, transport, use, and even after disposal.[6]

Just how powerful is institutional purchasing? Consider governments. In industrial countries, public purchasing accounts for as much as 25 percent of the gross domestic product (GDP). (See Figure 6–1.) Government procurement in the European Union (EU) alone totaled more than $1 trillion in 2001, or roughly 14 percent of GDP. In North America, it reached $2 trillion, or about 18 percent of GDP. This purchasing occurs at all levels of government: in 2002, the U.S. government spent roughly $350 billion on goods and services (excluding military spending), while the country's state and local authorities spent more than $400 billion.[7]

Corporations, universities, religious bodies, and other large institutions also have significant purchasing power. Many businesses, for instance, buy not only myriad finished products, such as pens and computers, but also raw materials, packaging, and other goods as inputs into the manufacturing process. By one estimate, aggregate spending by companies along their "supply chains" far outweighs the consumption of the finished products by individuals. As production becomes ever more global, manufacturers can play an important role in influencing the environmental behavior of suppliers in other countries, including in the developing world.[8]

Meanwhile, universities spend billions of dollars each year on everything from campus buildings to cafeteria food. In the United States, roughly 3,700 colleges and universities collectively bought some $250 billion in goods and services in 1999—equivalent to nearly 3 percent of the nation's GDP and more than the GDP of any country but the

PURCHASING FOR PEOPLE AND THE PLANET

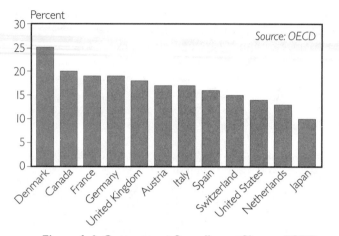

Figure 6–1. Government Spending as Share of GDP in Selected Countries, 1998

world's 18 largest. Religious institutions have similar clout, managing vast numbers of schools and houses of worship around the world. And international institutions like the United Nations and the World Bank buy large quantities of goods and services to run their headquarters and operations in industrial countries as well as to sustain their field offices and activities in the developing world—giving them a unique opportunity to help build sustainable markets worldwide. The United Nations alone bought nearly $4 billion worth of goods and services in 2000.[9]

But the sheer volume of their purchasing is only one reason institutions can be powerful agents for positive environmental change. "Unlike many individuals, large institutions take a very systematic approach to their purchasing," notes Scot Case of the Maryland-based Center for a New American Dream. "Purchases are clearly defined in detailed contracts that specify nearly every aspect of the product or service being bought." With structured methodologies already in place, inserting environmental considerations into institutional purchasing can be a relatively straightforward process,

with significant payoffs.[10]

Most organizations that make high-value purchases engage in some form of competitive bidding, meaning that they open their procurement up to many potential suppliers when awarding contracts. "Greening" procurement means that in addition to specifying basic requirements for quantity, price, function, or safety, institutional buyers make environmental demands on their suppliers as well. For instance, they may ask that products meet certain standards for energy efficiency or have a specific recycled content, or even that the suppliers themselves have green credentials. (See Box 6–1.) (Purchasers may also stipulate certain criteria for social justice, though this is not very common. The government of Belgium, for instance, is considering barring public contracts with companies whose production conditions support anti-democratic regimes or do not respect human rights.)[11]

For many common institutional purchases, alternatives now exist that are less resource-intensive, less polluting, and less harmful to human and environmental health than their conventional counterparts. By buying paper with even a small percentage of recycled content, for instance, institutions can divert significant amounts of waste from landfills. They can also save energy, wood, and other resources: the New York–based group Environmental Defense estimates that if the entire U.S. catalog industry switched its publications to just 10-percent recycled content paper, the savings in wood alone would be enough to stretch a 1.8-meter-high fence across the United States seven times.[12]

Buying greener products can also bring health benefits to employees and other build-

---

BOX 6–1. GREENING PURCHASING CONTRACTS

In contracts with suppliers, purchasers can ask that:

- Products display one or more positive environmental attributes, such as recycled content, energy or water efficiency, low toxicity, or biodegradability.

- Products generate less waste, including by having less packaging or being durable, reusable, or remanufactured—the city of Santa Monica in California asks its vendors to supply cleaning products in concentrated form to save packaging.

- Products meet certain environmental criteria during manufacturing or production, such as that paper be processed chlorine-free or be made out of timber from a sustainably managed forest.

- Suppliers reclaim or take back items such as batteries, electronics equipment, or carpeting at the end of their useful lives—some U.S. federal agencies now use "closed-loop" contracts, requiring contractors to pick up used oil products, tires, and toner cartridges for disposal.

- Suppliers themselves have environmental credentials—some government purchasers in Switzerland give preference to companies that have or are putting in place environmental management systems.

SOURCE: See endnote 11.

---

ing occupants. Many common purchases, including paints used on walls and furniture, pesticides for buildings and grounds maintenance, and products for custodial cleaning, contain toxic ingredients like heavy metals and volatile organic compounds. These substances can pollute indoor air and accumulate in living tissue, endangering human and environmental health. The Janitorial Products

Pollution Prevention Project reports that 6 out of every 100 janitors in Washington state have lost time from their jobs as a result of injuries related to the use of toxic cleaning products, particularly glass and toilet cleaners and degreasers.[13]

Many institutions are finding that green purchasing saves money as well. Some green products are cheaper right away than their conventional alternatives. Recycled toner cartridges for printers and copiers, for instance, can sell for a third of the price of new cartridges. Other items, such as low-flush toilets or compact fluorescent lamps, bring considerable cost savings over their lifetimes. Although CFLs cost up to 20 times more than incandescent bulbs, they last 10 times longer and use a quarter of the electricity to produce the same amount of light. Buying goods that are durable, remanufactured, or recyclable can lower the costs of product maintenance, replacement, or disposal. Meanwhile, cleaners and other products that are less toxic can reduce the insurance and workers' compensation expenses associated with certain workplace injuries.[14]

Perhaps most important, rising institutional demand can play a key role in building larger markets for greener goods and services. If these consumers increasingly seek out products and services that are more beneficial to the environment, producers will have a greater incentive to design and produce them. As markets for these items grow, propelled by the forces of competition and innovation, the resulting economies of scale will eventually drive down prices, making greener purchases more affordable for everyone.

Pioneers in Green Purchasing

Most institutions that buy green are targeting smaller purchases, like paper and office supplies, that are easy to shift without significant changes in the organization's practices. But some others have begun to fundamentally restructure the way they do business.

At the corporate level, green purchasing pioneers include companies in virtually all sectors of the economy, including banks, hotels, automakers, clothing retailers, and supermarkets. (See Table 6–1.) Many of these businesses are motivated by enlightened self-interest: they are finding that by embracing energy efficiency measures and other relatively small-scale changes in their internal operations, they can reduce their environmental impacts as well as improve their profitability. In 1992, the Business Council for Sustainable Development endorsed this approach as "eco-efficiency." L'Oreal, the world's largest cosmetics manufacturer, cut its greenhouse gas emissions 40 percent between 1990 and 2000 while increasing production 60 percent, largely by installing energy-efficient lighting in its facilities and introducing a recycling program to cut back on waste incineration. Anheuser-Busch and IBM are among the many other companies that have saved millions through improvements in water and energy efficiency.[15]

Even in cases where green purchasing does not lead to direct savings, it can bring overall business benefits. In a recent study for the Center for Advanced Purchasing Studies, Craig Carter and Marianne Jennings found that increased corporate social responsibility is generally correlated with higher revenues, healthier and safer work environments, and improved relationships with customers and suppliers—factors that can more than outweigh any potential monetary costs. Green purchasing can also be a way for companies to win "PR" points with their supporters and critics (though some environmental groups maintain that this is just a form of "greenwashing").[16]

Many companies also realize that they may

lose competitiveness by pursuing resource-intensive or environmentally destructive methods. The sportswear manufacturer Nike, for instance, has boosted the organic cotton content of its clothing because it worries about the potential health and environmental liabilities associated with conventional cotton production, which requires high inputs of chemical pesticides and fertilizers. "It's the only intelligent way to do business," says Heidi McCloskey, global sustainability director at Nike Apparel. "By managing and designing out every harmful product, Nike won't be at risk of paying higher costs in the future." In 2001, more than a third of the cotton garments the company produced contained at least 3-percent certified organic cotton.[17]

Nike is in the vanguard of companies that now hope to take a leading role in pushing wider markets for green products. In 2001 it helped launch Organic Exchange, a network of 55 businesses that intends to expand the use of organic cotton in manufacturing significantly over the next 10 years. Other companies, including Texas Instruments, Levi Strauss, and Ford Motor Company, have joined the Recycled Paper Coalition, founded in 1992 to use bulk corporate purchasing power to boost the supply and quality of recycled paper products (and to wean companies off virgin paper before regulations require it). The coalition's 270 members bought nearly 150,000 tons of recycled paper in 2002, with an average postconsumer content of 29 percent.[18]

But balancing green purchasing with the corporate profit motive can be a delicate process. Because companies are ultimately responsible to their bottom lines and beholden to shareholder and supplier relationships, in some cases it can still be a competitive disadvantage to do the right thing. Jeffrey Hollender, CEO of Seventh Generation, a manufacturer of environmentally

Table 6–1. Examples of Green Purchasing in Selected Companies

Bank of America	Boosted recycled paper purchases by 11 percent in 2001 to 54 percent of paper bought. Reuses and refurbishes office furniture and carpeting and uses recycled materials in fixtures and bank service counters. Aims to include supplier environmental requirements in all future contracts.
Boeing	By 1999, had retrofitted more than half of its floor space with efficient lighting, cutting energy costs by $12 million annually and saving energy that could power some 16,000 homes.
Canon	Gives priority in its global purchasing to nearly 4,600 company-approved green office supplies. Now working to green its procurement for plants in Japan, Asia, and North America. Outreach to suppliers has led to high rates of compliance with existing policies.
Federal Express	In 2002, pledged to replace all 44,000 fleet vehicles with diesel-electric trucks that would increase fuel efficiency by half and cut smog- and soot-causing emissions by 90 percent.
Hewlett-Packard	In 1999, committed to buying paper only from sustainable forest sources. Gives preference to suppliers that provide green products and have green business practices. Restricts or prohibits use of certain chemicals in manufacturing and packaging.
IKEA	Gives preference to wood from forests that are either certified as being sustainably managed or in transition to these standards. Buys wood via a four-step process that encourages suppliers to seek forest certification.
McDonald's	Spent more than $3 billion on recycled-content purchases between 1990 and 1999, including trays, tables, carpeting, and packaging. In 2001, adopted compostable food packaging made from reclaimed potato starch and other materials. Has installed energy-efficient lighting in restaurants.
Migros	In 2002, the Swiss supermarket became the first European retailer to stop buying palm oil supplies from ecologically unsound sources in Malaysia and Indonesia. Audits its suppliers for compliance with environmental criteria and labels products that "protect tropical forests."
Riu Hotels	By switching to bulk purchases of breakfast items, the German chain was able to cut waste by 5,100 kilograms annually, saving 24 million items of individual packaging and an average of 5 million plastic garbage bags each year.
Staples	In 2002, pledged to boost the average recycled content in its paper goods to 30 percent and to phase out purchases from endangered forests. Uses energy-efficient lighting and roofing material in its buildings. By end of 2003, aims to buy only recycled paper products for internal operations and to boost green power purchases to 5 percent.
Starbucks	Since November 2001, has offered financial incentives and supplier preference to coffee farmers who meet certain environmental, social, economic, and quality standards. In 2002, 28 percent of paper fiber used was post-consumer and 49 percent contained unbleached fiber.
Toyota	In 2001, switched some 1,400 office supply items and 300 computers and other equipment to green alternatives. Achieved 100-percent green purchasing in these areas in 2002. In fiscal year 2001, bought 500,000 kilowatt-hours of wind power and aims to boost this to 2 million kilowatt-hours per year.

SOURCE: See endnote 15.

sound household goods, notes that his company is constantly weighing the urge to boost the recycled content of its products against the higher cost of doing so. "At the final analysis, it's far better to make a slightly less environmentally benign product for a time than it is to find yourself out of business and unable to make any difference at all," Hollender says. "The trick is achieving a balance between moving too fast and not moving fast enough."[19]

Over the past decade or so, calls for greener government procurement have escalated as well. Most recently, delegates attending the 2002 World Summit on Sustainable Development in Johannesburg reiterated the need to "promote public procurement policies that encourage development and diffusion of environmentally sound goods and services." Many governments also increasingly recognize the value of greening operations as a way to streamline costs and achieve wider environmental policy goals, such as reducing waste and meeting targets for energy efficiency.[20]

Although some governments took steps to green their purchasing as early as 20 years ago, most activity only dates from the 1990s. (See Table 6–2.) Several countries—including Austria, Canada, Denmark, Germany, Japan, and the United States—now have strict national laws or policies requiring government agencies to buy green (though this does not mean they always do so). In most other countries where green purchasing now occurs, either policies "recommend" that purchasers consider environmentally preferable options or no specific policy exists, although purchasers are able to consider environmental variables in their buying. As at the industry level, most government activity has focused on buying recycled or energy-efficient products, although interest in renewable energy and other green purchases is rising as well.[21]

There has also been a flurry of green purchasing by city, state, and regional governments. Christoph Erdmenger, coordinator of eco-procurement activities at the International Council for Local Environmental Initiatives (ICLEI), notes that in most countries with strong green purchasing activities, local authorities have been the forerunners. In Europe, 250 municipal leaders from more than 30 countries pledged in the Hannover Call of 2000 to use their purchasing power "to direct development toward socially and environmentally sound solutions." Kolding in Denmark set an ambitious goal in 1998 of incorporating environmental considerations into 100 percent of its framework purchasing by 2002. By May 2001, roughly 70 percent of its purchasing requests had specified and integrated environmental demands, primarily in the areas of food, office equipment, cleaning products, information technology, and health care supplies.[22]

The United States has also seen greater progress in greening government purchases at the state, county, and city levels than at the national level. In 1999, Santa Monica, California, became the first U.S. city to buy 100 percent of its municipal power from renewable sources, including geothermal and wind energy. The state of Minnesota now has some 110 different contracts for green products and services, including alternatively fueled vehicles, low-toxicity cleaning supplies, energy-efficient computers, and solvent-free paint. Other local pioneers include the states of Massachusetts, Vermont, and Oregon; the city of Seattle, Washington; and Kalamazoo County in Michigan.[23]

In the developing world, Taiwan is one of a few countries that have formalized green public purchasing, stating a preference for approved green products in a 1998 presidential decree. Other governments have implemented legislation to support recycling programs, yet initiatives to actively promote

Table 6–2. Examples of Government Green Purchasing

Austria	Local-level activities date to the late 1980s. Federal laws from 1990 and 1993 require public authorities to insert environmental criteria in product specifications. (The 1993 law has been adopted by eight of nine provinces.) Since 1997, the Ministry of Environment has helped municipalities and other ministries buy green. In 1998, the government approved basic green buying guidelines in areas like office equipment, building, cleaning, and energy.
Canada	A strong national legislative and policy framework for green purchasing exists. Goals include making 20 percent of federal power purchases from green sources by 2005 and, where cost-effective and operationally feasible, running 75 percent of federal vehicles on alternative fuels by April 2004. Environment Canada's policy directs purchasers to consider product lifecycle impacts, to use ecolabeled products, and to adopt recycling, energy efficiency, and other green criteria when purchasing.
Denmark	A world leader in green purchasing. A 1994 law requires all national and local authorities to use recycled or recyclable products, and all authorities must also have a green purchasing policy. As of 2000, 10 of 14 counties had policies. At least half of the municipalities also attest to having or developing policies.
Germany	Federal waste law requires public institutions to give preference to green products in purchasing. State and municipal directives also require inclusion of environmental criteria in requests for bids, though economic criteria take priority in evaluation.
Japan	Another world leader in green purchasing, starting with local government activity in the early 1990s. A 2001 law requires national and local governmental organizations to develop policies and buy specified green products. As of early 2003, authorities in 47 prefectures and 12 major municipalities were buying green, and nearly half the 700 municipalities had policies. Most progress is in the areas of paper, office supplies, information technology equipment, cars, and home appliances.
United Kingdom	Rules allow purchasers to use environmental criteria in purchasing as long as this does not prevent fair competition. Authorities can choose how much weight will be given to environmental and other criteria when awarding contracts. Government departments were required to obtain at least 5 percent of their energy from renewable sources by March 2003, which must increase to 10 percent by 2008.
United States	A wide array of laws and policy directives requires federal agencies to buy green items, including recycled-content and energy-efficient products and alternatively fueled vehicles. Agency-wide coordination and implementation has been poor but is improving. Among the states, 47 of 50 boast "buy-recycled" policies, some dating to the late 1980s. At least a dozen states have broadened these to include other green purchases.

SOURCE: See endnote 21.

and buy recycled products have been slower to take off. There has been talk about getting green purchasing into government policy in Brazil, Iran, Mexico, and Thailand, and the government of Mauritius is moving toward greater use of recycled plastic and paper and has introduced more-efficient neon bulbs for street lighting.[24]

In most cases, it is too early to judge the overall impact of these government pioneers.

A few notable successes, however, point to the tremendous ability of green purchasing to influence markets. For instance, ICLEI attributes the ascendance of recycled paper to a standard office supply in many European countries to the cumulative demands of public authorities, which have given this product a competitive edge. A similar shift happened when the U.S. government boosted the recycled content standard for federal paper purchases to 30 percent in 1998. In 1994, only 12 percent of the copier paper bought by federal agencies had recycled content, and it was only 10-percent recycled material. By 2000, however, 90 percent of the paper purchased by the government's two leading paper buyers had 30-percent recycled content. The spike in government demand not only boosted the overall market standard for recycled content, it also helped elevate the standing of the government's main recycled paper supplier, Great White.[25]

Government green purchasing can be particularly effective in pushing markets where public buying accounts for a significant share of overall demand or where a technology is changing rapidly, as in the case of computer equipment. The U.S. government, the world's single largest computer purchaser, buys more than 1 million machines annually—roughly 7 percent of new computers worldwide. In 1993, President Bill Clinton issued an executive order requiring federal agencies to buy only computer equipment that meets the efficiency requirements described under the government's Energy Star program. Today, largely as a result of this increased demand, 95 percent of all monitors, 80 percent of computers, and 99 percent of printers sold in North America meet Energy Star standards. Analysts have linked a similar jump in the environmental performance of Japanese electronics to that country's preeminence in the green purchasing of these items.[26]

Pressures and Drivers

Institutions of all kinds increasingly face a wide range of regulatory and consumer pressures to buy green. For instance, many governments now offer rebates, tax breaks, and other economic incentives to encourage businesses, schools, individuals, and other consumers to invest in everything from energy-efficient appliances to alternatively fueled vehicles. In 2002, the U.S. Archdiocese of Los Angeles received thousands of dollars in local rebates when the Cathedral of Our Lady of the Angels became the city's first religious edifice to install solar panels on its roof, generating enough energy to power both the building and more than 60 residences.[27]

Governments are also using their regulatory authority to essentially force institutions to make greener purchases. New laws or regulations that require manufacturers to meet certain standards for energy efficiency or recyclability influence the way many companies now design and make their products. Vehicle manufacturers, for instance, have had to rethink both their sourcing and use of materials in order to meet the terms of a new European Union directive on end-of-life vehicles, which aims to reduce the portion of old cars that ends up in landfills. By 2007, 85 percent by weight of every new vehicle must be made from recyclable components (currently, vehicle recycling is limited to the 75 percent by weight that is metallic). DaimlerChrysler hopes to exceed this standard and achieve 95 percent recyclability by 2005, in part by boosting its use of recovered plastic and other materials. If widely adopted, this recycling process could save the world automotive industry $320 million per year.[28]

Governments are not the only ones pushing institutions to buy green. Around the world, individual consumers are beginning to translate personal concerns about their health, the environment, and social justice into greener buying at the household level. Today, some 63 million U.S. adults, or approximately 30 percent of households in the country, do some form of environmentally or socially conscious buying, according to a survey by LOHAS Consumer Research. In the United Kingdom, ethical purchases by individuals—in areas ranging from organic foods to renewable energy—increased 19 percent between 1999 and 2000, six times faster than the overall markets in the various sectors.[29]

These consumers increasingly expect better environmental performance from the institutions that guide them, the businesses they support, and the products they buy. Manufacturers in the United States report a growing volume of consumer requests for environmental information about their products, such as whether they contain recycled content. The 1999 Millennium Poll on Corporate Social Responsibility found that some 60 percent of consumers in 23 countries now expect companies to tackle key environmental and social issues through their businesses, in addition to making a profit and generating jobs.[30]

Consumer pressure was instrumental in getting municipal authorities in Ferrara, Italy, to introduce organic food into local school cafeterias. After a group of concerned parents drew attention to the poor quality of food being served in kindergartens in 1994, the town established a commission to study the possibility of switching its food procurement. Within four years, Ferrara had systematized the purchasing of organic food into a special procurement request, and by 2000, 80 percent of the food served in the city's kindergartens was organic.[31]

Growing numbers of concerned consumers, shareholders, and nongovernmental organizations (NGOs) are also participating in boycotts and other direct actions to pressure companies into shifting their buying practices. In recent years, advocacy groups like the Rainforest Action Network (RAN) and ForestEthics have organized public actions in the United States and around the world to get leading retailers like Home Depot to stop buying wood products originating in old-growth forests. (See Box 6–2.) Michael Marx, executive director of ForestEthics, notes that a key factor behind the success of these campaigns has been their focus on the private sector: "It's important to target corporate customers because they have an image. The goal is to raise the cost of doing business in an environmentally unfriendly way." Marx believes that boycotts and other public shaming actions can be much more effective tools for pushing environmental change than, for example, lobbying for regulatory action, which could take years or even decades.[32]

> **Today, some 63 million U.S. adults, or approximately 30 percent of households, do some form of environmentally or socially conscious buying.**

In other cases, NGOs are actively partnering with leading corporations to help them redirect their significant purchasing power toward environmental ends. The Alliance for Environmental Innovation, a project of the nonprofit Environmental Defense, is working with companies like Citigroup, Starbucks, Bristol-Myers Squibb, and Federal Express to shift industry purchases of paper, vehicles, and other products. And the new Climate Savers Program, a joint initiative of the World Wide Fund for Nature and the Virginia-based Center for Energy and Climate Solutions, works with global companies like

---

BOX 6–2. HOME DEPOT'S COMMITMENT TO SUSTAINABLE WOOD PRODUCTS

In the mid-1990s, the San Francisco–based Rainforest Action Network launched a high-profile campaign to pressure Home Depot, the world's largest home improvement retailer, to improve its wood buying practices. The Atlanta-based chain sells more than $5 billion worth of lumber, doors, plywood, and other wood products annually in its 1,450 stores worldwide.

RAN used consumer boycotts, in-store demonstrations, ad campaigns, and shareholder activism to draw public attention to Home Depot's practice of buying wood products that originate in highly endangered forests in British Columbia, Southeast Asia, and the Amazon. In August 1999, largely in response to this pressure, the company announced that it would phase out all purchases of old-growth wood by the end of 2002. As of January 2003, it had reduced its purchases of Indonesian *lauan* (a tropical hardwood used in door components) by 70 percent and shifted more than 90 percent of its cedar purchasing to second- and third-generation forests in the United States. Today, the company claims to know the original wood source of roughly 8,900 of its products.

Home Depot also pledged to give preference to products certified as coming from sustainably managed forests. (Currently, roughly 1 percent of wood sold worldwide is certified.) Between 1999 and 2002, the number of its suppliers selling wood approved by the Forest Stewardship Council (FSC), a leading forest certification body, jumped from only 5 to 40, and the value of its certified wood

purchases soared from $20 million to more than $200 million.

The company's decision had a significant ripple effect on the wider home improvement and home building markets. Within a year of the shift in policy, retailers accounting for well over one fifth of the wood sold for the U.S. home remodeling market, including leading competitors Lowe's and Wickes, Inc., announced that they too would phase out endangered wood products and favor certified wood. Two of the nation's biggest homebuilders also pledged not to buy endangered wood.

These policy shifts have raised the overall standard for the timber industry. With many companies now scrambling to get FSC approval, it could soon be a liability for other wood producers not to get certified. Michael Marx of ForestEthics notes that "one statement from Home Depot did more to change British Columbia's logging practices than 10 years of environmental protest."

But critics worry that Home Depot has not gone far enough in using its market power to influence its suppliers. One obstacle has been the higher cost to vendors of buying certified wood or producing synthetic alternatives, though Home Depot has agreed to absorb any price increases. Another challenge has been weaning consumers away from environmentally unsound options. According to the company, few customers are specifically asking for certified wood.

SOURCE: See endnote 32.

Johnson & Johnson, IBM, Nike, and Polaroid to boost their energy efficiency and use of green power. Similarly, the World Resources Institute is recruiting leading companies to help meet its goal of developing corporate markets for 1,000 megawatts of new green power by 2010—enough capacity for 750,000 American homes.[33]

Overcoming Obstacles

For the past few years, economist Julia Schreiner Alves has been trying to push her employer, the state of São Paulo in Brazil, to green its government purchasing. Home to 30 million people, São Paulo is second among all Brazilian states in purchasing power. But

Alves is one of the only people in her agency calling for greener purchasing, and she says that many of her colleagues, particularly in the purchasing department, are simply insensitive to the potential of green buying to generate positive environmental change.[34]

At a practical level, the success of green purchasing often comes down to the role of the professional purchaser. An institution's procurement department wields considerable power. In the United States, government purchasing departments supervise 50–80 percent of total buying. When purchasing is highly centralized, a single decision made by just one or a handful of buyers can have a tremendous ripple effect, influencing the products used by hundreds or even thousands of individuals. As a result, the buying activities of institutional purchasers often have far greater consequences for the planet than the daily choices of most household consumers.[35]

Unfortunately, many purchasers are not yet harnessing their tremendous power to leverage change. In some cases, they simply are not aware of the influence they could have. But they also face serious legal, political, and institutional obstacles at all stages of their work—from establishing a green purchasing program to selecting green products. Unless these barriers are addressed and the gap between good intentions and practical results is narrowed, today's pioneering green purchasing initiatives could be swallowed up in the rising swell of consumption.

Because of the often complex legal framework surrounding procurement, inserting green demands into the purchasing process is usually more easily said than done. Procurement rules can vary depending on the volume, value, or type of purchase, making it difficult for purchasers to determine whether environmental considerations are compatible with existing procedures. Green purchasing can be particularly challenging

in the developing world, where environmental or product standards are often so weak that buyers make poor-quality, or even dangerous, purchases. Corruption and weak enforcement in procurement offer buyers little incentive to make the most efficient purchases, much less buy environmentally preferable products.[36]

As governments worldwide update their procurement procedures and close loopholes that allow for inefficiency, waste, and corruption, this could either lead to more-restrictive regulations that make it harder to buy green or offer new opportunities. The European Commission, for instance, is now exploring the legal possibilities for green purchasing under the European Union Procurement Directives, which have historically made no reference to environmental concerns. A July 2001 communication examined how green criteria might be integrated into different stages of EU procurement, from product specification to supplier selection.[37]

In addition to ongoing legal uncertainties, green purchasing faces formidable political challenges. As markets for environmentally preferable products grow, industries that have an ongoing stake in conventional production (such as oil companies, fertilizer companies, and other manufacturers of less environmentally sound products) will likely lose out. These interests are using their significant influence over institutional purchasing decisions to prevent product alternatives from gaining ground. For years, Tom Ferguson of Perdue AgriRecycle has attended trade shows and garnered support from government buyers for his organic fertilizer product, which is derived from recycled poultry litter. But he is not breaking into the market, he says, because federal specification codes do not allow purchasers to buy alternatives like the one he sells. He notes that powerful industry groups, such as the Fertilizer Institute, will protect chemical agribusiness contracts at all costs. And "if the product or service is not in the government spec, then the hands of the government procurer—no matter how well meaning he is—are tied."[38]

Many green purchasing efforts fall short because organizations do not set strict targets for the activity and there is no system of accountability.

Even when the political climate is more receptive, purchasers face other obstacles to buying green. In most institutions, rules require them to buy the product or service that best meets their needs at the lowest price, which can rule out more expensive environmentally preferable products. Luz Aída Martínez Meléndez of Mexico's Ministry of Environment and Natural Resources complains that finding affordable product alternatives is one of the biggest barriers to green purchasing in her agency.[39]

But some institutions are finding innovative ways to address price concerns. In 2001, the city of Chicago and 48 suburbs pooled their jurisdictional resources to buy a larger block of electricity at a reduced rate, and they will use the savings to meet at least 20 percent of group power needs with renewable sources by 2006. Kansas City and Jackson County in Missouri have agreed to pay a premium of 15 percent more for alternative fuels, cleaning products, and other products they consider environmentally preferable. Other institutions allow purchasers to compare products based on the lifetime cost of ownership, rather than simply the purchase cost—which often reveals the green choice to be cheaper.[40]

There are also internal institutional obstacles to buying green. Because many organizations have no history of environmental responsibility, getting employees to recognize the benefits of adopting more environmentally sound practices can take time. Purchasers, managers, and product end-users are often accustomed to the status quo and resistant to new methods that may complicate their work. Moreover, skepticism about the functionality of many green products persists. For instance, many purchasers still avoid buying recycled paper because they believe it to be of substandard quality, even though these types of performance problems have largely been overcome.

Selecting a focus for green purchasing can be challenging as well. Should an institution target smaller, off-the-shelf commodities like cleaning products, office furniture, and paper or bigger-ticket items like buildings and transport? Ideally, the initiative would focus on changes that make the greatest difference overall, in terms of environment benefits and market influence. But this usually is not the case. Stuttgart, Germany, for instance, focuses its green buying primarily on paper, cleaning products, and computer equipment, even though 80 percent of municipal spending goes for electricity, heating, and the construction and renovation of buildings.[41]

Ultimately, the target may depend on an institution's environmental priorities, financial and legal constraints, and the overall ease or likelihood of adopting the changes. The city of Santa Monica, California, kicked off its green purchasing effort in 1994 with less-toxic cleaning products because a large body of knowledge about product alternatives already existed. Without doing too much additional research, buyers were able to replace traditional cleaners with less-toxic options in 15 of 17 product categories, saving 5 percent on annual costs and avoiding the purchase of 1.5 tons of hazardous materials per year. Japan's Green Purchasing Network, which encourages consumers of all kinds to buy green, is thought to be particularly successful because it focuses mainly on office supplies and electronics. (Some of the group's members, which include local governments, corporations, and NGOs, have achieved 100-percent green procurement of these items.)[42]

One important way to institutionalize green purchasing is by establishing an explicit written policy or law that reinforces the activity. Copenhagen's strategy, which went into effect in 1998, specified that within two years all office supplies had to be PVC-free, all photocopiers had to use 100-percent recycled paper, all printers had to use double-sided printing, and all toner cartridges had to be reused.[43]

But having a policy on the books does not always guarantee that the activity will take place. The United Kingdom's forward-thinking timber procurement rule is a case in point. In 2000, in response to rising worldwide concern about illegal logging, the central government adopted a policy requiring all departments and agencies to "actively seek" to buy wood products certified as coming from sustainably managed forests. A Greenpeace investigation in April 2002, however, revealed that authorities were clearly flouting this law when they refurbished the Cabinet Office in London with endangered sapele timber from Africa. Following up on the incident, the House of Commons Environmental Audit Committee confirmed that there has been "no systematic or even anecdotal evidence of any significant change in the pattern of timber procurement."[44]

Many U.S. green buying laws, too, have failed to live up to expectations. Under the 1976 Resource Conservation and Recovery Act and subsequent updates, federal agencies are required to consider the use of certain recycled, bio-based, and other environmen-

tally preferable products in their procurement and contracting above a specified dollar limit. But two recent reports, by the U.S. Environmental Protection Agency (EPA) and the U.S. General Accounting Office, found that not only were few federal agencies meeting the green requirements, but the majority of agency purchasers were not even aware of the rules.[45]

Julian Keniry, director of the Washington-based National Wildlife Federation's Campus Ecology program, says that many green purchasing efforts fall short because organizations do not set strict targets for the activity and there is no system of accountability. "Policies alone aren't enough," she says. "They need to be coupled with the goal-setting process. Otherwise, they're just words on paper." The more specific and quantifiable an institution's goals, the greater the likelihood that green buying will actually happen.[46]

In some cases, institutions do not impose any sanctions for noncompliance, giving purchasers little incentive to abide by the rules. The 2000 EPA survey, for instance, attributed lax federal compliance with U.S. "buy-recycled" laws to weak enforcement; even if purchasers were aware of the rules, they did not always perceive them to be mandatory. To encourage compliance, Vorarlberg in Austria now holds a regional contest to reward the most environmentally friendly town hall for its buying practices, while in the United States, Massachusetts gives prizes to the most successful state, municipal, and business green purchasers.[47]

At the same time, most institutional accounting systems are not set up to track purchases of recycled or green products, making it difficult to monitor activity. The ongoing decentralization of many government, university, and other institutional purchasing operations compounds the accountability problem. Canada's large government agencies now issue some 35,000 individual credit cards, allowing employees to select and charge their own supplies up to a fixed dollar limit, while more than half of U.S. federal purchases are charged on government bankcards.[48]

Some institutions are tackling the monitoring problem the old-fashioned way: tallying green purchasing receipts by hand. But others are developing more sophisticated systems. Kolding, Denmark, is creating an electronic form for recording green purchases, and the U.S. government has made headway in getting an automated tracking system for green products into the federal procurement system. Other institutions are avoiding the responsibility altogether, putting the onus on suppliers. U.S. renewable energy retailer Green Mountain Energy, for example, requires its paper supplier, Boise Cascade, to provide summary reports on all Green Mountain purchases of recycled paper.[49]

Identifying Green Products

An added challenge in "buying green" is knowing exactly what to look for. Relatively little is known about the environmental characteristics of most products and services on the market today, making it tough for buyers to compare products effectively. Tracing a product's origins up the chain of production can be particularly difficult. A purchaser may unwittingly buy paper originating from virgin forests in Southeast Asia (where forests are rapidly being cleared for agriculture and other purposes) because it has been repackaged and sold under so many different brand names that even most vendors cannot confirm its source. Without the time or scientific background to extensively research green product offerings, many purchasers simply prefer to be told what to buy.[50]

The absence of sound environmental infor-

mation has left manufacturers, environmentalists, and others confused about what exactly constitutes a "green" product or service. Should an "environmentally sound" paper, for instance, contain a maximum percentage of recycled content? Come from a sustainably harvested forest? Be processed chlorine-free? Or some combination of the above? For many green products, widely recognized environmental standards or specifications do not yet exist. In some cases, green products are so innovative that only a handful of companies produce them, or they undergo such a high rate of technological change that standards or specifications simply have not been developed. Yet without agreement on what's really "green," many manufacturers remain reluctant to invest in more environmentally sound technologies.[51]

Fortunately, sophisticated tools are being developed to help both manufacturers and purchasers evaluate the environmental performance of products. One particularly promising technique, lifecycle assessment (LCA), offers a methodology for identifying and quantifying the inputs, outputs, and potential environmental impacts of a given product or service throughout its life. (See Box 6–3.) Volvo, for instance, now uses lifecycle considerations to provide detailed information on the various environmental impacts that arise during vehicle manufacturing and use. And the U.S. Department of Commerce's new BEES software (Building for Environmental and Economic Sustainability) uses lifecycle data to help buyers compare and rate the environmental and economic performance of building materials based on their relative impacts in areas like global warming, indoor air quality, resource depletion, and solid wastes.[52]

Agreement is also emerging, at least among some stakeholders, on how to define certain green goods, such as paper and cleaning

---

BOX 6–3. THE LIFECYCLE APPROACH

A lifecycle approach allows us to look for the unintentional consequences of our actions throughout the entire life of products—from the extraction of the raw materials to the disposal of the product. By offering more complete information about everything from our transport systems to our energy sources, it can help us redirect consumption in a more sustainable direction. "Consumers are increasingly interested in the world behind the products they buy," notes Klaus Töpfer, Executive Director of the U.N. Environment Programme (UNEP). "Lifecycle thinking implies that everyone in the whole chain of a product's lifecycle, from cradle to grave, has a responsibility and a role to play."

In 2001, in response to a call from governments for a lifecycle economy, UNEP and the Society of Environmental Toxicology and Chemistry jointly initiated a Life Cycle Initiative. Through its three main programs—on Life Cycle Management, Life Cycle Inventory data, and Life Cycle Impact Assessment—the initiative works to develop and disseminate practical tools for evaluating the opportunities, risks, and trade-offs associated with products and services over their whole lifecycle. The Initiative is governed by an International Life Cycle Panel, which also serves as a key global forum for stakeholders and lifecycle experts from around the world.

The Initiative contributes as well to the wider 10-year framework of programs to promote sustainable consumption and production patterns called for at the World Summit on Sustainable Development in Johannesburg in 2002. The Johannesburg Plan of Implementation stressed the need for "policies to improve the products and services provided, while reducing environmental and health impacts, using, where appropriate, science-based approaches, such as life cycle analysis."

—*Guido Sonnemann, Division of Technology,*
Industry, and Economics, UNEP

SOURCE: See endnote 52.

---

products. In November 2002, some 56 environmental groups across North America adopted a set of common environmental criteria for environmentally preferable paper and released detailed guidance to advise paper buyers about their choices. That same year, government purchasers, industry representatives, and environmental groups joined forces under a new North American Green Purchasing Initiative to develop uniform criteria and contract language for green purchases of energy, paper, and cleaning products. One working group scored a big victory by agreeing on a single set of criteria for identifying green cleaning products in government contracts (previously, purchasers had used up to 17 different types of contract language).[53]

Many purchasers (and other consumers) also seek guidance from national, regional, and global ecolabeling initiatives. Ecolabels are seals of approval used to indicate that a product has met specified criteria for environmental soundness during one or more stages of its lifecycle. Though the range of products and services that carry ecolabels is relatively small, labels can already be found on everything from green electricity to wood products. Certifiers include government agencies, NGOs, professional or private groups, and international accreditation bodies. (See also Chapter 5.)[54]

Some institutions allow their purchasers to specifically call for ecolabeled items in their contracts. The city of Ferrara, Italy, for instance, seeks to buy paper that carries the

Nordic Swan ecolabel. But many purchasers (particularly those from government) hesitate to endorse specific ecolabeled products, asking instead that their suppliers satisfy the labels' underlying criteria. The state of Pennsylvania has indicated a desire to buy only cleaning products and paints that meet criteria set by Green Seal, a U.S. nonprofit organization that has developed rigorous environmental standards in some 30 product categories. A leading concern is that singling out specific labeled products could create an unfair barrier to trade under the rules of the World Trade Organization, by discriminating against smaller suppliers that may not be able to meet the costs of qualifying for and complying with the labels. (See Chapter 7.)[55]

Japan's Green Purchasing Network now boasts some 2,730 member organizations, including Sony, Toyota, and Canon.

There has been significant industry opposition to ecolabeling as well, particularly in the United States. Green Seal president Arthur Weissman explains that manufacturers like Procter & Gamble, a leading producer of household goods, have used a variety of tactics—from legal arguments to extensive government lobbying—to prevent green certified products from entering the U.S. marketplace. "The way they see it, it interrupts the relationship to the consumer," says Weissman. "A third party interferes with the brand."[56]

In some cases, global manufacturers with multiple product lines have resisted efforts to single out any of their products as environmentally preferable out of a concern that this might make their more conventional offerings look bad. "Once companies begin identifying some of their products as environmentally preferable, customers will want to know what's wrong with the other products,"

explains Scot Case, director of procurement strategies for the Center for a New American Dream. "A company could face legal liabilities if customers suddenly learn that many of its cleaning products, for example, are a toxic witches' brew of known carcinogens and reproductive toxins."[57]

Several other tools also now help purchasers more easily identify environmentally preferable products and services. Many organizations publish green purchasing guidelines or product lists for their purchasers to reference or provide detailed training manuals to lead buyers through the process. The U.S. EPA, for example, offers recommendations for purchasing some 54 different recycled-content products, including traffic cones, toner cartridges, plastic lumber, garden hoses, and building insulation. The city of Göteborg, Sweden, holds training seminars, lectures, and workshops to inform purchasers and other stakeholders about legal requirements, specific tools, and best practices for green procurement. By 2000, 80–90 percent of municipal staff there (both purchasers and end-users) had been trained in green purchasing.[58]

Spreading the Movement

For many years the effort to publicize and promote greener purchasing practices was scattershot, with much duplication of effort and little cross-fertilization of ideas. But this is beginning to change. Today initiatives at the international, regional, and local levels seek both to address the obstacles to green purchasing and to accelerate its adoption. And as more institutions recognize the benefits, they are beginning to share information and to learn from each other's successes and failures.

A number of organizations and networks, mainly in Europe, North America, and Japan, now publish green purchasing information, collect success stories, and publicize trends. They primarily target institutions with significant purchasing power, such as governments and large corporations, though many of their strategies are also applicable at a smaller scale. Some of these groups partner directly with industry leaders and government officials to encourage greener purchasing. Others rally the grassroots to boycott or otherwise pressure manufacturers or other institutions to change their buying practices. Many also use their resources to promote public debate and generate media interest in the green purchasing movement.

ICLEI's Eco-Procurement Program, launched in 1996, is a leader in promoting green purchasing among governments, businesses, and other institutions across Europe. More than 50 cities and other local governments in 20 countries now belong to the group's Buy-It-Green Network, which helps members exchange information and experiences, join forces, and make joint green purchases. The organization also holds yearly conferences and publishes a magazine that is distributed to more than 5,000 purchasers in Europe. And in one of the first efforts of its kind, ICLEI is working on a project to quantify the environmental savings associated with green purchasing, in order to determine how best to strategically combine the purchasing power of cities and to spread green purchasing across Europe. For instance, the project has found that replacing the 2.8 million desktop computers that EU governments buy annually with energy-efficient models could reduce European emissions by more than 830,000 tons of carbon dioxide equivalent.[59]

In North America, a leading proponent of institutional green purchasing is the U.S. EPA's Environmentally Preferable Purchasing program, created in 1993 by presidential executive order. The program offers support and information in such areas as construction, office products, conferencing and printing services, cleaning products, cafeteria procurement, and electronics. EPA has also launched several pilot projects, including partnerships with the Department of Defense (to green military operations and installations) and with the National Park Service (to help parks both green their purchasing and educate visitors about consumption). Through its central database of information, the EPA also serves as a clearinghouse on more than 600 environmentally preferable products and services, including links to 130 local, state, and federal green contract specifications, to 523 product environmental performance standards, and to 25 lists of vendors and products that meet these standards.[60]

The Maryland-based Center for a New American Dream helps large purchasers, particularly state and local governments, incorporate environmental considerations into their purchasing decisions. Its Procurement Strategies Program was an active driver behind the new North American Green Purchasing Initiative in 2002, which aims to generate a critical mass for green purchasing on the continent. The group also hopes to serve as a central clearinghouse for green purchasing information for manufacturers, purchasers, and suppliers.[61]

Japan's Green Purchasing Network (GPN) now boasts some 2,730 member organizations, including more than 2,100 businesses (among them Panasonic, Sony, Fuji, Xerox, Toyota, Honda, Canon, Nissan, and Mitsubishi); 360 local authorities in places like Tokyo, Osaka, Yokohama, Kobe, Sapporo, and Kyoto; and 270 consumer groups, co-ops, and other NGOs. GPN holds countrywide seminars and exhibitions on green purchasing, publishes purchasing guidelines and environmental data books on different products and services, and offers awards to exemplary organizations.[62]

In higher education, more than 275 university presidents and chancellors in over 40 countries have signed on to the 1990 Talloires Declaration, a 10-point action plan that, among other things, encourages universities to establish policies and practices of resource conservation, recycling, waste reduction, and environmentally sound operations. In the hospitality industry, the International Hotels Environment Initiative, a global nonprofit network of more than 8,000 hotels in 11 countries, sponsors a Web-based tool to help hotels improve their environmental performance (and save money) through purchases of everything from energy-efficient lighting to environmentally preferable flooring materials, refrigerators, and minibars.[63]

Since 1992, DaimlerChrysler has tapped Brazil's rainforests for environmentally sound coconut fiber and natural rubber, which it now uses in car seats and headrests.

There are also efforts to bring greater media attention to green purchasing. In February 2001, the Danish Environmental Protection Agency launched an intensive television, newspaper, and leaflet campaign to raise interest in products carrying ecolabels. Japan's Green Purchasing Network has worked hard to feature green purchasing prominently on television, in newspapers, and at government or corporate seminars. Organizations of all kinds also now use the Internet to inform their buyers about green purchasing, offering procedural tips and links to alternative products and services. King County, in the state of Washington, uses its comprehensive Web site and e-mail bulletins to disseminate success stories and other green purchasing developments.[64]

Fledgling efforts to spread green purchasing in the developing world are also under way, though considerable work remains to be done. ICLEI's ERNIE (Eco-Responsible Purchasing in Developing Countries and Nearly Industrialized Economies) program, supported by the Global Environment Facility, is working with local authorities in several cities—including São Paulo, Brazil; Durban, South Africa; and Puerto Princessa, Philippines—to develop green purchasing pilot projects. The initiative focuses mainly on buying energy-efficient appliances, and aims to address the various market and other barriers to green procurement, including the need to build capacity for local suppliers and manufacturers.[65]

One way institutions can help spread green purchasing to the developing world is by using their own procurements to strengthen local green markets. For instance, the United Nations, World Bank, donor agencies, and multinational corporations that operate in these countries can seek to buy a greater portion of their goods and services from local green suppliers, helping to build capacity for sustainable production. Since 1992, DaimlerChrysler has tapped Brazil's rainforests for environmentally sound coconut fiber and natural rubber, which it now uses in car seats, armrests, and headrests. In doing so, the automaker has not only eliminated the use of synthetic inputs in these vehicle parts, it has also boosted local markets for renewable materials and generated income and employment for farmers.[66]

In most cases, however, it is a big enough challenge simply to get international institutions to buy from developing countries, much less buy green. Although some of these institutions do try to buy locally, their procurement generally favors businesses in the industrial world. (In fact, most donor agencies tie their aid to purchases back home.) In 2000, only about a third of the procurements by the U.N. system went to the developing world. Occasionally, these institutions make socially responsible demands in their purchasing: UNICEF, for instance, seeks to develop sourcing policies and strategies that support national goals for boosting children's welfare. But so far they have rarely specified environmental criteria, in part because there is a risk that putting green specifications in contracts could alienate smaller suppliers that may not be able to meet them.[67]

By boosting green purchases in developing countries, international institutions can not only stimulate markets, they can also clean up their own acts in the face of mounting criticism about the environmental impacts of their activities. There is rising interest, for example, in inserting environmental criteria into the procurements associated with World Bank lending, as part of larger efforts to green the Bank's operations. The Bank is now working with a coalition of other multilateral development banks, U.N. agencies, and NGOs to stimulate green purchasing both within and outside member institutions. This interagency group hopes to also mainstream the idea of incorporating social justice criteria into members' procurement decisions.[68]

Clearly, the world's institutions have significant power to bring about environmental and social change through their purchases. But no matter how environmentally sound this purchasing is, it still uses resources and generates wastes. To truly mitigate the impacts of their consumption, institutions will need to find ways to meet their needs without buying new products—for instance, by eliminating unnecessary purchases and extending the lives of existing products. Pori, Finland, has implemented a citywide service for reusing goods that enables employees from any municipal department to trade or give away products they no longer need or use. And since 1994, the University of Wisconsin-Madison's SWAP (Surplus With a Purpose) project, which has subsequently been expanded statewide, has helped divert reusable office furniture, computers, and other goods away from landfills and to other users on campus and around the state.[69]

Green purchasing is not the only way to minimize the problems associated with excessive consumption. But it is an important step along the way to achieving a more sustainable world. As individuals, we will need to push the organizations we work for and rely on to join us in building such a world.

Paper

For most if its history, paper existed as a precious and rare commodity. Now it covers the planet. In modern times, we hardly notice most of the paper drifting through our lives. From the contents of our in-boxes to the currency in our wallets to the containers for our frozen dinners, paper is never far from reach.

Globally, paper consumption increased more than sixfold over the latter half of the twentieth century. The United States—at 331 kilograms per person per year, and roughly 30 percent of the world's total annual use—is the biggest paper user. On a per capita basis, the Japanese are next, at 250 kilograms each. Though invented as a tool to communicate, about half the paper in today's consumer society serves another purpose: packaging. From precious to disposable— paper now represents a big chunk of the modern waste stream, accounting for roughly 40 percent of the municipal solid waste burden in many industrial countries.[1]

While paper gets its name from "papyrus," a water reed plucked, pounded, and pressed into service by ancient Egyptians to record hieroglyphs, fibrous paper as we know it was invented in China less than 2,000 years ago. Over the next two millennia, rags and hemp were popular raw materials for paper. The Gutenberg Bible, the first and second drafts of the U.S. Declaration of Independence, and the original works of Mark Twain were all printed on hemp-based papers. It wasn't until 1850 that Friedrich Gottlob Keller of Germany devised a method of making paper from wood. It took several more decades for trees to become the raw material of choice, as others refined Keller's technique and found ways to mass-produce wood-based paper.[2]

In the twenty-first century, we hardly think of paper as derived from anything but wood. Indeed, 93 percent of today's paper comes from trees, and paper production is responsible for about a fifth of the total wood harvest worldwide. Newly cut trees account for 55 percent of the total supply, while 7 percent comes from non-tree sources and the remaining 38 percent is from recycled wood-based paper.[3]

Trees around the world feed the global paper supply. U.S. woodlands represent the largest contributor, at 30 percent of the total, but that share has been shrinking in recent decades as China and other developing countries increase production. Paper production has been moving within the United States as well. When logging was curtailed in ancient forests of the Pacific Northwest, paper production moved to lower-profile but biologically rich second-growth forests in the Southeast. These forests now supply one quarter of the world's paper. And second-growth forests supply 54 percent of the total paper derived from virgin wood worldwide. Tree plantations (often planted on newly deforested lands) are next at 30 percent, and old-growth, mostly boreal, forests account for the remaining 16 percent.[4]

The process of turning trees into paper begins at the chip mill, where a series of rotating blades reduces logs to poker-chip-sized pieces. The chips are carted to pulp mills, which may be thousands of kilometers away, where they are mixed with chemicals in vast pressure cookers and digested into a wet paste the consistency of oatmeal. Washed and bleached several times, this mix is finally pressed and dried, emerging as large rolls of paper for public consumption. In the end, a piece of writing paper might contain fibers from hundreds of different trees that have collectively traveled thousands of kilometers from forest to consumer.

Making paper is extremely resource-intensive. A ton of paper requires two or three times its weight in trees, along with a great deal of water and energy. Worldwide, the pulp and paper industry is the fifth largest industrial consumer of energy, and uses more water to produce a ton of product than any other industry. Paper mills can be obnoxious neighbors, emitting foul odors and generating large quantities of air and water pollution and solid waste. Though paper mills in the industrial world have taken some steps to clean up, elsewhere mills continue to spew appalling amounts of untreated toxic waste into the air, land, and waterways.[5]

Individuals and large institutions can help reduce the paper burden in a number of ways—from being more mindful in the office about paper use to being more diligent about recycling. Recycling saves more than trees. Using recycled content rather than virgin fibers to produce paper creates 74 percent less air pollution and 35 percent less water pollution.[6]

Large institutions, in particular, can play a key role in driving the market for recycled papers. In 2002, the 270 members of the U.S.-based Recycled Paper Coalition—an organization of major industries, nongovernmental organizations, and government agencies formed to use bulk purchasing power to foster the market for recycled paper—bought nearly 150,000 tons of recycled paper, with an average postconsumer content of 24 percent.[7]

Tackling packaging waste can also reap big dividends. Germany is a pioneer in this area, passing an ordinance in 1991 that required packaging producers and distributors to take back and reuse or recycle packaging materials, including paper. In the following three years, wastepaper recycling in Germany shot up to 54 percent, after stagnating at 45 percent for nearly 20 years. In 2003, the European Union Parliament adopted a law that would require member governments to set waste paper recycling goals of 60 percent by 2008.[8]

Paper is also returning, on a limited scale, to its nonwood roots. Several alternative fibers are now on the market—from the old standby, hemp, to kenaf (a leafy member of the hibiscus family), agricultural residues (cereal straws, cotton linters, banana peels, coconut shells, and others), and even denim scraps. Many "agrifibers" yield more pulp-per-acre than forests or tree farms, and they require fewer pesticides and herbicides. Fewer chemicals and less time and energy are needed to pulp agricultural fibers because they contain less lignin, a glue-like substance that makes plants and trees stand erect. In the future, some of these nonwood sources might once again become significant sources of paper.

—*Dave Tilford,*
Center for a New American Dream

CHAPTER 7

Linking Globalization, Consumption, and Governance

Hilary French

In May 2003, a delegation of indigenous leaders from the Ecuadorian and Peruvian Amazon visited Washington, D.C., to tell people about the environmental and social toll of oil extraction by U.S.-based corporations on their lands. Following meetings in Washington, the delegation went to Houston, Texas, to meet with Burlington Resources, a company that holds two oil development concessions covering 400,000 hectares of their ancestral territory.[1]

On behalf of 100,000 Shuar, Achuar, and Kichwa people living on approximately 1.6 million hectares of pristine rainforests, the delegation delivered a letter to the Chief Executive Officer of Burlington Resources calling on the company to cease all oil activities in the area and to leave the territory immediately. Citing the toxic contamination and forest destruction left behind by previous oil development operations elsewhere in the Amazon, the president of the Independent Federation of Shuar Peoples declared emphatically that "the Shuar and Achuar people of the Ecuadorian Amazon want it to be known

that the position of our communities is no to oil exploration, no to dialogue and negotiation, no to deforestation, no to contamination, and no to all oil activities."[2]

These indigenous leaders provided a vivid reminder of the great but often hidden toll that consumption in the world's richest countries can take on distant peoples and places. The delegation's visit put a human face on the tendency of today's global economy to insulate consumers from the various negative impacts of their purchases by stretching the distance between different phases of a product's lifecycle—from raw material extraction to processing, use, and finally disposal. While sales of sport-utility vehicles have skyrocketed in the United States over the last decade, for instance, few if any of the new owners stop to ponder the connection between their recent purchase and the fate of indigenous peoples whose lives and livelihoods have been torn asunder in the push for petroleum.[3]

Although the Amazonian delegation's visit was in some ways a cautionary tale, it also offered grounds for hope, as it demonstrated how the environmental and social challenges accompanying economic globalization are spurring innovative forms of political mobilization across international borders. To shift to environmentally sustainable patterns of consumption and production worldwide, we need to strengthen such alliances in pursuit of the new ground rules needed to forge a global economy based on protecting rather than plundering the planet's natural wealth.

The Spread of "McWorld"

Benjamin Barber's *Jihad vs. McWorld*, first published in 1995, was unusually prescient in describing our complicated world, in which two seemingly contradictory scenarios play out simultaneously: one "in which culture is pitted against culture, people against people, tribe against tribe" and a second in which "onrushing economic, technological, and ecological forces...demand integration and uniformity and...mesmerize peoples everywhere with fast music, fast computers, and fast food..., one McWorld tied together by communications, information, entertainment, and commerce."[4]

The global spread of "McWorld" is rapidly bringing the consumer society of the West to the rest of the world. Shortly after the fall of the Berlin Wall in 1989, billboards for western tobacco and liquor brands began to appear widely in Eastern Europe and the former Soviet Union, sometimes in the same central squares that once featured the busts of Communist leaders. And visitors to some of the most remote outposts of the developing world often find Coca-Cola kiosks at the end of the road. McDonalds itself currently operates 30,000 restaurants in 119 countries, while the German-based enterprise Siemens is represented in 190 countries, where it sells mobile phones, computers, medical supplies, lighting, and transporta-tion systems. (See Table 7–1.)[5]

The rapid globalization of the consumer economy over the 1990s was closely linked with a general economic boom that saw rapid growth in the movement of goods, services, and money across international borders. The value of world trade in goods increased by nearly 50 percent during the decade, climbing from $4.22 trillion to $6.25 trillion. Exports of commercial services such as banking, consulting, and tourism expanded even faster. (See Figure 7–1.) Foreign direct investment (FDI) also surged dramatically, reaching a peak of $1.4 trillion in 2000. The FDI explosion was spurred in part by a frenzy of corporate mergers, although that trend has dramatically reversed during the last few years in response to the global economic slowdown and a general weakening of business confidence following the terrorist attacks on the United States in September 2001. The growth of global trade and investment of recent decades has contributed to lower costs for many consumer goods, such as clothing, computers, and toys—a phenomenon that is hailed by traditional economists while decried by critics of the global consumption binge. (See Chapter 5.)[6]

One subcomponent of the broader overall expansion in world trade has been rapid growth in trade in a range of particularly environmentally sensitive commodities, such as minerals, forest products, fish, and agricultural produce. (See Figure 7–2.) The value of world trade in forest products, for instance, climbed fourfold between 1961 and 2001, reaching a peak of $148 billion in 2000, before dropping to $132 billion in 2001. At the same time, Earth's overall forest cover has steadily declined. Trade is by no means the only factor in this, but it has played a significant role. Similarly, the value of world fish exports nearly tripled between 1976 and 2001, reaching $56 billion in 2001. At the same time, the

Table 7–1. The Spread of "McWorld"

Corporation	Global Presence
Hennes & Mauritz	The Sweden-based clothing company employs 39,000 people in 17 European countries and the United States. It operates 893 stores and plans to open 110 more, expanding into Canada, in 2003. Turnover was $6.8 billion in 2002. H&M uses suppliers in Europe and Asia.
Levi Strauss	This American company sells clothing in more than 100 countries, and its trademark is registered in 160 countries. It employs 12,400 people worldwide. It reported total sales of $4.1 billion in 2002, and a net income of $151 million in 2001.
Tata Group	The Tata Group operates in seven industry sectors, including transportation, energy, chemicals, and communications services. Developed in India, it now has business partners in 11 countries across the world. It reported a turnover of $2.9 billion in 2001–02, more than double that of the previous year.
Altria Group, Inc.	The Altria Group is the parent company of Kraft Foods, the second largest branded food company in the world, and Philip Morris, the most profitable international tobacco company. The Altria Group had net revenues of $80.4 billion in 2002, including $28.7 billion from the international tobacco market. It employs 169,000 people in 150 countries.
Siemens	This German company employs 426,000 people and is represented in 190 countries. It sells mobile phones, computers, medical supplies, lighting, and transportation systems. In 2002, Siemens' net sales amounted to $96.4 billion, of which 79 percent were international. One million people hold shares in the company.
Yum! Brands	Formerly part of PepsiCo, this company and its six subsidiaries—KFC, Pizza Hut, Taco Bell, A&W, All-American Food Restaurants, and Long John Silvers—had global sales of over $24 billion in 2002. It operates 32,500 restaurants in more than 100 countries and employed 840,000 people in 2002. Yum! Brands opened 1,000 restaurants outside of the United States in 2001, nearly 3 per day. China now has 800 KFCs and 100 Pizza Huts.
McDonald's Corp.	McDonald's serves 46 million customers each day. It operates 30,000 restaurants in 119 countries. Its total revenue was $15.4 billion in 2002. On opening day in Kuwait City, the line for the McDonald's drive-through was over 10 kilometers long.
Domino's Pizza	Domino's opened its 7,000th store in 2001 and is active in 60 countries. Sales for all countries added up to $4 billion in 2002. Its delivery people drive over 14 million kilometers each week in the United States alone. Domino's uses 149 million pounds of cheese each year and 26.8 million pounds of pepperoni.
Coca-Cola	Coca-Cola sells more than 300 drink brands in over 200 countries. More than 70 percent of the corporation's income originates outside of the United States, and its net revenues reached $19.6 billion in 2002. Coca-Cola employs 60,000 people in Africa alone.

SOURCE: See endnote 5.

LINKING GLOBALIZATION, CONSUMPTION, AND GOVERNANCE

world saw a deterioration in the health of the world's fisheries, with the U.N. Food and Agriculture Organization estimating that 75 percent of the world's fish stocks are now fished at or beyond their sustainable limits.[7]

In a rather different kind of global exchange, countries whose ecological footprints exceed their available ecological capacity often import goods from countries enjoying surpluses, leading to ecological trade deficits. (See Chapter 1 for a discussion of the ecological footprint accounting system, which measures the amount of productive land an economy requires to produce the resources it needs and to assimilate its wastes.) Nations vary greatly in the size of these deficits; countries as diverse as Japan, the Netherlands, the United Arab Emirates, and the United States are all major importers of ecological capital. (See Figure 7–3.) Although there are times when this sort of global transfer makes ecological and economic sense, in effect it is enabling countries to live beyond their ecological means.[8]

The growing globalization of the world economy also serves to shield consumers and producers from the wastes generated in making, using, and ultimately disposing of the multitude of goods and gadgets that characterize the consumer economy. The resulting "out of sight, out of mind" mentality has the effect of shifting these burdens to others and

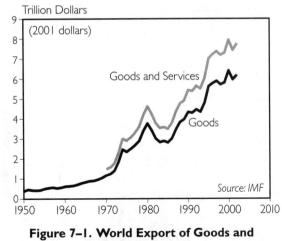

Figure 7–1. World Export of Goods and Services, 1950–2002

reducing the incentive to address unsustainable consumption patterns at their roots. (See Box 7–1.) The world's attention was first focused on the waste export problem in a significant way in the mid-1980s, when a series of highly publicized incidents—such as Philadelphia's wandering "garbage barge,"

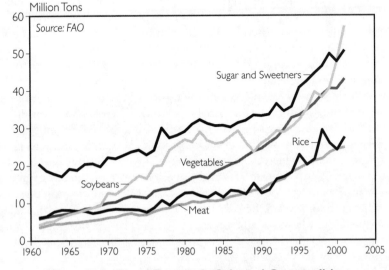

Figure 7–2. World Exports in Selected Commodities, 1961–2001

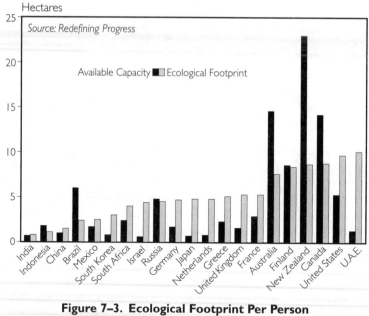

Figure 7–3. Ecological Footprint Per Person in Selected Nations, 1999

a ship loaded with toxic ash that was turned away from three states and five nations over 16 months—put the spotlight on the growing international commerce in hazardous and solid wastes. More recently, the profusion of electronic wastes created by the information age has led to a flourishing international trade in discarded computers, televisions, telephones, and the like.[9]

With markets for many consumer goods becoming saturated in industrial countries, corporate strategies increasingly rely on visions of rapid growth in developing countries, leading to increases in purchases of all manner of goods, from cars and televisions to paper and fast food. This trend is particularly pronounced in the Asia-Pacific region, which is now home to an estimated 684 million members of the global consumer class—more than in Western Europe and North America combined. (See Chapter 1.) While it is ethically problematic to suggest that

developing countries are not entitled to have the same options for material consumption that have long been taken for granted by western consumers, the global adoption of industrial country–style consumption patterns would place unbearable strains on the health of Earth's natural systems.[10]

In the face of this conundrum, some analysts from the developing world have in recent years begun to emphasize the opportunities rather than the downsides that await countries shifting to environmentally sustainable economies. The China Council for International Cooperation on Environment and Development, for instance, noted in a recent statement that "China's remarkably low per capita consumption pattern is an opportunity to avoid the mistakes of many other countries that have developed very high levels of material and energy consumption. Moving towards more sustainable consumption patterns could lead to more competitive domestic enterprises and greater access to international markets." The challenge is to develop strategies for leapfrogging directly to an economy where producers use cutting-edge green technologies on a widespread basis and where consumers practice sustainable purchasing as a matter of course. (See Chapters 5 and 6.)[11]

More than two years since terrorist attacks on New York and Washington brought the Jihad and McWorld paths onto a direct col-

BOX 7–1. FAIR TRADE AND THE CONSUMER

As poet and farmer Wendell Berry noted recently: "One of the primary results—and one of the primary needs—of industrialism is the separation of people and places and products from their histories." Each minute, practically, in the life of the modern consumer contains hidden interactions with people and portions of the planet hundreds or thousands of miles away. The global trade network allows consumers in large part to shed their dependence on immediate surroundings. The unfortunate consequence is that consumers are shielded from the profound effects their choices can have on the lives of people at the other end of the production and consumption line.

While the benefits of free trade flow to consumers and middle merchants, the burdens routinely flow elsewhere—out to the end points along the trade lines. We extract resources from and dump wastes into areas occupied by the poor and underrepresented. Although someone at the other end of the line is always there to accept the messy job of cleaning up after industrial-world consumers, those harmed frequently are not among those compensated, or the compensation is dwarfed by the damage to vital local resources.

The growing problem of electronic waste provides one vivid example. Consumers have little cause to contemplate what lurks within the plastic case of a computer or mobile phone, or what becomes of an electronic item scrapped for a newer model. To see what happens, they would have to travel to places like the Guiyu region of China's Guangdong Province. Hundreds of trucks rumble through there every day, carting spent computers, printers, and televisions from North America to dumping grounds scattered among the small villages of the region. For a dollar or two a day, unprotected migrant workers sift through mounds of electronic waste—burning plastics, cracking apart cathode-ray tubes, and pouring acid over circuit boards to extract precious metals and other valuable materials within.

Carcinogenic smoke fills the air around the dumps. The region's water has become so polluted that drinking water must be trucked in from 30 kilometers away.

Consumer desires and the welfare of those caught up in the process of meeting those desires can be intertwined in complex ways. Local people may depend financially on the industry producing the goods even as they suffer the harm caused by the industry. The harm is seen as an unfortunate but unavoidable side effect. But the damage to local resources and the adverse conditions under which people toil are typically byproducts of efforts to keep prices low for the end-consumer. To cite one example, the banana industry in Panama employs a whopping 70 percent of the population. To boost production, bananas are grown in huge, monoculture plantations that are heavily dosed with pesticides applied directly by unprotected workers or through indiscriminate aerial spraying. The chemicals pollute the local water supply and have been linked with increased cancer rates in communities near the plantations. In essence, the welfare of the workers and communities has been left out of the final price.

Better practices depend on consumers understanding the issues and supporting better systems. In the banana industry, activist groups have begun to call attention to the plight of workers, inspiring some companies to change their practices. Dole, for instance, is expanding its efforts to farm organic bananas, grown without pesticides. And Chiquita's Latin American plantations are now 100-percent certified by the Rainforest Alliance's Better Banana Project, which inspects plantations to verify that they are using sustainable practices that are healthier for banana consumers and for the environment, as well as beneficial for employees.

"Fair trade" is emerging in connection with some commodities as a tool to provide farmers and independent producers with more control over the sale of their products and a closer

---

BOX 7–1. (continued)

connection with end-consumers. Under fair-trade systems, small producers band together to form cooperatives, selling directly to retailers for a guaranteed minimum price. At present, coffee is the most visible example. In recent years, coffee prices paid to farmers plummeted to all-time lows while profits for big coffee retailers remained substantial. In Central America, well over a half-million coffee workers lost their jobs. Once-thriving villages turned into ghost towns, their former inhabitants crowded into dangerous shanty-towns on the outskirts of urban centers.

Through fair trade, many farmers are able to stay afloat. Members of the Oromiya Cooperative in Ethiopia get more than twice what their neighbors receive selling coffee on the open market. Fair trade has environmental benefits as well. Farmers with more-stable markets gain the breathing room to adopt a long-term focus. Members of the Miraflor Cooperative in Nicaragua—like many fair-trade coffee cooperatives—grow shade-grown, organic coffee in areas once heavily dosed with pesticides.

Though still a small share of the market, fair-trade coffee sales grew by 12 percent in 2001, compared with overall growth in coffee consumption of just 1.5 percent.

Consumers have the power to make the global trading system more just and sustainable. Outrage over unjust conditions and market demand for more socially responsible products can help change the way companies do business and create a better atmosphere for those at the ends of production and consumption lines. When the effects of consumption are hidden from view, social and environmental costs tend to be left off the balance sheets and reforms are difficult to enact. Greater awareness on the part of consumers and a willingness to act on that awareness, however, can reconnect consumer items to their histories and can counter the harm that often accompanies unconscious consumption.

—*Dave Tilford, Center for a New American Dream*

SOURCE: See endnote 9.

---

lision course, it appears increasingly clear that neither scenario will bring about a stable and secure future. Just prior to the first anniversary of September 11th, tens of thousands of people from around the globe gathered in Johannesburg, South Africa, for the World Summit on Sustainable Development. Conference participants implicitly rejected both the Jihad and McWorld paths while embracing the cause of building an environmentally sound and socially just global society. Sustainable development proponents worldwide now face the challenge of keeping both public attention and political will focused on the urgent need to breathe life into the many important international agreements forged in Johannesburg, including commitments to

transform unsustainable patterns of consumption and production.

Global Cooperation for Sustainable Consumption

The first concentrated international attention to consumption and production issues came a decade before the Johannesburg conference, when the United Nations convened the Earth Summit in Rio de Janeiro in 1992. At this historic gathering, governments officially recognized underlying consumption and production patterns as key driving forces for unsustainable development and highlighted the responsibility of nations to reverse them. Since then, consumption and production issues have been treated as two sides of the same coin in the world of international policy. This fusion reflects the inextricable links between the two phenomena: it is impossible to use sustainable products if nobody has produced them. But the linkage also reflects the reality that most governments find it more politically palatable to discuss the production side of the equation than controversial lifestyle issues.[12]

Agenda 21, the lengthy action plan for sustainable development that emerged from the Rio conference, highlighted the disparity between the "excessive demands and unsustainable lifestyles among the richer segments" and the inability of the poor to meet their basic needs for food, good health, shelter, and education. It also called on international institutions and national governments to undertake a number of initiatives to reverse unsustainable consumption and production patterns, such as promoting greater energy and resource efficiency, minimizing waste generation, encouraging environmentally sound purchasing decisions by both individuals and governments, and shifting toward pricing systems that incorporate hidden environmental costs. These commitments were particularly noteworthy in light of repeated pronouncements from U.S. officials that the American way of life was not negotiable at Rio.[13]

Responsibility for overseeing follow-up activities fell to the U.N. Commission on Sustainable Development (CSD), an intergovernmental body that meets annually to track efforts to implement the agreements reached in Rio. The CSD has provided a useful venue for several discussions of consumption and production issues among governments and nongovernmental observers over the last decade. But despite much talk, the deliberations have produced little in the way of concrete action.[14]

One exception has been a successful effort to revise the U.N. Guidelines for Consumer Protection. These guidelines are not binding, but nonetheless they offer a tool for governments to use in developing their own policies. The revised guidelines, adopted in 1998, encourage governments to implement a range of policy innovations to promote sustainable consumption, including conducting impartial environmental testing of products, strengthening regulatory mechanisms for consumer protection, and integrating sustainable practices into government operations. Unfortunately, a 2002 survey by the U.N. Environment Programme (UNEP) and Consumers International concluded that countries were making only slow progress in implementing the guidelines, with 38 percent of responding governments indicating they were not even aware of them.[15]

Several other international organizations were active on consumption and production issues in the decade following the Earth Summit. The Paris-based Organisation for Economic Co-operation and Development, a forum on economic and social policies for the world's major industrial countries, has sponsored a series of meetings and research papers aimed at encouraging governments to implement innovative policies on sustainable consumption and production, including ecolabeling systems that help consumers select environmentally sound products, "take-back" legislation requiring producers to retrieve packaging and discarded goods, reductions in government subsidies to environmentally harmful industries, and environmental taxes to internalize environmental costs into the price of products. (See Chapter 5.)[16]

UNEP is another active player in efforts to promote sustainable consumption on a global scale. This Nairobi-based U.N. program launched a Life-Cycle Initiative in 2002 to bring together industry leaders, academics, and policymakers to encourage the development and dissemination of practical tools for evaluating the environmental impacts of products over their entire lives. UNEP also cooperates with other U.N. agencies and the World Bank to encourage collaboration on the greening of procurement procedures at these institutions. It works with industries of particular importance to the quest for sustainable consumption, including the advertising, fashion, finance, and retail sectors, to encourage them to take steps to promote sustainable consumption and production. And it seeks to engage nongovernmental organizations (NGOs) in the shift toward sustainable consumption, including consumer and youth groups. (See Box 7–2.)[17]

The 1990s also saw governments make progress toward strengthening several international treaties that address threats to the global environment. These are more binding than the cooperative activities just described and thus form a key component of broader efforts to change unsustainable patterns of consumption and production. For example, nations participating in a 1995 accord on the cooperative management of international fisheries committed to developing national policies to restore fish stocks to healthy levels, thereby encouraging sustainable fish harvesting and consumption. Governments that signed the 2000 Cartagena Protocol on Biosafety to the U.N. Convention on Biological Diversity agreed to abide by a system of prior informed consent for international shipments of genetically modified organisms (GMOs) and products containing them, which gives importing countries greater control over whether to use such products domestically. Countries abiding by the 2000 Stockholm Convention on Persistent Organic Pollutants (POPs) pledge to regulate the production and use of 12 particularly harmful chemicals, including eliminating 9 of them altogether. And countries agreeing to the carbon dioxide emissions targets in the 1997 Kyoto Protocol to the U.N.'s Framework Convention on Climate Change need to shift toward less fossil-fuel-intensive energy use in order to meet them.[18]

Enough countries have now formally ratified both the fish and biosafety conventions to bring these treaties into legal force, making their provisions binding on participating nations. This is not yet the case for either the POPs Convention or the Kyoto Protocol, although many signatory countries are nonetheless in the process of changing their national policies accordingly.[19]

In addition to initiatives by international institutions and governments, the decade since the Rio conference also saw the development of new information tools such as international labeling and certification systems in response to a heightened sensibility on the part of individual consumers to the ties that bind them through global product chains to people and communities in distant lands. One example is the growing popularity of coffee, bananas, and other produce that meet the criteria for being labeled as either organic or traded on fair economic terms, or both. (See Chapter 4.) Another case in point is the impact of the Forest Stewardship Council (FSC), an independent body established in 1993 to set standards for sustainable forest production through a cooperative process involving timber traders and retailers as well as environmental organizations and forest dwellers. A decade later, the FSC had certified over 39 million hectares of commercial forest in 58 countries, more than six times as much area as in 1998, although still only some 1 percent of the world's forests.[20]

A Marine Stewardship Council (MSC) modeled on the FSC was created a few years later. Seven fisheries to date have been certified as meeting the MSC's environmental standard for being well managed and sus-

---

BOX 7–2. HARNESSING THE POWER OF YOUNG PEOPLE TO CHANGE THE WORLD

There are more than 1 billion young people between the ages of 15 and 24, according to the U.N. Population Fund, and more than 500 million young people will enter the labor force in developing countries over the next decade. These numbers point to the huge potential influence that young people can have in shaping a better future as a result of their lifestyle choices and their professional contributions. But the buying and decisionmaking powers of 1 billion young people today are far from homogeneous. Half of them live in poverty. At the other end of the spectrum, young people in affluent societies account for a growing share of total consumption and are under constant pressure to buy more. With the globalization of cinema, television, and advertising, there is a danger that the tendency of these media to glorify the materialistic youth lifestyles in the world's most affluent countries could have a negative impact on the attitudes and consumption patterns of other youngsters.

In response to these trends, the U.N. Environment Programme and UNESCO did a survey in 2000—called "Is the Future Yours?"—on attitudes toward consumption among people between the ages of 18 and 25. More than 8,000 people in 24 countries responded to the survey, providing important information about the aspirations and interests of youth, their awareness of environmentally and ethically responsible consumption, and their vision of their roles in improving the world for the future. The survey found that young people understand the environmental impact of their use and disposal of products, but that they are less aware of the impact of their shopping habits, particularly for food and clothing. It also found that young people consider environmental, human rights, and health issues to be major concerns for the future, but that they favor individual over collective action to address them.

While the survey found that young people do not generally make links between their personal behavior and global problems, there are nonetheless many examples of youth activists who are hard at work urging their communities and governments to promote sustainable consumption. For example, a 23-year-old Peruvian activist enlisted Shell in a project paying for the installation of solar panels in a remote mountain village, a youngster in Cameroon travels from village to village to teach other young people how to use water more safely and efficiently, young "supermarket rangers" in Sweden have opened a dialogue with supermarkets to ensure that sustainable products are easily available to consumers, and in the United States young people have developed a gift guide with fair-trade and enviromentally friendly gift suggestions. In response to the survey results and a subsequent workshop convened to discuss them, UNEP and UNESCO launched the YouthXchange Project to develop tools to help young people take action to promote sustainable consumption.

—Isabella Marras, U.N. Environment Programme

SOURCE: See endnote 17.

---

tainable, including the Alaska salmon fishery, the New Zealand hoki fishery, and the Western Australian rock lobster fishery; many more are currently in the assessment stages. Some 170 MSC-certified seafood products are now offered in 14 countries. But as with forest products, these still represent only a small share of total production. Tipping the balance so that sustainably harvested products are the rule rather than the exception will require new incentives and regulations to bring about a broader transformation in the global marketplace. (See Chapters 5 and 6.)[21]

Toward this end, several important initiatives to encourage global corporations to adopt more sustainable production techniques have unfolded over the last decade. In 2000, the United Nations launched the Global Compact, which calls on participating companies to integrate nine core values related to human rights, labor standards, and environmental protection into their operations. More than 1,200 companies from over 50 countries have signed on so far, although critics have charged that the program requires little in the way of specific actions and that it fails to provide for effective monitoring of implementation or compliance. More recently, 17 leading banks from 10 countries adopted the Equator Principles for managing environmental and social risks in lending operations. Participating banks agreed to require clients borrowing money for large projects such as dams and power plants to adhere to the World Bank's environmental and social standards, which are rapidly developing into de facto international baselines for both public and private investments.[22]

Despite these steps forward, the sobering reality is that the limited gains made since 1992 in shifting toward more-sustainable patterns of consumption and production have been largely overwhelmed by the continued global growth of the consumer society. Delegates spent many hours during the World Summit in Johannesburg debating what they might do to turn this situation around. The power of vested interests and institutional inertia translated into reluctance on the part of many governments to commit to a clear program of action toward this end. Nonetheless, the official Plan of Implementation that governments agreed to stipulates that all countries should promote sustainable consumption and production patterns and that governments, international organizations, the private sector, and NGOs, among others, should play important roles in bringing about the needed shifts. Among other things, the Plan of Implementation calls for increasing investments in cleaner production and eco-efficiency, enhancing corporate environmental and social responsibility, and promoting the internalization of environmental costs and environmentally sound public procurement policies. (See Box 7–3.)[23]

The Plan of Implementation also endorses the development of a 10-year framework of programs at the international level to support regional and national initiatives to accelerate the shift toward sustainable consumption and production, including offering a better range of products and services to consumers, providing them with more information about the health and safety of various products, and establishing programs of capacity building and technology transfer to help share these gains with developing countries. In June 2003 the United Nations convened an international experts meeting in Marrakech to jumpstart this process, and it has organized related regional meetings in both Asia and Latin America.[24]

In addition to the formal process just described, the World Summit also generated more than 230 partnership agreements in which diverse stakeholders pledged to take joint action to help achieve the broad range of targets related to sustainable development that were agreed to in Johannesburg. Several of these partnerships were specifically linked to the cross-cutting challenge of changing unsustainable patterns of consumption and production. (See Table 7–2.) For instance, a bicycling refurbishing project led by the Dutch NGO Velo Mondial and supported by the bicycle manufacturer Shimano plans to collect bikes for repair and distribution in Africa. The partners in the initiative expect to gather 12,500 bicycles (one container) a week in the first year, growing to up to daily shipments by 2006, as the market requires. And the U.S.-based Collaborative Labeling and Appliance

---

BOX 7–3. HIGHLIGHTS FROM THE JOHANNESBURG PLAN OF IMPLEMENTATION

The Johannesburg Plan of Implementation is one of two negotiated documents from the 2002 World Summit on Sustainable Development. It encourages countries to fulfill their commitments from the 1992 Rio Earth Summit by participating in a 10-year framework of programs on sustainable consumption and production. Broad expectations and goals for this framework include the following:

- Have industrial countries take the lead in promoting sustainable consumption and production.

- Through common but differentiated responsibilities, make sure that all countries benefit from the process of shifting toward sustainable consumption and production.

- Make sustainable consumption and production crosscutting issues and include them in sustainable development policies.

- Focus on youth, especially in industrial countries. Use consumer information tools and advertising campaigns to communicate issues of sustainable consumption and production to youth.

- Promote implementation of the polluter pays principle, which internalizes environmental

costs and incorporates the financial burden of pollution into the price of a product.

- Incorporate lifecycle analysis into policy, in order to track a product from production to consumption and disposal. Use this approach to increase product efficiency.

- Support public procurement policies that encourage the development of environmentally sound goods and services.

- Develop cleaner, more efficient, and more affordable energy sources to diversify the energy supply. Phase out energy subsidies that inhibit sustainable development.

- Encourage voluntary industry initiatives that promote corporate environmental and social responsibility, especially among financial institutions. Examples include codes of conduct, certification, ISO standards, and Global Reporting Initiative Guidelines.

- Collect cost-effective examples of cleaner production and promote cleaner production methods, especially in developing countries and among small and medium-sized enterprises.

SOURCE: See endnote 23.

Standards Program, an initiative involving more than 36 governments as well as several international organizations and NGOs, will work to cut residential and commercial energy use by 5 percent by designing energy efficiency standards and labels and providing technical assistance to 35 developing countries. Many of these partnerships show considerable promise, but it will be important for NGOs and other advocates of sustainable development to monitor their implementation efforts so that laudable commitments are not forgotten as the momentum generated by Johannesburg fades.[25]

From Johannesburg to Cancún and Beyond

A year after the World Summit on Sustainable Development, world attention was again focused on a major international meeting, albeit one of a rather different complexion: the ministerial meeting of the World Trade Organization (WTO) in Cancún, Mexico, in September 2003. The WTO is predicated on a fundamentally different worldview than the sustainable development philosophy that underpins the Rio and Johannesburg agreements, yet its provisions stand to have a large

Table 7–2. Selected Partnerships on Consumption and Production Connected with the World Summit on Sustainable Development

Arab Civil Union for Waste Management

Leader: Children and Mothers Welfare Society (Bahrain)
Others: Council of Arab Ministers Responsible for the Environment, Kuwait Environmental Protection Society, Gulf Net for Environment NGOs Societies, Ajial Youth Group

Six Arab governments are working with the United Nations and regional NGOs to create a regional strategy that will facilitate civil society involvement in community-based solid waste management projects. They will actively involve women and youth and will initiate relevant technology transfer.

Awareness Raising and Training on Sustainable Consumption and Production

Leader: UNEP Division of Technology, Industry, and Economics
Others: Governments of the Netherlands and Sweden, Consumers International, National Cleaner Production Centres

The United Nations is working with two governments and several international NGOs to increase awareness of sustainable consumption and production among government and small and medium-sized enterprise officers in 30 countries. The groups are also working to increase the share of governments implementing the U.N. Guidelines for Consumer Protection from 20 to 50 percent over three years.

The Cement Sustainability Initiative

Leader: World Business Council for Sustainable Development
Others: Government of Portugal, UN University (in Japan), 12 leading cement companies, WWF International, 25 other sponsors in 15 countries

Initiated in 1999, this partnership identifies and facilitates cement companies' efforts to implement sustainable practices. Incorporating national governments, companies, and NGOs, it creates dialogue with cement companies on issues like climate change management, use of raw materials, employee health, and internal business processes. Companies representing one third of the global cement capacity are now involved, and three companies have already implemented a carbon dioxide protocol.

Introduction to Social Standards in Production

Leader: Federal Ministry of Economic Cooperation and Development (Germany)
Others: German Agency for Technical Cooperation, Faber Castell, German Metalworkers' Union, public authorities in Asia

The German government is coordinating efforts to implement a "social charter" at Faber Castell's Indian supply companies. The German Metalworkers' Union has developed the charter, per International Labour Organization standards. An implementation workshop with various partners was convened in 2002, and company partners have undergone a first inspection.

Selling of Responsible Products via Big Retail Chains in Europe: Best Practices and Dialogue

Leader: Reseau de Consommateurs Responsables Asbl
Others: European Commission; European Network for Responsible Consumption; CSR Europe; Oxford Centre for the Environment, Ethics & Society; Die Verbraucher Initiative; consumer groups in Italy and Denmark

Several European consumer NGOs, supported by the European Commission and university representatives, held a conference in June 2003 on the Distribution of Ethical Products via Large Retail Chains in the EU. They also compiled a database of 20 retail case studies in the EU and worked to involve various stakeholders in a dialogue about the best ways to make environmentally and socially responsible products more available in European supermarkets.

Table 7–2. (continued)

Youth Dialogue on Consumption, Lifestyles, and Sustainability

Leader: Federation of German Consumer Organizations
Others: Governments of Germany, Mexico, and Peru; UNEP; UNESCO; Consumers International; national consumer groups and youth groups; Massachusetts Institute of Technology; Media Ecology Technology Association

With the support of three governments, various consumer NGOs are increasing awareness about consumption issues among youth through idea exchange online and in workshops. They are creating a network, based in Europe and Mexico, to educate young consumers about the impact of consumption on sustainable development.

Certification for Sustainable Tourism

Leader: Costa Rican Tourist Board
Others: Governments of Belize, Guatemala, Honduras, El Salvador, Nicaragua, and Panama; Sustainable Tourism Accreditation Commission; Central American Integration System

The Costa Rican Tourist Board is working with five Central American governments and tourism associations to transfer a successful Costa Rican sustainable tourism program to eight countries in Central America by 2006. They will promote the use of local agricultural products and handicrafts while integrating economic, environmental, and sociocultural concerns into business models.

SOURCE: See endnote 25.

impact on the ability of both consumers and governments to promote sustainable business practices worldwide. But the WTO negotiations broke down at Cancún, providing reform-minded governments and activists with an opportunity to push for bringing future trade negotiations into better balance with sustainable development concerns.[26]

When the WTO was created in 1995, trade specialists argued that legislators were passing disingenuous environmental laws that lacked a scientific rationale and that primarily sought to keep foreign products off their shelves. Many governments shared these concerns about "green protectionism," particularly developing-country governments worried that the growth of environmental regulations in the industrial world would be a significant barrier to their own products. Environmental analysts, in contrast, tended to view the laws at issue not as disguised trade barriers at all but as legitimate measures aimed at

protecting the environment and human health. In many cases, they had been passed only after painstaking political battles against vested interests at home.[27]

The accord that created the World Trade Organization nonetheless included several provisions that imposed new restrictions on the ability of governments to make laws designed to protect human, animal, and plant health. Trade officials maintained that the new restrictions were aimed at ferreting out disguised trade barriers, not at preventing countries from undertaking policies legitimately motivated by environmental or health and safety concerns. But the new WTO restrictions set the stage for a series of high-profile disputes between trade and environmental policies, such as conflicts over U.S. laws that restrict imports of tuna fish caught in ways that harm dolphins and of shrimp harvested by methods harmful to sea turtles. (See Table 7–3.) Although the legal reasoning that WTO dispute panels have used in their rulings has become more sensitive to environmental concerns in recent years, fundamental differences remain between international trade rules and emerging environmental practices that could impede efforts to promote more-sustainable patterns of consumption and production.[28]

Some of these differences are well exemplified in a long-running dispute between the European Union (EU) and the United States over the sale of meat produced with the use of growth hormones. A European law forbidding this was first passed in the late 1980s in response to widespread concern among consumers that eating hormone-treated beef might cause cancer and reproductive health problems. The measure always applied equally to domestically raised and imported livestock, and it thus passed the WTO's bedrock test of nondiscrimination. But the ban posed a threat to the hormone-hooked U.S. livestock industry, as it blocked hundreds of millions of dollars worth of U.S. beef exports.[29]

The U.S. beef industry convinced the U.S. government to take up its cause at the WTO, where the government argued that the law was not scientifically justified and was not based on an adequate risk assessment. The European Commission, though, maintained that the law was consistent with the precautionary principle, an emerging tenet of inter-

---

Table 7–3. Key Trade Conflicts Related to Sustainable Consumption and Production

---

Beef Hormones (European Union and United States)

The European Union banned the import of beef from the United States after growth-promoting hormones were found in the meat, because it deemed the hormones a health risk. The United States filed a dispute with the WTO Dispute Settlement Body, arguing that the ban constituted an unfair trade barrier. In 1998, the WTO Panel ruled that the EU ban was inconsistent with WTO rules. The EU refused to lift the ban. In 1999, the United States imposed $117 million per year in retaliatory trade restrictions against the EU. Following the release of new studies showing that growth hormones pose a risk to human health, in October 2003 the EU issued a new directive that refined its bans on various growth hormones found in meat. Claiming that this new directive follows WTO recommendations, the EU hopes that the United States will lift its trade sanctions.

Tuna-Dolphin (United States and Mexico)

Following the domestic Marine Mammal Protection Act, the United States imposed an embargo on Mexican tuna that were caught using a controversial fishing technique called "setting nets on dolphins." Mexico argued that this embargo created an unfair trade barrier, and it lodged a complaint under General Agreement on Tariffs and Trade (GATT) law. In September 1991, the GATT panel concluded that the United States could not embargo Mexican tuna imports because the embargo addressed the way in which the tuna was produced, not the quality or content of the product.

Shrimp-Turtle (India and United States)

India, with other Asian countries, filed a complaint with the WTO when the United States banned imports of specific shrimp and shrimp products. Under the Endangered Species Act, the United States required shrimp boats to use "turtle excluder devices" to prevent endangered sea turtles from being caught in shrimp nets. The United States lost the ruling because it discriminated against the Asian countries by not providing them with adequate technical assistance to protect the turtles. Though a WTO Appellate Body ruled against the United States, it made clear that a country does have the right to take trade action to protect its domestic environment.

national environmental law that holds that "where there are threats of serious or irreversible damage, lack of full scientific certainty shall not be used as a reason for postponing cost-effective measures to prevent environmental degradation." But a WTO appeals panel ruled in February 1998 that the European law did in fact violate WTO rules, paving the way for the U.S. government in July 1999 to retaliate by slapping WTO-authorized 100-percent tariffs on $117 million worth of European imports, including fruit juices, mustard, pork, truffles, and Roquefort cheese. Four years later, the European law still stands and the sanctions remain in place, although the EU is calling for them to be lifted now that it has completed a risk assessment that the EU says validates its law.[30]

Meanwhile, the EU, the United States, and other countries are now embroiled in another major agricultural trade controversy—one that has important implications for the right of consumers to make their own choices about the possible health and environmental impacts of their purchasing decisions. The issue this time is a European Union moratorium on granting approvals for planting or importing many varieties of genetically modified seeds and crops. After complaining about the situation for several years, in May 2003 the U.S. government joined forces with Argentina and Canada to file a formal WTO case in

Table 7–3. (continued)

Swordfish (Chile and European Union)

As early as 1991 Chile, fearing that its swordfish stock was being depleted, stopped allowing Spanish ships to dock at its ports or secure new swordfishing licenses. Overexploitation and bans aimed at renewing the fishery caused Chile's annual swordfish catch to drop in half between 1994 and 1999. The EU alleges that Chile's ban on Spanish ships, which hinders the transport of goods from factory ships to export vessels, violates WTO agreements on the free movement of goods. Chile argues that the UN Convention on the Law of the Sea allows it to protect its marine resources. In November 2000, the EU requested a WTO panel to resolve the dispute. The panel was suspended, however, when the EU and Chile reached a settlement in January 2001. The settlement allowed some EU ships to dock at Chilean ports and provided for multilateral scientific monitoring of the fishery in question.

Asbestos (France and Canada)

Canada challenged a French ban on chrysotile asbestos, a carcinogenic mineral found in many products. Canada alleged that a complete ban violated the WTO's requirement to use the "least trade-restrictive" means of dealing with the health issue. A WTO panel upheld France's ban on asbestos. Reaffirming the fact that asbestos is a carcinogen, in February 2001, the panel held that safer alternatives exist. Civil society hailed this decision as "the first time in its five-year history" that the WTO had ruled for public health.

Genetically Modified Organisms (United States and European Union)

The European Union argues that there may be health and ecological risks associated with genetically modified organisms and has been slow to approve the use or import of products containing them. The U.S. government does not view GMOs as a health risk, and Canada has stated that there is no scientific basis for the EU regulations. Under GMO legislation adopted by the European Parliament in July 2003 and expected to take effect in early 2004, the EU would mandate that all food and animal feed products containing more than 0.9 percent GMOs be labeled as such and that all genetically modified foodstuffs must be traceable. In August, the United States, Canada, and Argentina asked the WTO to form a dispute panel regarding the EU ban on GMOs.

SOURCE: See endnote 28.

protest of this EU policy. A few months later, the European Parliament passed EU legislation that paves the way for food containing genetically modified organisms to be sold in Europe, as long as it is clearly labeled as such and a system is in place to trace genetically modified foodstuffs from the port to the supermarket. EU officials hope that the labeling legislation will render the recent U.S. trade challenge moot, but U.S. government officials are skeptical, arguing that the labeling legislation might itself pose an unfair barrier to trade.[31]

Trade negotiations provide opportunities to push for policy reforms needed to promote more-sustainable consumption and production.

As in the beef hormone case, the U.S. government maintains that restrictions on GMOs violate WTO rules because hard scientific evidence of adverse health and ecological effects is lacking. The EU and most consumer and environmental groups, on the other hand, see the labeling initiative as a reasonable solution to the impasse, in that it would allow some trade in products with GMOs to take place while protecting the right of consumers to make informed risk calculations for themselves. Labeling initiatives enjoy broad public support in both Europe and the United States, with more than 90 percent of consumers in favor of such programs.[32]

In the background of the current trade controversy over GMOs is a larger issue: What should be done when international trade law is on a collision course with the international environmental treaties needed to encourage consumers and producers to shift to more environmentally sound practices? Although no country has thus far lodged a formal WTO challenge against the provisions of an environmental treaty, arguments about WTO consistency often arise during negotiations. These tensions were much in evidence during negotiations on the Cartagena Protocol on Biosafety of 2000, an agreement forged under the umbrella of the U.N. Convention on Biological Diversity that endorses the need for governments to sometimes take precautionary steps to prevent the possibility of irreversible environmental harm in the face of scientific uncertainty. In the current U.S.-European dispute over GMOs, the question could arise of whether WTO rules should trump the provisions of the biosafety protocol or vice versa. An international coalition of NGOs recently launched a campaign soliciting signatories to a "Citizen's Objection" statement that calls for the WTO to dismiss the complaint against the EU and for the dispute to be settled under the Cartagena Protocol instead. So far, 184 organizations from 48 countries have signed on.[33]

Despite the many possible clashes between international trade law and environmental goals and priorities, trade negotiations also provide opportunities to push for policy reforms needed to promote more-sustainable consumption and production. For example, WTO rules and negotiations could be used to encourage countries to reduce and reform governmental subsidies to environmentally sensitive industries, such as agriculture, fossil fuels, fishing, and forestry. Or they could be used to provide preferable trade treatment to "green consumer goods" such as energy-efficient lightbulbs, recycled paper, organic produce, and certified forest and fish products.[34]

Both the desire to minimize clashes between trade and environmental rules and the possibility of promoting synergies led governments to agree in Doha, Qatar, in November 2001 to initiate talks on selected environmental issues as part of a Doha Mandate for a new round of international trade talks. Among other commitments, trade ministers decided to enter into negotiations on the trade implications of environmental labeling requirements, on the relationship between WTO rules and trade measures contained in multilateral environmental agreements, and on the effect of environmental measures on market access. They also agreed to work on strengthening WTO restrictions on fishing subsidies and to discuss the reduction of tariff and non-tariff barriers to trade in environmental goods and services.[35]

Many other subjects slated for discussion under the Doha Mandate could also have important implications for efforts to promote more-sustainable patterns of consumption and production. Efforts to reduce or redirect agricultural subsidies, for instance, could be a powerful boost to more environmentally and socially sound food systems. Negotiations on transparency in government procurement, for instance, would have a bearing on green purchasing initiatives. And proposed talks on reducing restrictions on international investment and on trade in services could limit the latitude that governments have to implement and enforce environmental regulations.[36]

Yet the breakdown of negotiations at the WTO meeting in Cancún in September 2003 raised fundamental questions about where the organization—and any new round of trade talks—is headed. In preparations for the meeting, little progress was made on any of the specifically environmental issues on the table. But it was disputes over issues such as investment and government procurement and simmering tensions over agricultural trade subsidies that finally brought the talks to a halt. Reactions to the stalemate were mixed, even among NGOs. Some felt that the breakdown revealed a failure of political will to confront pressing development concerns; others viewed Cancún as a pivotal turning point in which developing-country governments joined forces in a powerful new coalition, backed up by a strengthened civil society.[37]

In the months ahead, governments and civil society organizations alike will be stepping back to contemplate the larger lessons of recent events. The way forward is not immediately clear, with the situation complicated by the need to forge a consensus among a great diversity of interests from around the world. Still, recent events suggest that the terms of the debate are shifting as people worldwide come to understand that our current unsustainable course threatens both human well-being and ecological health. Although the powerful forces of both Jihad and McWorld continue to sweep the globe, hope for the future comes from the growing number of people who reject each of these paths but support the development of a global community based on respect for people and nature.

Cotton T-shirts

In 1913 the U.S. Navy issued an all-white cotton undershirt to all hands—the first recorded appearance of a T-shirt. Then in 1938 the U.S. retail giant Sears introduced a line of T-shirts for civilian use. But it wasn't until the 1950s that the garment really became popular, thanks to rebel heartthrobs Marlon Brando, James Dean, and Elvis Presley.[1]

Nowadays a T-shirt is a comparatively cheap way for consumers around the world to sport the logo of a favorite company, sports team, or designer. But even when they are made of "natural" cotton, T-shirts come at a high cost to workers and the environment.

Cotton is the world's best-selling fiber, and each year farmers from Texas to Turkey harvest more than 19 million tons of it. Yet the crop packs an environmental wallop. Farmers use nearly $2.6 billion worth of pesticides on cotton a year worldwide—more than 10 percent of the global total, according to Pesticide Action Network North America. The World Health Organization has classified many pesticides commonly used on cotton as "extremely hazardous," including organophosphates like parathion and diazinon that are particularly dangerous to the nervous systems of infants and children.[2]

Cotton pesticides sicken and kill farm workers as well. Between 1997 and 2000, cotton fields were the sites of 116 reported farmworker acute pesticide poisoning cases in California. And more than 500 cotton farmers in India's cotton-producing state of Andhra Pradesh are believed to have died in 2001 as a result of pesticide exposure. In many cases, farmers are unaware of or fail to use proper safety procedures when handling and disposing of chemicals: in one survey in Benin, West Africa, 45 percent of cotton farmers said they used pesticide containers to carry water, while 20–35 percent used them to hold milk or soup. People have also been harmed in factories and communities where cotton pesticides are produced: the infamous toxic gas release at a Union Carbide facility in Bhopal, India, in 1984 killed 8,000 people.[3]

Over the past decade, ecologists have recorded devastating harm to birds, fish, and other wildlife from chemicals used on cotton. Before harvest, farmers often use herbicides to defoliate the cotton plants and to allow easier access to the boll, a pod containing seeds and linty fibers—a practice that can destroy wildlife habitat.[4]

Cotton pesticides can pollute local water bodies as well, endangering human and ecosystem health. Aldicarb, a compound that has been found to cause immune system abnormalities at even low levels of consumption, has been detected in groundwater in seven American states, according to Cornell University Cooperative Extension. And in 1998 the U.S. Geological Survey reported surface water contamination from herbicides and insecticides used on cotton in the South. Meanwhile, halfway around the world, the diversion of water to irrigate cotton—a thirsty crop—has shrunk Uzbekistan's Aral Sea to one fifth its original size.[5]

After a cotton field is harvested, the bolls are "ginned" to separate the fibers from the seeds. The fibers are then packed into bales of about 500 pounds (roughly 225 kilos) each. (U.S. textile makers use about 11 million bales of cotton a year.) A spinning factory cleans the fibers and twists them into yarn, which is woven into cloth on mechanized looms. The shipping from farm to factory requires energy, typically from fossil fuels; so does the production of thread and cloth, as spinneries are no longer powered by mules.[6]

After a T-shirt is produced, it is usually dyed and treated with fabric finishes. Chemical dyes, and even some natural dyes, often contain copper, zinc, and other heavy metals, which are toxic and can pollute water through factory runoff. Fabric finishes, such as those to repel stains, wrinkles, and water, may contain petrochemicals such as formaldehyde, a carcinogen.[7]

This does not mean that concerned consumers should choose synthetic fabrics instead. Polyester fibers are made from petroleum, a nonrenewable resource whose extraction and shipping damages the environment, most visibly in oil spills. According to one estimate, if the oil used in production and transport is included, a cotton T-shirt blended with polyester can release approximately one quarter of its weight in air pollutants and 10 times its weight in carbon dioxide.[8]

China is the world's top cotton producer, followed by the United States and India. And the United States is the leading exporter of the fiber, shipping more than 10.5 million bales a year worldwide, predominantly to Asia and Mexico. (Other top exporters are the countries of the former Soviet Union and Australia.) Poorer countries seeking to sell their cotton on the world market increasingly find themselves undercut by subsidies and trade barriers that aid U.S. cotton farmers.[9]

China is also the world's top producer of T-shirts, supplying about 65 percent of the total—most of which are sold to the United States and Europe. (Americans spent $6.2 billion on some 478 million T-shirts in 2002.) As in many developing countries, garment factory workers in China earn low wages and work long hours. Apparel manufacturers from industrial countries commonly use sweatshops in Central America and Southeast Asia, where they can operate under laxer labor and environmental regulations than at home.[10]

What's a T-shirt wearer to do? The most ecological choice, apart from buying used clothing, is a T-shirt made from certified organic cotton, grown without synthetic pesticides and fertilizers. At one Egyptian farm project, organic cultivation has boosted cotton yields by more than 30 percent, and the fiber is processed into textiles without any synthetic chemicals. The best choice in terms of worker welfare is garments certified by the Fair Trade Federation. In a positive trend, organic cotton and fair-trade garment makers are joining hands to both protect the environment and promote social justice.[11]

—*Mindy Pennybacker, The Green Guide*

CHAPTER 8

Rethinking the Good Life

Gary Gardner and Erik Assadourian

Bogotá, the capital of Colombia, is commonly associated with civil war and violence. But in the late 1990s, the city's reputation began to change as Mayor Enrique Peñalosa led a campaign to improve the quality of life there. School enrollments increased by 200,000 students—some 34 percent—during Peñalosa's tenure. His administration built or totally rebuilt 1,243 parks—some small, some very large—which are now used by 1.5 million visitors annually. An effective rapid transit system, accessible to all, was planned and constructed. And the city's murder rate fell dramatically: today, there are fewer murders per capita in Bogotá than there are in Washington, D.C.[1]

By any standard, the city's advance is a developmental success. Yet Bogotá's transformation was achieved in a rather unorthodox way. When Peñalosa took office, consultants proposed building a $600-million elevated highway, a standard transportation solution in many car-bound cities. Instead, the mayor created a cheaper yet more effective rapid transit system using the city's existing bus lines. The system carries 780,000 passengers daily—more than the costlier Washington, D.C., subway does—and is so good that 15 percent of the regular riders are car owners. Peñalosa also invested in hundreds of kilometers of bike paths and in pedestrian-only streets. And he strengthened the city's cultural infrastructure by building new public libraries and schools, connecting them with a network of 14,000 computers. Together with the rehabilitated parks, the transportation and cultural improvements advanced a strategic goal for Bogotá: to orient urban life around people and communities.[2]

Peñalosa uses an unusual yardstick to evaluate his development strategy. "A city is successful not when it's rich," he says, "but when its people are happy." That statement deflates decades of development thinking in rich and poor countries alike. After all, most governments make ongoing increases in gross domestic product (GDP) a chief priority of domestic policy, under the assumption that wealth secured is well-being delivered. Yet undue emphasis on generating wealth, especially by encouraging heavy consumption, may be yielding diminishing returns. Overall quality of life is suffering in some of the world's richest countries as people experience greater stress and time pressures and less satisfying social relationships and as the natural environment shows more and more signs of distress. Meanwhile, in poor countries quality of life is degraded by a failure to meet people's basic needs.[3]

Rethinking what constitutes "the good life" is overdue in a world on a fast track to self-inflicted ill health and planet-wide damage to forests, oceans, biodiversity, and other natural resources. By redefining prosperity to emphasize a higher quality of life rather than the mere accumulation of goods, individuals, communities, and governments can focus on delivering what people most desire. Indeed, a new understanding of the good life can be built not around wealth but around well-being: having basic survival needs met, along with freedom, health, security, and satisfying social relations. Consumption would still be important, to be sure, but only to the extent that it boosts quality of life. Indeed, a well-being society might strive to minimize the consumption required to support a dignified and satisfying life.

Wealth and Well-being

Wealth and well-being are less like antagonists and more like long-lost siblings. After all, the word "wealth" is rooted in "weal"—a synonym for well-being that traditionally had a community orientation. Yet wealth is now used to mean material goods and financial holdings, primarily of individuals—a far more narrow usage than its roots would imply. Building a society of well-being essentially involves recapturing the original, broad-based understanding of the term wealth.[4]

The idea of well-being as a personal and policy goal is increasingly commonplace, appearing everywhere from popular magazines to official publications of multinational organizations, such as *The Well-being of Nations* by the Organisation for Economic Co-operation and Development in 2001 and *Ecosystems and Human Well-being* by the Millennium Ecosystem Assessment in 2003. Even the Canadian House of Commons picked up the term in legislation passed in June of 2003 entitled the Canada Well-Being Measurement Act.[5]

Definitions of the concept vary, but tend to coalesce around several themes:
• the basics for survival, including food, shelter, and a secure livelihood;
• good health, both personally and in terms of a robust natural environment;
• good social relations, including an experience of social cohesion and of a supportive social network;
• security, both personal safety and in terms of personal possessions; and
• freedom, which includes the capacity to achieve developmental potential.[6]

In shorthand form, the term essentially denotes a high quality of life in which daily activities unfold more deliberately and with less stress. Societies focused on well-being involve more interaction with family, friends, and neighbors, a more direct experience of nature, and more attention to finding fulfillment and creative expression than in accumulating goods. They emphasize lifestyles that avoid abusing your own health, other people, or the natural world. In short, they yield a deeper sense of satisfaction with life than many people report experiencing today.

What provides for a satisfying life? In recent years, psychologists studying measures of life satisfaction have largely confirmed the old adage that money can't buy happiness—at least not for people who are already affluent. The disconnection between money and happiness in wealthy countries is perhaps most clearly illustrated when growth in income in industrial countries is plotted against levels of happiness. In the United States, for example, the average person's income more than doubled between 1957 and 2002, yet the share of people reporting themselves to be "very happy" over that period remained static. (See Figure 8–1.)[7]

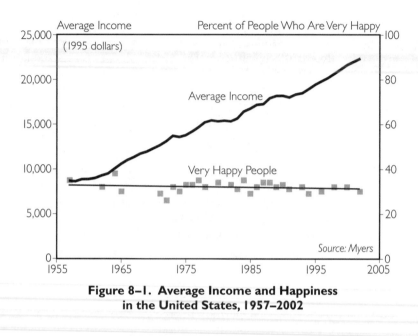

Figure 8–1. Average Income and Happiness in the United States, 1957–2002

Not surprisingly, the relationship between wealth and life satisfaction is different in poor countries. There, income and well-being are indeed coupled, probably because more of a poor person's income is used to meet basic needs. (See Chapter 1.) Findings from the World Values Survey, a set of surveys of life satisfaction in more than 65 countries conducted between 1990 and 2000, indicate that income and happiness tend to track well until about $13,000 of annual income per person (in 1995 purchasing power parity). After that, additional income appears to yield only modest additions in self-reported happiness.[8]

If psychologists are clear about the limits of wealth for delivering happiness, they are equally clear in describing what does contribute to life satisfaction. Again and again, studies suggest that happy people tend to have strong, supportive relationships, a sense of control over their lives, good health, and fulfilling work. These factors are increasingly under stress in fast-paced, industrial societies,

where people often attempt to use consumption as a substitute for genuine sources of happiness. Yet there are at least some individuals, communities, and governments that are dissatisfied with life quality and are beginning to make an effort to build lives, neighborhoods, and societies of well-being.[9]

The Power of One

During the summer of 2003, some 50 million Americans signed up for a government-sponsored National Do Not Call Registry designed to prevent commercial telemarketers from phoning them. The outpouring of response to this new government program—in essence, an attempt by people to reclaim some of their time and privacy from increasingly aggressive marketing tactics—hints at the frustration many individuals feel when economic forces begin to dominate rather than serve them. Yet a small but growing number of consumers are questioning the way they shop, the amount of "stuff" crowding and complicating their lives, and the amount of time they spend at work. These dissatisfied consumers have not yet built a coherent movement, because their actions are mostly private ones occurring in unconnected pockets in many nations. Still, the spontaneous and grassroots nature of these activities may signal a deeply felt desire by many people to build a satisfying life for themselves and their families.[10]

Perhaps the most apparent expression of a desire for a higher quality of life is found in the growing numbers of people who shop with an eye toward well-being. In Europe, for example, demand for organically grown foods drove sales up to $10 billion in 2002, 8 percent above the previous year, as a public bruised by mad cow disease and other food scares increasingly sought assurances of the safety of its food supply. Market analysts estimate that 142 million Europeans are consumers of organics, although a "loyal" core of 20 million accounted for 69 percent of the expenditures on these products in 2001. And 150 million people in Europe are either vegetarians or have reduced their consumption of meat.[11]

Meanwhile, in the United States the group of consumers interested in shopping for better health and a better environment is large enough to have earned recognition by market researchers as a distinct demographic group. Dubbed LOHAS consumers—people who lead Lifestyles of Health and Sustainability— these shoppers buy everything from compact fluorescent lightbulbs and solar cells to fair-trade coffee and chocolate (products that pay a just wage to producers or that have a lighter environmental impact than mainstream purchases do). This group now includes nearly one third of adult Americans and in 2000 accounted for about $230 billion in purchases— some 3 percent of total U.S. consumer expenditures. Although this is a relatively low share of expenditures compared with the number of people identified as LOHAS consumers, this is probably due to the few options for healthy consumption available today.[12]

In many countries, people are joining consumer cooperatives to leverage their market power for a higher quality of life. In Japan, for example, the 250,000-member Seikatsu Club Consumers' Cooperative Union stocks foods free of agricultural chemicals and artificial additives and preservatives, along with household products free of toxins. The club puts its goods in reusable glass jars in order to help reduce the 60 percent of household waste that is packaging. In contrast to many supermarkets that stock tens of thousands of individual items, the Seikatsu Club co-ops carry just 2,000 items, mostly basic foodstuffs. The co-ops typically carry only one or two choices per item, but for members seeking to live a more satisfying life, the high quality, healthy foods, and reduced waste apparently compensate for the somewhat lessened choice. And Seikatsu members are not alone; some 50 million people belong to local co-ops that are affiliated with Consumer Coop International, a global body that helps facilitate training for local consumer co-ops.[13]

In some cases, individuals are turning to organizations for help in greening their consumption. A coalition of organizations in 19 countries known as the Global Action Plan offers training to families on reducing waste, lessening energy use, and switching to eco-friendly products. In the Netherlands, at least 10,000 households are working on redirecting their consumption; after training, these people cut their household waste on average by 28 percent. Six to nine months later, the figure was 39 percent. And in 2003, the French government launched a similar initiative, *la famille durable* (the sustainable family), that offers practical ways for people to live sustainably at home, school, and work and on vacation.[14]

And in the United States, the Center for a New American Dream urges people to live a life of "more fun, less stuff." Through its Turn the Tide program, the Center encourages people to follow a simple nine-step environmental conservation plan, involving such actions as switching to water-efficient faucets and eating less meat. The 14,000 members of this initiative report saving more than 500 million liters of water and preventing over 4 million kilograms of carbon dioxide from being released into the atmosphere.[15]

Beyond a shift in shopping habits, many consumers are trying to simplify their lifestyles in broader ways—a process sometimes called "downshifting." Analyst Cecile Andrews describes the motivation for these individuals: "A lot of people [are] rushed and frenzied and stressed. They have no time for their friends; they snap at their family; they're not laughing very much." Many, she says, "are looking for ways to simplify their lives—to rush less, work less, and spend less. They are beginning to slow down and enjoy life again."[16]

Estimates of the numbers of downshifters are imprecise, but interest in simplifying appears to be growing. In seven European countries, the number of people who have voluntarily reduced their working hours has grown at 5.3 percent each year over the past five years, for example. And the trend toward simplicity is expected to continue. The number of people in these same countries who could at least partially embrace a voluntary simplicity lifestyle is expected to grow from about 7 million in 1997 to at least 13 million in 2007.[17]

Meanwhile, two research surveys in the United States in the mid-1990s suggested that around a quarter of the population were working to simplify their lives, although the extent of course varied greatly from person to person. And the media have registered growing interest in the topic. Articles in U.S. news-

papers about simplifying lifestyles grew three- to fivefold between 1996 and 1998. In 1997, the Public Broadcasting System aired a documentary called *Affluenza*, which treated consumerism as a contagious disease and offered suggestions for inoculating yourself against it. The program was very popular and was later distributed in 17 countries.[18]

Yet individual initiatives are only part of what is needed to build a society of well-being. Individual efforts alone do not necessarily help to build strong, healthy communities (although they can free up time that could lead to greater community involvement), nor can they address the structural obstacles to genuine consumer choice—the lack of organic produce in the supermarket, for instance. Some critics even argue that, pursued in isolation, individual initiatives can be counterproductive. An "individualization of responsibility," as political and environmental scientist Michael Maniates notes, distracts attention from the role that such institutions as business and government play in perpetuating unhealthy consumption. Moreover, to the extent that individuals see their power residing primarily in their pocketbooks, they may neglect their key roles as parents, educators, community members, and citizens in building a society of well-being.[19]

The need for individuals to act collectively to improve their quality of life led a group in Norway in 2000 to launch a campaign entitled 07-06-05. Campaigners are rallying Norwegians to count down to June 7th, 2005, the one-hundredth anniversary of Norway's independence from Sweden, and to once again declare their independence—but this time from the "time poverty" that has accompanied the ascendancy of the consumer culture.[20]

In the United States, an alliance known as the U.S. Simplicity Forum is trying to mobilize the millions of Americans struggling with too much to do and too little time. They organized Take Back Your Time Day on October 24th, 2003, urging Americans to leave work early, arrive late, take longer than usual lunches, or even skip work altogether. Thousands joined events at neighbors' homes, local churches, meeting halls, and universities to discuss the time poverty facing the average American. The date was deliberately chosen—it was nine weeks before the end of the year—to remind Americans that they are some of the most overworked people in the industrial world, putting in 350 hours more on the job (that is, nine workweeks) each year than the average European.[21]

Organizers hope to use the energy of the American initiative to start a popular movement centered on reclaiming time for a higher quality of life. The campaign would seek to reform national vacation laws, working hours, and other measures that would free up time for the neglected elements of life, such as family, friends, and community. As Take Back Your Time Day coordinator and *Affluenza* producer John de Graaf explains, "The Time Movement is about looking beyond GDP as the measure of a good society and understanding that the real purpose of our economy is not material growth without end, but a balanced, fulfilling, and sustainable life for all."[22]

The Ties That Bind

Humans are social beings, so it is little surprise that good relationships are one of the most important ingredients for a high quality of life. Harvard Professor of Public Policy Robert Putnam notes that "the single most common finding from a half century's research on the correlates of life satisfaction…is that happiness is best predicted by the breadth and depth of one's social connections." Thus individual efforts to build a satisfying life are more likely to be successful if some of them involve family, friends, or neighbors. Fortunately, individual efforts and community efforts often work hand in hand. The person who works fewer hours each week finds more time for family, friends, and community. And community ties, which are strengthened, for example, when neighbors share tools or babysitting responsibilities, can reduce family expenses and help people lead simpler lives.[23]

People who are socially connected tend to be healthier—often significantly so. More than a dozen long-term studies in Japan, Scandinavia, and the United States show that the chances of dying in a given year, no matter the cause, is two to five times greater for people who are isolated than for socially connected people. For example, one study found that in 1,234 heart attack patients, the rate of a recurring attack within six months was nearly double for those living alone. And a Harvard study of health and mistrust in the United States concluded that moving to a state with a high level of social connections from a state where the level is low would improve a person's health almost as much as quitting smoking.[24]

A particularly impressive example of the relationship between social connectedness and health comes from a study of the town of Roseto, Pennsylvania, which caught the attention of researchers in the 1960s because its rate of heart attacks was less than half the rates in neighboring towns. The usual causes of such an anomaly—diet, exercise, weight, smoking, genetic predisposition, and so on—did not explain the Roseto phenomenon. In fact, people in Roseto scored worse on many of these risk factors than their neighbors. So the researchers looked for other possible explanations and found that the town had a tight-knit social structure that had produced community-initiated sports clubs, churches, a newspaper, and a Scout troop. Extensive informal socializing was the norm. Eventually researchers gave credit to the strong social ties of the residents—most were from the same village in Italy and worked hard to maintain their sense of community in the United States—for the higher levels of health. The sad postscript to the story is that starting in the late 1960s, as social ties weakened in this town and across the United States, the heart attack rate in Roseto rose, eventually surpassing that of a neighboring town.[25]

Strong social ties are especially helpful in promoting collective consumption, which often has social and environmental advantages.

Researchers offer various explanations for the link between social connectedness and lower risk of health problems. Some are quite practical: connected people have someone to depend on if they run into health problems, thereby reducing the likelihood that sickness will develop into a serious health condition. Social networks may reinforce healthy behaviors; studies show that isolated people are more likely to smoke or drink, for example. And cohesive communities may be more effective at lobbying for medical care. But the connection may run deeper. Social contact may actually stimulate a person's immune system to resist disease and stress. Laboratory animals, for example, are more likely to develop hardening of the arteries when isolated, while animals and humans in isolation both tend to experience decreased immune response and higher blood pressure.[26]

International development professionals also now acknowledge that strong social ties are a major contributor to a country's development. The World Bank, for instance, sees social connectedness as a form of capital—an asset that yields a stream of benefits useful for

development. Just as a bank account (financial capital) yields interest, social ties tend to build trust, reciprocity, or information networks, all of which can grease the wheels of economic activity. Trust, for example, facilitates financial transactions by creating a climate of confidence in contractual relationships or in the safety of investments. A World Bank study of social contacts among agricultural traders in Madagascar found that those who are part of an extensive network of traders and can count on colleagues for help in times of trouble have higher incomes than traders with fewer contacts. Indeed, the connected traders say that relationships are more important for their success than many economic factors, including the price of their traded goods or access to credit or equipment.[27]

A lack of social capital also seems to be connected with poor economic growth at the national level. Stephen Knack of the World Bank warns that low levels of societal trust may lock countries in a "poverty trap," in which the vicious circle of mistrust, low investment, and poverty is difficult to break. Knack and his colleagues tested the relationship between trust and economic performance in 29 countries included in the World Values Survey. They found that each 12-point rise in the survey's measure of trust was associated with a 1-percent increase in annual income growth, and that each 7-point rise in trust corresponded to a 1-percent increase in investment's share of GDP.[28]

The role of social glue in facilitating economic transactions is especially evident in microcredit initiatives such as the Grameen Bank of Bangladesh, which provides small loans to very poor women who lack the collateral to borrow from a commercial bank. Participating women organize themselves into borrowing groups of five, and each group applies to the Bank for loans, often of less than $100. The women count on knowledge of their neighbors' dependability when they extend invitations to join the group. This information function—something commercial banks spend money on when they compile an applicant's credit history—is an example of how social capital can lower the costs of financial activity. Social ties are also meant to serve as collateral for the loans. Because women are jointly responsible for repayment, and because a default puts all five in jeopardy of disqualification for future loans, each woman is subject to strong social pressure to repay.[29]

The economic payoff of these types of social connectedness has made microcredit successful in many parts of the world. The Grameen Bank claims that 98 percent of its microcredit loans are repaid, a better record than in most commercial banks. Grameen has inspired the spread of microcredit globally. An initiative known as the Microcredit Summit Campaign has set a goal of enrolling 100 million people in microcredit programs by 2005. By the end of 2002, they were more than halfway there, with 68 million people participating.[30]

Beyond improving health and facilitating economic security, strong social ties are especially helpful in promoting collective consumption, which often has social and environmental advantages. A good example of this is co-housing, a modern form of village living in which 10–40 individual households live in a development designed to stimulate neighborly interaction. Privacy is valued and respected, but residents share key spaces, including a common dining hall, gardens, and recreational space. Started in the late 1960s, more than 200 co-housing communities have been established in Denmark. The movement has spread to the Netherlands, Scandinavia, Australia, Canada, and the United States, where 50 new co-housing interest groups are established each year (although more than half of these do not

survive to see a community established, because of the steep challenges involved, including gaining permits and financing as well as building the community).[31]

In a co-housing community, houses often share common walls with neighboring homes and are clustered around a courtyard or pedestrian walkway. Cars are typically confined to the perimeter of the community property. The design means that these communities often use less energy and fewer materials than neighborhoods full of private homes. A study of 18 communities in the United States in the mid-1990s found that, compared with before they moved into co-housing, members owned 4 percent fewer cars, while their ownership of washers and dryers dropped by 25 percent and of lawnmowers by 75 percent. The average living space per household in the 18 communities—including each unit's share of the common room area—was about 1,400 square feet, two thirds as big as the average new U.S. home in the mid-1990s. Shared basement space for mechanical services and common entryways for adjoining dwellings reduce living space with little sacrifice of livability. And building in tight clusters allows yard space to be shared without a major loss of privacy. As a result of these features, the average co-housing community in the study used only half as much land per dwelling as in a conventional suburban U.S. development.[32]

But perhaps the greatest contribution of co-housing communities to a high quality of life is the social ties they create. The communities are self-managed, which encourages interactions and sharing. Children typically have many adults watching as they play, as well as an abundance of playmates and babysitters. Most of the communities offer two or more common meals per week, with on average 58 percent of members attending. Interestingly, in contrast to "time-saving" meals offered by food companies, which typically feature highly processed and packaged foods such as instant mashed potatoes or frozen pizza, the co-housing approach to common meals saves time without sacrificing on food quality. At the Nomad Cohousing Community in Colorado, for instance, where there are two community meals a week, residents spend 2.5–3 hours every five to six weeks helping with cooking and cleanup. Compared with cooking a family meal each day, this occasional sharing of effort frees up 9 hours of labor for every family over six weeks.[33]

In many developing countries, too, collective consumption is more feasible in communities with a strong social base. (See Box 8–1.) A World Bank study of 64 villages in Rajasthan, India, for example, found that conservation and development of watersheds was more successful in villages that exhibited strong levels of trust, informal networks, and solidarity than in villages that had fewer of these social assets. And in Bangladesh, cooperative garbage collection programs (where local government failed to provide it) were undertaken and successful in areas where certain forms of social capital—in this case, norms of reciprocity and sharing—were well developed.[34]

Creating Infrastructures of Well-being

When individuals or communities seek to enhance their quality of life, they may be handcuffed by the set of choices available to them. Organic produce, reusable beverage bottles, or mass transit obviously cannot be bought if they are not offered for sale. The rules and policies that determine the set of choices available, such as oil subsidies that make fossil energy cheaper than wind power, zoning laws that encourage sprawling development, or building codes that frown on the use of recycled building materials, are essentially the "infrastructure of consumption." Creating a higher quality of life requires us all—individuals and communities—to help create new political, physical, and cultural "infrastructures of well-being."[35]

Some governments are beginning to use their authority to help create a political environment conducive to well-being. The most basic of their initiatives is to properly assess community or societal health, as the city of Santa Monica is doing through a Sustainable City Plan. In place since 1994, the plan aims to decrease overall community consumption, especially the use of materials and resources that are not local, nonrenewable, not recycled, and not recyclable. It also seeks to develop a diversity of transportation options, to minimize the use of hazardous or toxic materials, to preserve open space, and to encourage participation in community decisionmaking. The plan uses 66 indicators to measure its progress, such as solid waste generation, cost of living, share of major streets with bike lanes, percent of tree canopy coverage, voting rates, share of residents who volunteer, greenhouse gas emissions, number of homeless, and crime rates. Many of Santa Monica's initial targets have been met or exceeded, according to the city, and more ambitious goals have been set for 2010.[36]

At the national level, the standard tool used to measure societal health, GDP, is much too narrow to serve as a yardstick of well-being because it sums all economic transactions, regardless of their contribution to quality of life. It also ignores entire swaths of nonmarket activity that contribute to individual and community well-being, such as the child care provided by a stay-at-home parent. Throughout the 1990s, researchers worked to develop alternative measures, such as the Ecological Footprint, the Genuine Progress Indicator, the Human Development Index, and the Living Planet Index, to com-

---

BOX 8–1. THE GAVIOTAS EXPERIENCE: MAKING WELL-BEING A PRIORITY

Gaviotas is a village of 200 people in rural Colombia with a global reputation for innovative development. Governing their approach is a strong concern for the quality of village life and for the natural environment. For starters, villagers ensure that basic needs are met: residents pay nothing for meals, medical care, education, and housing. All adults have work, whether in the various village enterprises that manufacture solar collectors and windmills, in organic and hydroponic agriculture, or in forestry initiatives.

Social needs are addressed as well, through the rhythm of daily activities. Members work together in village businesses and regularly eat together in the large refectory, even though each home has a kitchen. Music and other cultural events are a regular part of village life. With survival and social needs met in abundance, the atmosphere is peaceful: the community has had no police force, jail, or mayor in its 33-year history. Community norms are set by members and enforced through social pressure.

Gaviotas is known worldwide for its many inventions, including a water pump that village kids work as they ride their seesaw, windmills designed for the gentle breezes of the Colombian plains, a pressurized solar water heater, and a pedal-powered cassava grinder. The

technologies enhance the quality of life of these villagers, but also of other interested communities. As a matter of principle—and in line with their primary interest in advancing quality of life, not just in generating wealth—the villagers do not patent their inventions, which are made widely available. Thousands of the windmills have been installed by Gaviotas technicians across Colombia, and the design has been copied throughout Latin America.

For the villagers, well-being also means treading lightly on the environment. Gaviotas is now self-sufficient in energy, making ample use of solar and wind power and of methane produced from cattle manure. Its air-cooled and solar-heated former hospital (now a water purification center) was named by a Japanese architectural journal as one of the 40 most important buildings in the world. Its agricultural activities are organic. And it is the center of the largest reforestation project in Colombia, having converted tens of thousands of hectares of savannah to forest, from which villagers extract and sell only resin, even though logging would be more lucrative. The villagers believe that a healthy forest generating modest resources is better than a depleted one that yields a temporary bonanza.

SOURCE: See endnote 34.

---

plement the perspective of GDP. (See also Chapters 1 and 7.) One such effort, the Well-being Index developed by sustainability consultant Robert Prescott-Allen, is noteworthy for its comprehensiveness. (See Box 8–2.)[37]

In addition to recalibrating yardsticks for societal health, governments are using their extensive legislative and regulatory powers to shape the way people consume and the values a society internalizes regarding consumption. Eliminating perverse subsidies and adopting pollution taxes, for example, have already

proved useful in creating a cleaner environment and a higher quality of life in many European countries. (See also Chapter 5.)

And many governments in Europe are helping workers and families to carve out extra time each week. Belgium, Denmark, France, the Netherlands, and Norway now have 35- to 38-hour workweeks, which in addition to freeing up valuable time for workers often help to create new jobs. The Netherlands has two particularly creative approaches to paring back working hours. Employers

---

BOX 8–2. MEASURING WELL-BEING

The Wellbeing Index uses 87 indicators to measure human and ecological well-being—ranging from life expectancy and school enrolment rates to the extent of deforestation and levels of carbon emissions. The 87 indicators can help countries identify the areas in which their quality of life is suffering. Values from the array of indicators are standardized and summed into a single score for ease of comparison across 180 countries.

The results are revealing: some two thirds of the world's people live in countries with a bad or poor rating for human well-being. Only Norway, Denmark, and Finland receive the highest of the five rating levels. Meanwhile, countries with a poor or bad environmental rating cover almost half of Earth's land area. And no country receives a good environmental rating.

The Index's separate measures of human and environmental well-being help crystallize an ideal development goal: to improve people's lives with the least possible environmental impact. Indeed, the Index reveals that meeting people's needs can be done at a range of environmental price tags. The Netherlands and Sweden have roughly the same human well-being score, for example, but the Netherlands scores much lower on environmental health. This suggests that *how* a nation meets its development goals is as important as *whether* it meets them.

SOURCE: See endnote 37.

give the same benefits and promotion opportunities to part-time and full-time workers, making part-time work attractive for many. And the government encourages parents with small children to work the equivalent of no more than 1.5 jobs between the two of them,

so that more time is available to meet the heavy demands of caring for young children. In addition to reforms of the workweek, many countries provide generous paid family leave to new parents. Sweden, for instance, grants 15 months of leave per child at up to 80 percent of salary, compared with the 12 weeks of unpaid leave that is offered in the United States.[38]

Government interventions like these are likely to create a less stressful home environment. Finland, for example, has very strong policies supporting the employment of mothers, including paid parental leave, tax relief for child care, publicly funded child care, and other measures. (In one study, Finland ranked first among 14 nations in provision of these benefits.) A 2001 study of the psychological benefit to parents of these measures found that, in contrast to the United States, where parenting tended to be associated with poor psychological well-being because of the stress involved and lack of family support, parenting in Finland correlated either neutrally or positively with psychological well-being. For fathers, the results were strongly positive, but for mothers somewhat less so, indicating that support for them could be strengthened.[39]

Central to changing the legal and political infrastructure of well-being is achieving clarity about the importance of providing public services. The increased priority given to private consumption in many countries in recent decades has often given public services a bad name. But societies pay a social price when private consumption is pursued at the expense of public investment. A 2003 report by the Fabian Society in the United Kingdom demonstrates this. Privatizing public schools, the report noted, can lead to the best schools attracting the best students, while the worst schools get a disproportionate share of disciplinary cases. Privatized bus services can leave unprofitable routes unserved and profitable routes overcrowded, sending more people into their cars, as happened when U.K. local bus services were privatized.[40]

Of course, deciding which goods should be publicly provided is a knotty political problem, but one the public can and should be involved in. An inspiring example of public involvement in setting priorities for government funds comes from Porto Alegre, Brazil. Officials there have used a "participatory budget" process since 1989 to involve citizens directly in decisions on how to allocate the municipal budget. The process has produced greater governmental transparency and accountability, a reduction in the share of city revenues consumed by salaries, and a reduction in the share of contracts allocated on a patronage basis. It has also led to increases in the amount of money spent on education, basic services, and urban infrastructure—initiatives that have improved residents' quality of life. In addition, the process has mobilized more people each year, with 40,000 of the 1.3 million residents participating in the 1999 budget process. Most get involved by joining neighborhood meetings, so the process has helped to increase grassroots involvement, allowed new local leaders to emerge, and empowered some of Porto Alegre's poorer communities. Participatory budgeting has now spread to 140 communities—2.5 percent of Brazil's municipalities.[41]

Attention to the design of physical infrastructure is also critical to improving quality of life. Car-centered suburban dwellings, for example, have long been criticized for weakening community cohesion, in part because of the time required to commute to work. Social scientist Robert Putnam has noted that each additional 10 minutes of daily commuting time is associated with a 10-percent decline in involvement in community affairs. With the average American adult now spending 72 minutes a day behind the wheel, often alone, community cohesion is bound to suffer. In 2003, sprawling suburban developments were also criticized for their adverse effects on health. A U.S. study of more than 200,000 people in 448 counties found that those living in low-density suburban communities tended to spend less time walking and weighed 6 pounds more on average than those living in densely populated areas. Suburbanites were also found to be as likely as cigarette smokers to have high blood pressure.[42]

Meanwhile, urban design can deter—or attract—cyclists. Surveys in the United States indicate that a principal reason Americans give for not cycling is that they regard the practice as unsafe. And it is. Measured per kilometer of travel, cycling in the United States is more dangerous than any other form of transportation. Yet the accident rate for cyclists in the Netherlands and Germany is only one quarter the U.S. rate, largely because those nations invest in bike lanes, stoplights that favor cyclists, and other infrastructure developments that make cycling safe. The Netherlands has doubled the length of its network of bikeways in the past 20 years, and Germany has tripled its network.[43]

When they are well designed, cities can become attractive places to spend time, which encourages greater civic interaction. Both factors tend to boost quality of life. By converting streets into pedestrian thoroughfares, mixing housing and shops, creating plazas and parks, and taking other steps, city centers can be stimulating places to be. In Copenhagen, for example, outdoor cafes, public squares, and street performers attract the public in the summer, while skating rinks, heated benches, and gaslit heaters on street corners make winters enjoyable. And the city has gone out of its way to make cycling easy, not only by providing bike lanes, but also by making bicycles available for a modest deposit that is refunded when the bicycle is returned.[44]

Such design innovations happen when a city is serious about making quality of life a priority. One demonstration of such seriousness comes from Austin, Texas, which used an incentive program known as the Smart Growth Criteria Matrix to control where and how growth took place and to enhance quality of life. The city used a series of criteria to score proposed development projects, with high-scoring projects qualifying to have city fees waived. Analyst Guy Dauncey describes the incentive criteria this way:

> You can win more points for a downtown location, and for a location within one block of a transit stop or two blocks of a light rail station. There are points for…smaller setbacks, front porches, back lanes, narrow streets, and a community orientation. There are points for mixed residential, office and retail use, for residential units above commercial, and for encouraging street level pedestrian uses. The Matrix also offers points for being bicycle friendly, for traffic calming, for greenways and affordable housing, for using local contractors and architects, for water and energy efficiency, for incorporating a neighbourhood food market and other retail stores, for preserving heritage structures, and for re-using existing buildings. There are points for landscaping, streetscaping, for being consistent with local neighbourhood plans, and for local participation and support.[45]

Some businesses are also starting to recognize that they can make their own physical infrastructure more amenable to the well-being of employees. At the new world headquarters in Kansas for Sprint, a telecommunications firm, cars must park in garages at the edge of the corporate campus, requiring employees to walk some distance into work. Buildings feature slow elevators, which encourages people to use the stairs. And the eating area in the complex is located away from the offices rather than conveniently in the middle of them, so that employees must put some energy into getting to their food. This innovative design reflects an understanding that advancing well-being is not always synonymous with maximizing convenience or comfort.[46]

New political and physical infrastructures of consumption are being supplemented by a budding new cultural framework, particularly in promoting an ethic of consumption for well-being. In this regard, people are increasingly active in demanding a higher ethical standard of advertisers. In Sweden, all advertising is forbidden in programming directed at children, a particularly impressionable group. And in the United States, cigarette ads have been forbidden on television for decades. The European Union recently expanded its ban on ads for cigarettes on television to cover more media, including newspapers, magazines, radio, and the Internet by 2005, as well as sporting events by 2006. Setting boundaries for advertising is a sensitive topic, given concerns that such parameters might limit free speech, but these examples demonstrate that countries can strike a healthy balance between protecting speech and public health.[47]

Meanwhile, advertising itself is being used as a tool to fight the high number of consumption messages bombarding consumers. The Canadian group Adbusters sponsors TV "uncommercials" that encourage viewers to reduce consumption, leave their cars in their garages, or turn off their televisions. Some governments are placing ads or public service announcements on television and other media to encourage more-sustainable consumption, as the Thai government has done through humorous TV commercials urging consumers to use less energy and water. The

U.N. Environment Programme (UNEP) takes a different approach, working with advertisers to develop ads that encourage people to use sustainable products. (See Box 8–3.)[48]

Education is also important in reshaping culture for a higher quality of life. Australia and Canada now mandate a media education curriculum in their schools. These programs help make students aware of how the media and advertising shape their values and culture. And students are taught how to differentiate between reality and marketing hyperbole—whether in commercials or embedded in programming. Consumption education, in particular, may be a necessary corrective to advertising's incessant proclamations of the desirability of consumption. In Brazil, the nongovernmental group Instituto Akatu has worked with schools, businesses, and Scout troops to educate participants to "consume consciously." The organization uses a variety of tools—from the Internet to pamphlets, comic books, and games—to teach the environmental and social consequences of consumption and to tell people how to lobby governments to press for changes in policy that will help promote conscious consumption.[49]

Getting to the Good Life

Lurking beneath growing dissatisfaction with the consumer society is a simple question: What is an economy for? The traditional responses, including prosperity, jobs, and expanded opportunity, seem logical enough—until they become dysfunctional, that is. When prosperity makes us overweight, overwork leaves us exhausted, and a "you can have it all" mindset leads us to neglect fam-

---

BOX 8–3. ENCOURAGING ADVERTISERS TO PROMOTE SUSTAINABILITY

Marketing is a powerful tool that is often implicated in stimulating consumption—and, therefore, in undermining efforts to build a sustainable world. But the U.N. Environment Programme is trying to turn marketers into allies by enlisting them to promote sustainability. In 1999, a UNEP Forum on Advertising and Communication was established to raise awareness of "sustainable consumption"—consumption that improves life quality while minimizing social and ecological inequities—and to encourage advertisers and marketers to promote it.

Key business associations within the advertising and marketing industry have responded by developing pro-sustainability publications in cooperation with UNEP and by organizing special sessions on sustainable development at their international congresses. For example, the advertising agency McCann-Erickson published with UNEP a leaflet called "Can Sustainability Sell?" targeted at companies and marketing professionals to convince them that "far from depressing sales, sustainable principles could be essential to protect both brand health and future profitability." In partnership with Sustain-Ability and UNEP, the European Association of Communications Agencies prepared a guide for advertising agencies that describes the growing international market for sustainable consumption. And the World Association of Research Professionals has ordered a survey on consumer attitudes toward sustainability issues.

Moreover, UNEP is cooperating with specific industry sectors—notably, the automotive, tourism, and retail sectors—to help them develop innovative marketing strategies that would further promote sustainable options.

—*Solange Montillaud-Joyel,*
U.N. Environment Programme

SOURCE: See endnote 48.

---

ily and friends, people start to question more deeply the direction of their lives as well as the system that helps steer them in that direction. The signals emerging in some industrial countries—and some developing ones as well—suggest that many of us are looking for more from life than a bigger house and a new car. People long for something deeper: happy, dignified, and meaningful lives—in a word, well-being. And they expect their economies to be a tool to this end, not an obstacle to it.

Everyone will need to become practiced at wrestling with a key question: How much is enough?

Societies with a high quality of life are people-centered, with proper attention given to promoting interactions among human beings. Urban areas designed with attention to pedestrians, to leisure, and to human expression, for example, would bring people together in constructive and satisfying ways—for public concerts, festivals, or simply the informal interactions made possible in outdoor markets. Economies would have a local character, so that produce, talent, and goods unique to the region would be favored over imports from distant shores. By strengthening the web of relationships between farmer and city dweller, artisan and client, producer and consumer, local economies have a "human-scale" character that far-flung economies often lack.

Nurturing relationships requires time and may involve corralling many of the "time thieves" of modern life, starting with work. Experience in several European countries has demonstrated that the 40-hour workweek is clearly not sacrosanct, so that people can arrive home earlier or have longer weekends to spend with their children or friends. And housing that is not spread out in scattered sub-

urbs could prevent the daily commute that robs many people of astonishing amounts of time: a commute of more than an hour a day, the norm for many American suburbanites, means a worker spends the equivalent of six workweeks in transit each year. Society's focus on time-saving devices, the use of which has only led to more frenzied lives, needs to be replaced with simpler, time-saving lifestyles.[50]

A well-being society would offer consumers a sufficient range of genuine choices rather than a large array of virtually identical products. Businesses would be encouraged through economic incentives to deliver what consumers really seek—reliable transportation, not necessarily a car; or tasty, seasonal local produce rather than fruits and vegetables shipped in from another country; or strong neighborhood relationships in lieu of a large house with a big yard. Choice would be redefined to mean options for increasing quality of life rather than selections among individual products or services.

For individuals, genuine choice would likely include the choice not to consume. Everyone will need to become practiced at wrestling with a key question: How much is enough? Responses will vary from person to person, but a guideline worth considering is one from the Chinese philosopher Lao Tzu: "To know when you have enough is to be rich." Consumers who embrace this ancient wisdom take a large step toward escaping the tyranny of social comparison and marketing that drives so much of today's consumption.[51]

People in a well-being society would also develop close relationships with the natural environment. They would recognize the trees in their parks and the flowers in their yards as easily as they identify corporate logos. They would understand the environmental foundations of their economic activity: where their water comes from, where their garbage goes, and whether coal, nuclear, or renewable energy runs the power plant that generates their electricity. They would likely enjoy developing projects at home that help them to live more intimately with nature—a rain-catching cistern, for example, or a compost bin or vegetable garden. In short, they would learn to love nature and to become advocates for it. As the late Harvard biologist Stephen Jay Gould once said: "We must develop an emotional and spiritual bond with nature, for we will not fight to save what we do not love."[52]

Finally, a society focused on well-being would ensure that everyone in it has access to healthy food, clean water and sanitation, education, health care, and physical security. It is virtually impossible to imagine a society of well-being that does not provide for people's basic needs. And more than that, it is inconceivable that a well-being society would be satisfied with its own success if others outside its borders are suffering on a broad scale. Indeed, those societies that rank highest in the Wellbeing Index, especially in northern Europe, also have some of the world's most generous foreign aid programs.[53]

Making the transition to a society of well-being will undoubtedly be a challenge, given people's habit of placing consumption at the apex of societal values. But any move in this direction starts out with two strong advantages. First, the human family today has a base of knowledge, technology, and skills far surpassing anything previous generations have known. Ironically, this base is the product of an economic system oriented toward high levels of consumption. But our twentieth-century consumption-oriented development choices, however misguided, can be redeemed now by ensuring that today's stocks of knowledge and technology are invested in well-being rather than in continued material accumulation for its own sake.

A second advantage is simple but powerful: for many people, a life of well-being is preferred to a life of high consumption. Former Prime Minister Ruud Lubbers of the Netherlands captured this fundamental reality when he noted that in their effort to build a high quality of life, the Dutch work limited hours: "We like it that way. Needless to say, there is more room for all those important aspects of our lives that are not part of our jobs, for which we are not paid, and for which there is never enough time." The desire for a higher quality of life may be more imperfectly formed in other industrial societies, but the signals are there: workers who want free time more than a pay raise, shoppers who choose organic food and other "ethical" products, people who seek stronger family relationships. When the components of a well-being society are made available, the reception is often strikingly positive.[54]

By nurturing relationships, facilitating healthy choices, learning to live in harmony with nature, and tending to the basic needs of all, societies can shift from an emphasis on consumption to an emphasis on well-being. This could be as great an achievement in the twenty-first century as the tremendous advances in opportunity, convenience, and comfort were in the twentieth.

Notes

Preface

1. Gary Cross, *An All-Consuming Century: Why Commercialism Won in Modern America* (New York: Columbia University Press, 2000).

2. Matthew Bentley, *Sustainable Consumption: Ethics, National Indices and International Relations* (PhD dissertation, American Graduate School of International Relations and Diplomacy, Paris, 2003).

3. Lebow quote from Vance Packard, *The Waste Makers* (New York: David Mckay, 1960).

4. Fresh water from Igor A. Shiklomanov, *Assessment of Water Resources and Water Availability in the World* (St. Petersburg, Russia: State Hydrological Institute, 1996); Janet L. Sawin, "Fossil Fuel Use Up," in Worldwatch Institue, *Vital Signs 2003* (New York: W.W. Norton & Company, 2003), pp. 34–35.

5. Current number and growth calculated from "Car Sales Booming in China," *All Things Considered*, National Public Radio, 17 September 2003, from "China's Private Car Ownership Tops 10 Million," *People's Daily*, 14 June 2003, and from Liu Wei, "China's Demand of Cars to Exceed 4.2 Million in 2003," *People's Daily*, 30 July 2003.

6. Alan Durning, *How Much Is Enough?* (New York: W.W. Norton & Company, 1992).

State of the World: A Year in Review

October 2002. U.N. Environment Programme (UNEP), Finance Initiative Climate Change Working Group, *Climate Change and the Financial Services Industry* (Nairobi: 2002); Colin Woodard, "Popularity Burdens World's Favorite Coastline," *Christian Science Monitor*, 10 October 2002; International Food Policy Research Institute (IFPRI) and International Water Management Institute, *Global Water Outlook to 2025: Averting an Impending Crisis* (Washington, DC: IFPRI, 2002); "Pelo Menos 300 Mil Crianças Trabalham em Minas da Colômbia," *Reuters*, 31 October 2002.

November 2002. Coalition to Stop the Use of Child Soldiers, *Child Soldiers 1379 Report* (London: November 2002); UNESCO, "UNESCO Adds 18 New Sites to World Network of Biosphere Reserves," press release (Paris: 8 November 2002); "Galicia (Spain) Oil Spill, November 2002," *Earthnet*, 20 November 2002.

December 2002. World Food Programme, "Hunger Crisis Set to Worsen in 2003 Despite Fresh Donations," press release (Johannesburg: 30 December 2002); Gareth Harding, "EU Bans Tobacco Ads," *UPI Wire*, 2 December 2002; vehicles from Chang-Ran Kim, "The Future is Here—Japan Launches Fuel Cell Cars," *Reuters*, 3 December 2002, and from Deena Beasley, "Honda, Toyota Deliver Fuel Cell Cars in California," *Reuters*, 3 December 2002; "Wind Around the World," *Renew On Line*, November-December 2002, at www-tec.open.ac.uk/eeru/natta/renewonline/rol40/11.htm, viewed 8 October 2003; Human Rights Commission of Pakistan, "461 Honor Killings of Women Reported in Two Provinces," *UN Wire*, 12 December 2002; "Warm Arctic Summer Melted Much Ice," *Science News*, 21 and 28 December 2002.

January 2003. Natural Resources Defense Council, *Rewriting the Rules* (New York: 31 December 2002); David Karoly and James Risbey, *Global Warming Contributes to Australia's Worst Drought* (Sydney: World Wide Fund for Nature Australia, 15 January 2003); Clive James, "Global Status of Commercialized Transgenic Crops: 2002," press release (Manila: International Service for the Acquisition of Agri-biotech Applications, 16 January 2003).

February 2003. UNEP, "Power Stations Threaten People and Wildlife with Mercury Poisoning," press release (Nairobi: 3 February 2003); Clean Edge, Inc., *Clean Energy Trends 2003* (San Francisco: 20 February 2003); "UNICEF Official Cites 'Largest Slave Trade in History'," *UN Wire*, 20 February 2003; Department of Trade and Industry, *Our Energy Future—Creating a Low Carbon Economy*, Energy White Paper (London: February 2003); U.N. Population Division, *World Population Prospects: The 2002 Revision* (New York: February 2003).

March 2003. "World Bank Says Companies Paid $1 Billion in Extortion Money," *UN Wire*, 7 March 2003; Sid Perkins, "Warmer Climates Accelerate Life Cycles of Plants, Animals," *Science News*, 8 March 2003; Global Policy Forum, "Iraq Crisis," at www.globalpolicy.org/security/issues/irqindx.htm, viewed 8 October 2003; United Nations, *Water for People, Water for Life: The United Nations World Water Development Report* (Paris: UNESCO Publishing and Berghahn Books, 2003).

April 2003. J. Travis, "Moving On: Now The Human Genome is Really Done," *Science News*, 19 April 2003; "Sea Burial for Canada's Cod Fisheries," *Science News*, 17 May 2003; World Health Organization (WHO) and UNICEF, *Africa Malaria Report* (Geneva: WHO, April 2003); Foreign Policy Magazine and Center for Global Development, "New Ranking: Netherlands is Most Development-Friendly Nation," press release (Washington, DC: 28 April 2003).

May 2003. Ransom Myers and Boris Worm, "Rapid Worldwide Depletion of Predatory Fish Communities," *Nature*, 15 May 2003; WHO,

"WHO Framework Convention on Tobacco Control," at www.who.int/features/2003/08/en, viewed 8 October 2003; World Bank, "Multilateral Initiative to Manage South America's Largest Groundwater Reservoir Launched," press release (Montevideo, Uruguay: 23 May 2003).

June 2003. "Water Starts to Fill Three Gorges Dam," *UN Wire*, 2 June 2003; UNEP, "New Initiative to Combat Growing Global Menace of Environmental Crime," press release (Nairobi: 2 June 2003); Otto Pohl, "World Panel Will Now Act to Conserve the Whale Population," *New York Times*, 17 June 2003; McDonald's Corporation, "McDonald's Calls for Phase-Out of Growth Promoting Antibiotics in Meat Supply, Establishes Global Policy on Antibiotic Use," press release (Oak Brook, IL: 19 June 2003); "Deforestation of Amazon is on the Rise," *UN Wire*, 26 June 2003.

July 2003. WHO, "SARS Outbreak Contained Worldwide," press release (Geneva: 5 July 2003); Roger Thurow, "AIDS Fuels Famine in Africa: As Swaziland Farmers Die, Their Land Goes Unplanted," *Wall Street Journal*, 9 July 2003; "Europe Adopts Climate Emissions Trading Law," *Environmental News Service*, 22 July 2003.

August 2003. UNEP and World Conservation Monitoring Centre (WCMC), "Wild Forests 'Living Museums' of Virtually Extinct Species," press release (Cambridge: WCMC, 4 August 2003); "EU 'Regrets' US Action on GM Crops," *BBC World News*, 8 August 2003; Nancy Gibbs, "Lights Out," *Time*, 25 August 2003; International Council on Mining and Metals, "Landmark 'No-go' Pledge from Leading Mining Companies," press release (London: 20 August 2003); Denis Hémon and Eric Jougla, "Surmortalité liée à la Canicule d'Août 2003—Rapport d'étape" (Paris: National Institute for Health and Medical Research, 25 September 2003).

September 2003. "Northern Hemisphere Temperature Hits 2,000 Year High," *Environmental News Service,* 2 September 2003; UNEP, "World's Protected Areas Top 100,000, Exceed Size of India and China—UN Report," press release (Nairobi: 9 September 2003); Patrick Smith, "Poor

Nations Keep Heat on Trade," *Christian Science Monitor*, 30 September 2003; World Meteorological Organization, "Antarctic Ozone Hole Unusually Large," press release (Geneva: 16 September 2003); "Chile Indians End Protest Against Hydro-Power Dam," *Reuters*, 19 September 2003.

Chapter 1.
The State of Consumption Today

1. Wayne W. J. Xing, "Shifting Gears," *The China Business Review*, November-December 1997.

2. "China's Private Car Ownership Tops 10 Million," *People's Daily*, 14 June 2003; 11,000 a day is a Worldwatch calculation based on the data in Liu Wei, "China's Demand of Cars to Exceed 4.2 Million in 2003," *People's Daily*, 30 July 2003; auto sales from "Car Sales Booming in China," *All Things Considered*, National Public Radio, 17 September 2003; "150mn Chinese Families to Buy Cars in Next 15 Years," *People's Daily*, 12 March 2003; U.S. auto fleet from Ward's Communications, *Ward's Motor Vehicle Facts & Figures 2001* (Southfield, MI: 2001), p. 38; attitudes of Chinese from "Car Sales Booming," op. cit. this note.

3. Julie Chao, "Pacific Currents: China Trying to Cope With Burgeoning Car Culture," *Seattle Post-Intelligencer*, 8 September 2003; foreign investments from Clay Chandler, "China Goes Car Crazy: Suburbs, Drive-ins, Car Washes—This Revolution Has Wheels," *Fortune*, 11 August 2003.

4. Matthew Bentley, *Sustainable Consumption: Ethics, National Indices and International Relations* (PhD dissertation, American Graduate School of International Relations and Diplomacy, Paris, 2003).

5. Population projection from United Nations, *World Population Prospects, The 2002 Revision* (New York: 2003).

6. U.S. Department of Transportation, Bureau of Transportation Statistics, *National Household Travel Survey 2001 Highlights Report* (Washington, DC: 2003); SUVs from Oak Ridge National Lab-

oratory, *Transportation Energy Data Book*, Edition 22 (Oak Ridge, TN: September 2002), p. 7-1; house size from Joint Center for Housing Studies, *State of the Nation's Housing 2003* (Cambridge, MA: Harvard University, 2003), p. 32; household size from U.S. Department of Agriculture, Economic Research Service, *Race and Ethnicity in Rural America: Marital Status and Household Structure*, at www.ers.usda.gov/Briefing/RaceAndEthnic/familystructure.htm, updated 24 December 2002; obesity industry from Jui Chakravorty, "Catering to Obese Becoming Big Business," *Reuters*, 4 October 2003.

7. Private consumption expenditures (in 1995 dollars) are Worldwatch calculations based on World Bank, *World Development Indicators Database*, at media.worldbank.org/secure/data/qquery.php, viewed 2 June 2003. Box 1–1 from the following: United Nations, op. cit. note 5, p. 1; Americans' projected meat consumption and all population data from United Nations Population Division, online database, at esa.un.org/unpp, viewed 20 September 2003, and from U.N. Food and Agriculture Organization (FAO), *FAOSTAT Statistical Database*, at apps.fao.org, updated 30 June 2003; carbon emissions are Worldwatch calculations based on data in Molly O. Sheehan, "Carbon Emissions and Temperature Climb," in Worldwatch Institute, *Vital Signs 2003* (New York: W.W. Norton & Company, 2003), pp. 40–41; household size and energy use from Nico Keilman, "The Threat of Small Households," *Nature*, 30 January 2003, p. 489. Table 1–1 contains Worldwatch calculations based on population and private consumption expenditures data from World Bank, op. cit. this note; totals add up to 98 and 99 percent because data are unavailable for a few small countries.

8. Poverty numbers are World Bank estimates cited on Millennium Development Goals Web site, at www.developmentgoals.com/Poverty.htm.

9. Table 1–2 from Bentley, op. cit. note 4. Share of global consumer class is a Worldwatch calculation.

10. Table 1–3 from Bentley, op. cit. note 4; expenditures in sub-Saharan Africa (in 1995 dol-

lars) from World Bank, op. cit. note 7.

11. Bentley, op. cit. note 4.

12. U.N. Environment Programme (UNEP), "UNEP Urges Asia-Pacific Towards a Cleaner, Greener Development Path," press release (Nairobi: 19 May 2003); U.S. auto fleet from Ward's Communications, op. cit. note 2, p. 38; Matthew Bentley, "Forging New Paths to Sustainable Development," UNEP Background Paper, Asia Pacific Expert Meeting on Promoting Sustainable Consumption and Production Patterns, Yogyakarta, Indonesia, 21–23 May 2003.

13. Daily calories from FAO, op. cit. note 7; number of undernourished from idem, *The State of Food Insecurity in the World*, as cited in United Nations Statistical Division, *Millennium Indicators Database*, at unstats.un.org/unsd/mi/mi_series_xrxx.asp?row_id=640, viewed 23 October 2003; Table 1–4 from World Bank, *World Development Indicators 2000* (Washington, DC: 2000), pp. 222–24.

14. Calories from animal products from FAO, op. cit. note 7; meat consumption from Danielle Nierenberg, "Meat Production and Consumption Grow," in Worldwatch Institute, op. cit. note 7, p. 30; fast food in India from Saritha Rai, "Taste of India in U.S. Wrappers," *New York Times*, 29 April 2003, and from Seth Mydans, "Clustering in Cities, Asians Are Becoming Obese," *New York Times*, 13 March 2003; undernourished in India from U.N. Development Programme (UNDP), *Human Development Report 2003* (New York: Oxford University Press, 2003), p. 199.

15. Data on clean water and sanitation from UNICEF, *The State of the World's Children 2003* (New York: 2003), p. 95; definitions of "safe drinking water" and "adequate sanitation" from UNDP, op. cit. note 14, pp. 357–58.

16. Janet Abramovitz and Ashley Matoon, *Paper Cuts: Recovering the Paper Landscape*, Worldwatch Paper 149 (Washington, DC: Worldwatch Institute, December 1999), pp. 6, 11–12.

17. Table 1–5 from World Bank, op. cit. note 7;

households with televisions and number with cable service from International Telecommunication Union (ITU), *World Telecommunication Development Report 2002* (Geneva: 2002); average television habits from Robert Kubey and Mihaly Csikszentmihalyi, "Television Addiction Is No Mere Metaphor," *Scientific American*, February 2002, pp. 74–80.

18. Phones from ITU, op. cit. note 17; Internet users from idem, "Internet Indicators: Hosts, Users and Number of PCs," at www.itu.int/ITU-D/ict/statistics/at_glance/Internet02.pdf, viewed 9 October 2003.

19. Table 1–6 from the following: makeup and perfumes from "Pots of Promise," *The Economist*, 24 May 2003, pp. 69–71; pet food and ice cream from UNDP, *Human Development Report 1998* (New York: Oxford University Press, 1998), p. 37; ocean cruises from Lisa Mastny, "Cruise Industry Buoyant," in Worldwatch Institute, *Vital Signs 2002* (New York: W.W. Norton & Company, 2002), p. 122; additional annual investments needed from Michael Renner, "Military Expenditures on the Rise," in Worldwatch Institute, op. cit. note 7, p. 119, except for immunizing estimate from Erik Assadourian, "Consumption Patterns Contribute to Mortality," in Worldwatch Institute, op. cit. note 7, p. 108.

20. Materials increases from Gary Gardner and Payal Sampat, *Mind Over Matter: Recasting the Role of Materials in Our Lives*, Worldwatch Paper 144 (Washington DC: Worldwatch Institute, December 1998), p. 16; metals intensity from Payal Sampat, "Metals Production Climbs," in Worldwatch Institute, op. cit. note 19, pp. 66–67.

21. Fossil fuel use from Janet L. Sawin, "Fossil Fuel Use Up," in Worldwatch Institute, op. cit. note 7, pp. 34–35; metals from Payal Sampat, "Scrapping Mining Dependence," in Worldwatch Institute, *State of the World 2003* (New York: W.W. Norton & Company, 2003), p. 113.

22. Abramovitz and Mattoon, op. cit. note 16, p. 20.

23. Sampat, op. cit. note 21, p. 114; Organisation for Economic Co-operation and Development (OECD), *OECD Environmental Data Compendium 2002* (Paris: 2003), p. 14.

24. Data and FAO projections from Abramovitz and Mattoon, op. cit. note 16, pp. 20–21, 40, 52.

25. Peter Lunt, "Psychological Approaches to Consumption: Varieties of Research—Past, Present and Future," in Daniel Miller, ed., *Acknowledging Consumption* (London: Routledge, 1995), pp. 238–63.

26. Ibid.; John A. Bargh, "Losing Consciousness: Automatic Influences on Consumer Judgment, Behavior, and Motivation," *Journal of Consumer Research*, September 2002, pp. 280–85.

27. Tim Jackson and Nic Marks, "Consumption, Sustainable Welfare and Human Needs," *Ecological Economics*, vol. 28, no. 3 (1999), pp. 421–42.

28. J. R. McNeill, *Something New Under the Sun: An Environmental History of the 20th Century World* (New York: W.W. Norton & Company, 2001), p. 315; Toyota plant from Thomas L. Friedman, *The Lexus and the Olive Tree* (New York: Farrar, Strauss, Giroux, 1999), p. 26; semiconductor costs from "Cost per Megabit Trends," IC Knowledge Web site, at www.icknowledge .com/economics/productcosts2.html, viewed 2 September 2003.

29. Australians from "Freer Trade Cuts the Cost of Living," at WTO website at www.wto.org/ english/thewto_e/whatis_e/10ben_e/10b04_e .htm, viewed 17 October 2003; "Kingdom of Bahrain Joins WTO's Information Technology Agreement," 18 July 2003, at www.wto.org/ english/news_e/news03_e/news_bahrain_ita_18 jul03_e.htm, viewed 17 October 2003; cost reductions from Telecommunications Industry Association, "Information Technology Agreement Promises to Eliminate Tariffs on Most IT Products by the Year 2000," *PulseOnline Newsletter Archive*, October 1997, at www.tiaonline.org/media/ pulse/1997/pulse1097-3.cfm, viewed 17 October 2003.

30. International Labour Organization, Committee on Employment and Social Policy, *Employment and Social Policy in Respect of Export Processing Zones* (Geneva: November 2002); Jim Lobe, "Unions Assail WTO for Ignoring Worker Rights," *OneWorld US*, 8 September 2003; International Confederation of Free Trade Unions, *Export Processing Zones—Symbols of Exploitation and a Development Dead-End* (Brussels: September 2003).

31. Ransom A. Myers and Boris Worm, "Rapid Worldwide Depletion of Predatory Fish Communities," *Nature*, 15 May 2003, pp. 280–83; mining from Craig B. Andrews, *Mineral Sector Technologies: Policy Implications for Developing Countries*, Industry and Energy Division Note No. 19 (Washington, DC: World Bank, 1992), and from Craig Andrews, discussion with Claudia Meulenberg, Worldwatch Institute, 22 September 2003; Dogwood Alliance, "Paper and Chipboard Production," at www.dogwoodalliance.org/ chipmill.asp#chipboard, viewed 22 October 2003.

32. Oil prices from Organization of the Petroleum Exporting Countries, *OPEC Annual Statistics Bulletin 2001* (Vienna: 2001), p. 119; air freight rates from David Hummels, "Have International Transportation Costs Declined?" draft (Chicago: University of Chicago, November 1999), pp. 4–5; division of labor from Nathan Rosenberg, "Technology," in Glenn Porter, ed., *Encyclopedia of American Economic History*, vol. 1 (New York: Charles Scribner's Sons, 1980), pp. 294–308.

33. ITU, *Telecommunications Indicators in the World, 2000*, at www.rtnda.org/resources/wired web/text.html; Moore's Law from Webopedia, at www.webopedia.com/TERM/M/Moores_Law .html, viewed 9 October 2003.

34. Semiconductor costs from Joseph I. Lieberman, *White Paper: National Security Aspects of the Global Migration of the U.S. Semiconductor Industry* (Washington, DC: Office of Senator Lieberman, June 2003); fixed costs from Rosenberg, op. cit. note 32.

35. Global, U.S., and Chinese advertising spending from Bob Coen, *Universal McCann's Insider's Report on Advertising Expenditures*, June 2003, at www.mccann.com/insight/bobcoen.html, viewed 9 October 2003; Figure 1–1 from ibid. and from Bob Coen, *Estimated World Advertising Expenditures*, at www.mccann.com/insight/bobcoen .html, viewed 9 October 2003; newspapers and mail from John de Graff, David Wann, and Thomas Naylor, *Affluenza: The All-Consuming Epidemic* (San Francisco, CA: Berrett-Koehler Publishers, Inc., 2001), p. 149; "Television Clutter in Prime Time and Early Morning Reach All Time Highs," press release (New York: American Association of Advertising Agencies and Association of National Advertisers, Inc., 12 April 1999).

36. Teen smoking from Madeline A. Dalton et al., "Effect of Viewing Smoking in Movies on Adolescent Smoking Initiation: A Cohort Study," *The Lancet*, 10 June 2003, pp. 281–85; ban and increase from Stanton A. Glantz, "Smoking in Movies: A Major Problem and a Real Solution," *The Lancet*, 10 June 2003, p. 258; 85 percent and revenue from foreign sales from James D. Sargeant et al., "Brand Appearances in Contemporary Cinema Films and Contribution to Global Marketing of Cigarettes," *The Lancet*, 6 January 2001, pp. 29–32; three times more prevalent from A. R. Hazan, H. L. Lipton, and S. A. Glantz, "Popular Films Do Not Reflect Current Tobacco Use," *American Journal of Public Health*, vol. 84, no. 6 (1994), pp. 998–1000; World Health Organization (WHO), *'Bollywood': Victim or Ally? A WHO Study on the Portrayal of Tobacco in Indian Cinema* (Geneva: February 2003).

37. Consumer credit growth from Lizabeth Cohen, *A Consumer's Republic: The Politics of Mass Consumption in Postwar America* (New York: Alfred A. Knopf, 2003), pp. 123–24; 61 percent from Robert D. Manning, "Perpetual Debt, Predatory Plastic: From the Company Store to the World of Late Fees and Overlimit Penalties," *Southern Exposure*, summer 2003, p. 51; per capita income from World Bank, *World Development Indicators 2003* (Washington, DC: 2003), pp. 14–16.

38. Credit growth and quote from Joshua Kurlantzick, "Charging Ahead: America's Biggest

New Export—Credit Cards—Could Bring Down the World Economy," *Washington Monthly*, May 2003, pp. 28–29.

39. Economic subsidies from OECD, *Towards Sustainable Consumption: An Economic Conceptual Framework* (Paris: Environment Directorate, June 2002), p. 41; suburban home subsidies from Cohen, op. cit. note 37; Scott Bernstein, Center for Neighborhood Technology, Chicago, discussion with Gary Gardner, 20 August 1998.

40. Paul Hawken, Amory Lovins, and L. Hunter Lovins, *Natural Capitalism: Creating the Next Industrial Revolution* (Boston: Little, Brown, and Company, 1999), pp. 57–60.

41. Sampat, op. cit. note 21, p. 117.

42. OECD, *OECD Environmental Data 2002* (Paris: 2002), p. 11; "Norway—Household Waste Increases More Than Ever," *Warmer Bulletin*, 28 June 2003.

43. Andrew Balmford et al., "Economic Reasons for Conserving Wild Nature," *Science*, 9 August 2002; Living Planet Index from WWF International, UNEP, and Redefining Progress, *Living Planet Report 2002*, at www.panda.org/ news_facts/publications/general/livingplanet/ind ex.cfm, p. 21. Table 1–7 from various editions of Worldwatch Institute's *Vital Signs* as follows: fossil fuel use, carbon emissions, and sea level rise from *Vital Signs 2003*; forests, wood use, cropland, rangelands, food production, and water deficit from *Vital Signs 2002*; wetlands from *Vital Signs 2001*; groundwater from *Vital Signs 2000*; in addition, large predator fish depletion from Myers and Worm, op. cit. note 31. Figure 1–2 from WWF International, UNEP, and Redefining Progress, op. cit. this note, and from Angus Maddison, *The World Economy: A Millennial Perspective* (Paris: OECD, 2001), pp. 272–321, with updates from International Monetary Fund, *World Economic Outlook Database* (Washington, DC: December 2002).

44. WWF International, UNEP, and Redefining Progress, op. cit. note 43; Mathis Wackernagel et al., "Tracking the Ecological Overshoot of the

Human Economy," *Proceedings of the National Academy of Sciences*, 9 July 2002, p. 9268.

45. Deaths from Majid Ezzati and Alan D. Lopez, "Estimates of Global Mortality Attributable to Smoking in 2000," *The Lancet*, 13 September 2003, pp. 847–52; $150 billion from "Annual Smoking-Attributable Mortality, Years of Potential Life Lost, and Economic Costs— United States, 1995–1999," *Morbidity and Mortality Weekly Report*, 12 April 2002, p. 303; revenue from Judith Mackay and Michael Eriksen, *The Tobacco Atlas* (Geneva: WHO, 2002), p. 50; overweight and obesity from WHO and FAO, *Joint WHO/FAO Expert Consultation on Diet, Nutrition and the Prevention of Chronic Diseases* (Geneva: 2002); National Center for Health Statistics, *Health, United States, 2003* (Hyattsville, MD: 2003); U.S. Department of Health and Human Services, *The Surgeon General's Call to Action to Prevent and Decrease Overweight and Obesity, 2001* (Washington, DC: 2001).

46. The Fordham Institute for Innovation in Social Policy, *The Social Report 2003* (New York: 2003); UNDP, op. cit. note 14, pp. 248–49.

47. OECD, *The Well Being of Nations: The Role of Human and Social Capital* (Paris: 2001), pp. 99–103.

48. Robert Putnam, *Bowling Alone: The Collapse and Revival of American Community* (New York: Simon & Schuster, 2000), pp. 189–246.

49. Michael Bond, "The Pursuit of Happiness," *New Scientist*, 4 October 2003, pp. 40–47.

50. Anders Hayden, "Europe's Work-Time Alternatives," in John de Graaf, ed., *Take Back Your Time* (San Francisco: Berrett Koehler, 2003), p. 204.

Plastic Bags

1. Film and Bag Federation, "Great Moments in Plastic Bag History," at www.plasticbag.com/ environmental/history.html.

2. Howard Rappaport, Director of Global Plastics and Polymers, Chemical Market Associates International, discussion with author, 2 September 2003; National Economic Development and Labour Council (Nedlac), *Socio-Economic Impact of the Proposed Plastic Bag Regulations*, FRIDGE Study No. 29 (Cape Town, South Africa: October 2001).

3. Total production, 80 percent of use in North America and Western Europe, and one quarter from Asia estimates from Rappaport, op. cit. note 2; 100 billion from L. J. Williamson, "It's Not My Bag, Baby!" *OnEarth*, summer 2003, and from Chaz Miller, "Plastic Film," *Waste Age*, November 2002.

4. Franklin Associates Inc., *Resource and Environmental Profile Analysis of Polyethylene and Unbleached Paper Grocery Sacks* (Prairie Village, KS: 1990); Nedlac, op. cit. note 2; Fehily Timoney & Company, "Consultancy Study on Plastic Bags," at www.fehilytimoney.com.

5. Jeremia Njeru, "Managing the Problem of Plastic Bag Waste in Nairobi, Kenya: Demand-side and Supply-side Considerations," presentation at Association of American Geographers, 99th Annual Meeting, New Orleans, 5–8 March 2003; Beijing from Xu Zhengfeng, "Putting an End to a Plastic Plague," *Inter Press Service*, 17 August 1999; "national flag" from Shawn Pogatchnik, "Ireland, Pioneer of the Plastic-bag Tax, Plans Fees on Three Other Litter Sources," *Associated Press*, 16 July 2003; "national flower" from "South Africa Bans Plastic Bags," *BBC News*, 9 May 2003.

6. Biodegradable Products Institute, at www.bpiworld.org; Steve Mojo, Executive Director, Biodegradable Products Institute, discussion with author, 15 September 2003.

7. Ladakh from Tsewang Rigzin, "Leh's Successful Plastic Ban," *Ladags Melong*, 17 June 2002; Moazzem Hossain, "Bangladesh Bans Polythene," *BBC News*, 1 January 2002.

8. South Africa from Toby Reynolds, "South Africa Moves to Curb Flimsy Plastic Bag Scourge," *Reuters*, 1 October 2002; Ireland from Sean Federico-O'Murchu, "Irish Take Lead with Plastic

Bag Levy," *MSNBC*, 4 August 2003; other national policies from Pogatchnik, op. cit. note 5, and from John Roach, "Are Plastic Grocery Bags Sacking the Environment?" *National Geographic*, 2 September 2003.

9. James Watts, manager, Weaver Street Market, discussion with author, 8 September 2003.

Chapter 2.
Making Better Energy Choices

1. "Mountaintop Mining," *Morning Edition*, National Public Radio, 25 June 2003; Environmental Media Services, "Mountaintop Removal Strip Mining," 7 May 2002, in Mining the Mountains Series, *The Charleston Gazette Online*, at www.wvgazette.com/static/series/mining; Penny Loeb, "The Coalfield Communities of Southern West Virginia: Mining's Impact on Communities," March 2003, at www.wvcoalfield.com.

2. Efficiency of coal plants from U.S. Department of Energy (DOE), Office of Fossil Energy, "DOE Launches Project to Improve Materials for Supercritical Coal Plants," press release (Pittsburgh, PA: 16 October 2001).

3. Oil use calculated by Worldwatch with data from BP, *Statistical Review of World Energy 2003* (London: June 2003), p. 38.

4. Increases 1850–1970 from John Holdren, "The Transition to Costlier Energy," in Lee Schipper et al., *Energy Efficiency and Human Activity: Past Trends, Future Prospects* (Cambridge, U.K.: Cambridge University Press, 1992), p. 7; population from Molly O. Sheehan, "Population Growth Slows," in Worldwatch Institute, *Vital Signs 2003* (New York: W.W. Norton & Company), p. 67; increase in fossil fuels by 2002 calculated by Worldwatch with data from Janet L. Sawin, "Fossil Fuel Use Up," in ibid, p. 35, and from BP, op. cit. note 3, p. 38; 28 percent from DOE, Energy Information Administration (EIA), *Energy in Africa* (Washington, DC: Office and Energy Markets and End Use, 1999), p. 8.

5. Energy savings from Howard Geller, *Energy Revolution: Policies for a Sustainable Future* (Wash-

ington, DC: Island Press, 2003), p. 133; Amory B. Lovins, "U.S. Energy Security Facts (For a Typical Year, 2000)," fact sheet (Snowmass, CO: Rocky Mountain Institute (RMI), 18 April 2003).

6. U.S. losses for 2000 from DOE, EIA, "Production and End-Use Data," in *Annual Energy Review 1999* (Washington, DC: 2000) (note that DOE acknowledges using generous efficiency assumptions); Amory Lovins, "Twenty Hydrogen Myths," fact sheet (Snowmass, CO: RMI, 2003), p. 11.

7. Eastern Europe, former Soviet states, industrial countries, and share of oil use from BP, op. cit. note 3; U.S. increase from DOE, EIA, *Annual Energy Review 2001* (Washington, DC: 2002), with updates from *International Petroleum Monthly*, July 2002; U.S. share of reserves from Stacy C. Davis and Susan W. Diegel, *Transportation Energy Data Book: Edition 22* (Oak Ridge, TN: Oak Ridge National Laboratory, September 2002), pp. 1–6.

8. Energy use ratio and lack of access from United Nations, WEHAB Working Group, *A Framework for Action on Energy*, prepared for the World Summit on Sustainable Development (New York: 2002), p. 7; 2.5 billion from "Facts About Energy," Johannesburg Summit, 26 August–4 September 2002; average American calculated by Worldwatch using energy data from BP, op. cit. note 3, and from International Energy Agency (IEA), "Renewables in Global Energy Supply," fact sheet (Paris: November 2002). Energy data include traditional biomass use. Table 2–1 from the following: total energy (excluding noncommercial biomass) and electricity are 2000 data, and carbon dioxide (CO_2) emissions are 1999 data, from World Bank, *World Development Indicators 2003* (Washington, DC: 2003), pp. 144–46, 148–50, 294–96; oil data are for 2002 from BP, op. cit. note 3, and from www.nationmaster.com, except for Ethiopia, which was calculated by Worldwatch with 2001 estimated data from the U.S. Central Intelligence Agency, *The World Factbook: Ethiopia* (Washington, DC: 2003).

9. Petroleum use quadrupling from BP, op. cit. note 3; growth in Indian consumer class from

National Council of Applied Economic Research (NCAER), cited in Sunil Jain and Nandini Lakshman, "The Return of the Consumer," rediff.com (India), 7 June 2003; homeless families from Office of the Registrar General, India, "Tables on Houses, Household Amenities and Assets: India," *Census of India 2001*, April 2003, and from idem, "India at a Glance: Number of Households and Household Types," *Census of India 1991*.

10. Consumption and rising incomes from Manoj Kumar, "Tryst with Developing World Consumers: A Case Study of India," *The ICFAI Journal of Marketing Management*, November 2002; oil use rankings from DOE, EIA statistics, at www.nationmaster.com. Box 2–1 from the following: DOE, EIA, *World Primary Energy Consumption, 1992–2001* (Washington, DC: 2003); IEA, *World Energy Outlook 2002* (Paris: Organisation for Economic Co-operation and Development (OECD), 2002), p. 27; DOE, EIA, *China Country Brief* (Washington, DC: 2003); Lester R. Brown, *Plan B* (New York: W.W. Norton & Company, 2003), p. 11; NCAER, op. cit. note 9; Sylvester Research Ltd., *World Wave*, cited in Kumar, op. cit. this note; Neha Kaushik, "Durables Ownership Set to Rise–NCAER Says Mid-Income Households Will Grow Rapidly," *The Hindu Business Line*, 27 February 2003; Lee Schipper, "Energy and Life: Indicators of the Link Between Energy, the Economy, and Lifestyles," p. 6, e-mail to author, 5 August 2003; Oleg Dzioubinski and Ralph Chipman, *Trends in Consumption and Production: Household Energy Consumption*, Department of Economic & Social Affairs Discussion Paper No. 6 (New York: United Nations, 1999), pp. 9–10; growth in car sales in 2002 from "Car Sales Booming in China," *All Things Considered*, National Public Radio, 17 September 2003; 2000–05 growth in fleet calculated with data from ibid., from "China's Private Car Ownership Tops 10 Million," *People's Daily*, 14 June 2003, and from Liu Wei, "China's Demand of Cars to Exceed 4.2 Million in 2003," *People's Daily*, 30 July 2003.

11. China calculated with world oil production and population data from DOE, EIA, *International Energy Outlook 2003* (Washington, DC: 2003), Tables D2 and A15, and from idem, *International Energy Annual 2001* (Washington, DC:

2003), Table 1.2; "Analysts Claim Early Peak in World Oil Demand," *Oil & Gas Journal Online*, 12 August 2002.

12. Transportation share from IEA, *World Energy Outlook 2000* (Paris: OECD, 2000), p. 25; share of oil use from European Union (EU), "The EU and the World Summit on Sustainable Development: Partnerships for Sustainable Energy," *European Union Online*, at europa.eu.int/comm/ environment/wssd/energy_en.html; U.S. share of world transport energy from U.N. Environment Programme (UNEP), *North America's Environment: Thirty-Year State of the Environment and Policy Retrospective* (Washington, DC: 2002), p. xv; rapid increase from Intergovernmental Panel on Climate Change (IPCC), *Climate Change 2001: Mitigation* (Cambridge, U.K.: Cambridge University Press, 2001), p. 368.

13. Shift to more-intensive modes from EU, "Transport Overview—Market Overview and Trends: Past and Current Trends," ATLAS Project, *European Commission Online*, at europa.eu.int/comm/energy_transport/atlas/ httmlu/, viewed 16 May 2003, and from Molly O'Meara Sheehan, "Making Better Transportation Choices," in Worldwatch Institute, *State of the World 2001* (New York: W.W. Norton & Company, 2001), p. 107; passengers from ibid., p. 106; freight and air travel from Jean-Paul Rodrigue, "Transportation and Energy," Hofstra University, at people.hofstra.edu/geotrans/eng/ ch8en/conc8en/ch8c2en.html, viewed 9 May 2003; trucks from Joseph Romm with Arthur Rosenfeld and Susan Herrmann, *The Internet Economy and Global Warming* (Washington, DC: Center for Energy and Climate Solutions, 1999), Chapter V, p. 9.

14. Lee Schipper, *Indicators of Energy Use and Efficiency: Understanding the Link Between Energy and Human Activity* (Paris: OECD/IEA, 1997), p. 18; passenger vehicle statistics from Michael Renner, "Vehicle Production Inches Up," in Worldwatch Institute, op. cit. note 4, p. 56; annual additions from Lester Brown, "Paving the Planet: Cars and Crops Competing for Land," Alert 12 (Washington, DC: Earth Policy Institute, February 2001).

15. U.S. share of cars and warming contribution from Renner, op. cit. note 14; share of oil use from Karl H. Hellman and Robert M. Heavenrich, *Light-Duty Automotive Technology and Fuel Economy Trends: 1975 Through 2003* (Washington, DC: U.S. Environmental Protection Agency (EPA), April 2003), Executive Summary, p. 1; distance traveled from Dean Anderson, *Progress Towards Energy Sustainability in OECD Countries*, Rio+5 Report (Paris: Helio International, 1997); Americans opting to drive from Schipper, op. cit. note 14, p. 103; quote from Alan Pisarski, cited in Lisa Rein and Robin Shulman, "The Rise of the Multi-Car Family," *Washington Post*, 19 July 2003; licenses from Jane Holtz Kay, *Asphalt Nation* (New York: Crown Publishers, 1997), p. 271; cars per household from Patricia S. Hu and Jennifer R. Young, *Summary of Travel Trends: 1995 Nationwide Personal Transportation Survey* (Washington, DC: U.S. Department of Transportation, Federal Highway Administration, 1999), pp. 9, 28.

16. Comparisons to U.S. ownership from Renner, op. cit. note 14; causes from EU, op. cit. note 13; Japan from Anderson, op. cit. note 15; Poland from Environment Policy Committee (EPC), Working Party on National Environmental Policy, *Sustainable Consumption: Sector Case Study Series—Household Food Consumption: Trends, Environmental Impacts and Policy Responses* (Paris: OECD, December 2001), p. 6; Asia and Pacific (includes two- and three-wheeled two-stroke vehicles) from "Making Polluters Pay," *ADB Review*, May-June 2002. Table 2–2 from the following: 1950–90 from American Automobile Manufacturers Association, cited in Lynn Price et al., "Sectoral Trends and Driving Forces of Global Energy Use and Greenhouse Gas Emissions," in *Mitigation and Adaptation Strategies for Global Change*, vol. 3, no. 2/4 (1998), pp. 263–319; 1999 from Ward's Communications, *Ward's Motor Vehicle Facts & Figures 2001* (Southfield, MI: 2001), pp. 50–53.

17. U.S. trends from Hellman and Heavenrich, op. cit. note 15, p. 1, and from Danny Hakim, "Fuel Economy Hits 22-Year Low," *New York Times*, 3 May 2003; Model T from "Sierra Club Challenges Ford's Fuel Economy at 100," *Reuters*, 5 June 2003; top speed from Ford Motor Company, "Model 'T' Facts," press release (Dearborn, MI: 22 May 2003); half of U.S. sales from Hellman and Heavenrich, op. cit. note 15, p. 3; trends elsewhere from EU, op. cit. note 13, from Anderson, op. cit. note 15, and from Neha Kaushik, "More Car Per Car Maker," *The Hindu Business Line*, 10 April 2003; David Healy, "The Number of Motor Vehicles in Use Worldwide Will Grow from 625 Million Today to 1 Billion," *Purchasing Magazine Online*, 15 January 1998.

18. U.K. data from Anderson, op. cit. note 15; EU, "Performance by Mode of Transport, EU15: 1970–2000," ATLAS Project, *EU Energy and Transport in Figures, European Commission Online*, at europa.eu.int/comm/energy_transport/atlas/httmlu/; U.S. calculated with data from Davis and Diegel, op. cit. note 7, p. 11-8.

19. Deaths from Sheehan, op. cit. note 13, p. 110, and from "Study: Greenhouse Gas Cuts Could Aid Health," *USA Today*, 22 July 2002; external costs of road transport from Sheehan, "Sprawling Cities Have Global Effects," in *Vital Signs 2002* (New York: W.W. Norton & Company, 2002), pp. 152–53; and social inequities and one third of U.S. population from Kay, op. cit. note 15, p. 33.

20. United States led in public transit from Transportation Research Board (TRB), National Research Council, *Making Transit Work: Insight from Western Europe, Canada, and the United States*, Special Report 257 (Washington, DC: National Academy Press, 2001), p. 2; rail from Kay, op. cit. note 15, pp. 166, 192; postwar trends from ibid., pp. 224–33, and from TRB, op. cit. this note, p. 4; commuter subsidies from Matthew Daly, "Congressman Considers Tax Breaks for Cyclists," *Anchorage Daily News*, 22 March 2003; drop in U.S. ridership from Sheehan, op. cit. note 13, p. 117; Michael Powell, "Licensed to Drive? Fuhgeddaboutit! Most New Yorkers Do Without Wheels," *Washington Post*, 19 August 2003; Denver from Surface Transportation Policy Project, "Transit Grows Faster than Driving for Fifth Year in a Row," press release (Washington, DC: 17 April 2002).

21. Traffic policies and user fees from TRB, op.

cit. note 20; postwar trends from "Reducing the City's Footprint," in UN HABITAT, *The State of the World Cities Report 2001* (Nairobi: 2001), and from Kay, op. cit. note 15, pp. 233, 318–19; Tokyo from UN HABITAT, op. cit. this note; Japanese from Lee Schipper, Scott Murtishaw, and Fridtjof Unander, "International Comparisons of Sectoral Carbon Dioxide Emissions Using a Cross-Country Decomposition Technique," *Energy Journal*, vol. 22, no. 2 (2001), pp. 35–75; U.S. and West European share from TRB, op. cit. note 20, p. 1; Canada share and fuel use ratio from Robert J. Shapiro, Kevin A. Hassett, and Frank S. Arnold, *The Benefits of Public Transportation: Conserving Energy and Preserving the Air We Breathe*, at www.publictransporta tion.org/pdf/preserving_air.pdf.

22. Most rapid growth from IEA, *Energy Efficiency Initiative Volume I* (Paris: OECD, 1998), p. 17, and from EU, "Transport Overview—Overall Market Drivers," ATLAS Project, *European Commission Online*, at europa.eu.int/comm/ energy_transport/atlas/htmlu/tomarpast.html, viewed 16 May 2003; tax credits from Lee Douglas, "Oregon Moves to Claw Back Bush's Big SUV Tax Break," *Reuters*, 27 June 2003 (technically, this is a subsidy for agriculture that contains a loophole for SUVs).

23. Today's engine efficiency from National Clean Bus Project, "Fact Sheet on Clean Buses: Protecting Public Health, the Environment and Providing Greater Energy Security," distributed at Environmental and Energy Study Institute briefing, Washington, DC, 13 May 2003, p. 5; Danny Hakim, "Hybrid Cars Are Catching On," *New York Times*, 28 January 2003; U.S. sales from "Toyota, Nissan Join Hands for Hybrid Cars," IndiaCar.net, 3 September 2002. Box 2—2 from the following: near-term fuel economy potential from Chris Baltimore, "US Must Cut Auto Greenhouse Gases-Research Group," *Reuters*, 2 June 2003; triple fuel economy from Robert U. Ayres, "The Energy We Overlook," *World Watch*, November/December 2001, p. 35; efficiency increasing consumption from EU, op. cit. note 22, and from Horace Herring, "Does Energy Efficiency Save Energy? The Debate and Its Consequences," *Applied Energy*, July 1999, pp. 209–26.

24. Rosa Moreno, Greenpeace Chile, discussion with author, 27 June 2003.

25. "Bogota Car Free Day Creates New Model for Organising Transportation in World Cities," at www.challenge.stockholm.se/feature_right.asp?Id Nr=5; Tooker Gomberg, "How Bogotá Beat Cars," Greenspriation!Articles (undated); Bogotá today from "Car Free City," *Habitat Newsletter*, June 2001; other cities from UN HABITAT, *Cities in a Globalizing World: Global Report on Human Settlements 2001* (Sterling, VA: Earthscan Publications, 2001), p. 142.

26. London drivers' speed and other cities from Transport for London (TFL), at www.tfl.gov.uk/ tfl/cc_intro.shtml; 25 percent from TFL Press Centre, press releases (London: 17 February 2003 through 1 April 2003), and from TFL, *Central London Congestion Charging Scheme: Three Months On* (London: Congestion Charging Division, June 2003).

27. "ACCESS—EUROCITIES for a New Mobility Culture: Introduction," at www.access-EUROCITIES.org; "Zermatt—General Information: Zermatt—The Village without Cars," Tourism Office of Zermatt, at www.zermatt.ch/ e/in_general; Freiberg and choosing car-free from Sam Tracy and Mark Peterson, "How & Why To Be Auto-Free," *The Twin Cities Green Guide*, at www.thegreenguide.org; German communities from "Bremen: A Car Free City," at www.epe.be/workbooks/tcui/example7.html.

28. IPCC, op. cit. note 12, p. 91; fastest growth in homes from United States Energy Association, *USEA Climate Change Mitigation Option Handbook*, Version 1.0 (Washington, DC: June 1999), Chapter 7.

29. Consumption from IEA, *Cool Appliances: Policy Strategies for Energy Efficient Homes* (Paris: OECD, April 2003), p. 11; different shares from Dzioubinski and Chipman, op. cit. note 10, pp. 1–3, 5, 9–10.

30. One fourth from "Housing," in UN HABITAT, op. cit. note 21; U.S. housing growth and size according to National Association of Home

Builders, cited in Elizabeth Chang, "And How Do We Heat those Starter Castles?" *Washington Post*, 16 February 2003; European house size from ENERDATA/Odyssee, cited in European Environment Agency, "Indicator Fact Sheet Signals 2001—Chapter Households" (Copenhagen: 2001); Japan Information Network, "Social Environment: Housing," at www.jinjapan.org/today/society; *Japan Lumber Journal*, cited in U.S. Department of Agriculture, Foreign Agricultural Service, *Forest Products Trade Policy Highlights—June 2000* (Washington, DC: 2000); Africans based on an average floor space of 8 square meters per person in cities, found in "Housing," op. cit. this note.

31. Increased energy due to shrinking households from Schipper, op. cit. note 10, p. 11.

32. IEA, "IEA Study Shows How to Save Energy and Reduce Harmful Emissions by Using More Efficient Domestic Appliances," press release (Paris: 16 April 2003); saturation from World Energy Council (WEC), "Labeling and Efficiency Standards," in *Energy Efficiency Policies and Indicators* (London: 2001); refrigerator size from idem, "Annex 1—Case Studies on Energy Efficiency Policy Measures: United States of America," in ibid.; cooling systems from DOE, EIA, "Changes in Energy Usage in Residential Housing Units," in *The 1997 Residential Energy Consumption Survey—Two Decades* (Washington, DC: 1997). Box 2–3 from the following: "Cold Water Poured on IT's Environmental Pluses," *Environment Daily*, 11 June 2003; N. Cohen, "The Environmental Impacts of E-Commerce," in L. M. Hilty and P. W. Gilgen, eds., *Sustainability in the Information Society, 15th International Symposium on Informatics for Environmental Protection* (Marburg, Germany: Metropolis Verlag, 2001); Klaus Fichter, "E-Commerce: Sorting Out the Environmental Consequences," *Journal of Industrial Ecology*, vol. 6, no. 2 (2003); John A. Laitner, "Information Technology and U.S. Energy Consumption: Energy Hog, Productivity Tool, or Both?" *Journal of Industrial Ecology*, vol. 6, no. 2 (2003); "Computer Related Electricity Use Overestimated," *Environment News Service*, 5 February 2001; I. Greusing and S. Zangl, "Comparing Print- and Online-Mail Order Catalogues, Con-

sumer Acceptance, Environmental and Economical Analysis," IZT-Discussion Paper No. 44 (Berlin: Innovation Center for Telecommunication Technology, 2000); Dawn Anfuso, "Readers Prefer Paper to Online," *iMediaConnection.com*, 25 April 2002; E. Heiskanen et al, *Dematerialization: The Potential of ICT and Services* (Helsinki: Finnish Ministry of the Environment, 2001); D. Takahashi, "Power Integrations' Chip Cuts Appliance Power Waste," *Dow Jones Newswires*, 25 September 1998; DOE, Office of Energy Efficiency and Renewable Energy, *Technology Snapshots Featuring the Toyota Prius* (Washington, DC: 2001); Patrick Mazza, *The Smart Energy Network: Electricity's Third Great Revolution* (Olympia, WA: Climate Solutions, 2003).

33. Projections and standby share from IEA, op. cit. note 29, p. 12; additional capacity calculated by Worldwatch assuming 75 percent capacity factor, 7 percent transmission and distribution losses, using 2020 electricity demand (final end-use) projections for OECD from IEA, op. cit. note 10, p. 414, and assuming that one ton of oil equivalent equals approximately 12 megawatt-hours of electricity as noted in BP, op. cit. note 3; CO_2 emissions calculated by Worldwatch with data from IEA, op. cit. note 10, p. 417.

34. Developing-country building needs from Dzioubinski and Chipman, op. cit. note 10, pp. 5, 9–10, and from Sujay Basu, *Report on India Energy Scene*, Rio+5 Report (Paris: Helio International, 1997); three fourths of Indians from Office of the Registrar General, "Tables on Houses, Household Amenities and Assets: India," op. cit. note 9, and from idem, "India at a Glance," op. cit. note 9. Table 2–3 from the following: Schipper, op. cit. note 14, p. 144; Scott Murtishaw and Lee Schipper, "Disaggregated Analysis of U.S. Energy Consumption in the 1990s: Evidence of the Effects of the Internet and Rapid Economic Growth," *Energy Policy* 29 (2001), p. 1347; IEA, cited in EPC, op. cit. note 16, p. 28; National Bureau of Statistics, *China Statistical Yearbook* (Beijing: National Statistical Press, various years); IEA, op. cit. note 28, p. 35.

35. Demand growth since 1990 calculated by Worldwatch with data from IEA, op. cit. note 10, pp. 410–11, 458–59; energy consumption relative to income from Dzioubinski and Chipman, op. cit. note 10, p. 4; tripling of energy use from Mark Levine et al., *Energy Efficiency Improvement Utilising High Technology: An Assessment of Energy Use in Industry and Buildings* (London: WEC, 1995); televisions from Matthew Bentley, "Forging New Paths to Sustainable Development," Background Paper, Asia Pacific Expert Meeting on Promoting Sustainable Consumption and Production Patterns, Yogyakarta, Indonesia, 21–23 May 2003, p. 3; Indian refrigerators from Jain and Lakshman, op. cit. note 9; projections from IEA, op. cit. note 10, p. 29.

36. U.S. energy savings from American Council for an Energy-Efficient Economy (ACEEE), "Energy Efficiency Progress and Potential," fact sheet (Washington, DC: undated); financial savings from Stephen Meyers et al., *Realized and Prospective Impacts of U.S. Energy Efficiency Standards for Residential Appliances* (Berkeley, CA: Lawrence Berkeley National Laboratory, June 2002), pp. 21–41; Europe from WEC, op. cit. note 32; potential improvements over next decade from IEA, op. cit. note 32; improvements by 2030 from IEA, op. cit. note 29, p. 17.

37. Potential savings and barriers from Dzioubinski and Chipman, op. cit. note 10, p. 3; Thai savings from Jas Singh and Carol Mulholland, "DSM in Thailand: A Case Study," Joint U.N. Development Programme (UNDP)/World Bank Energy Sector Management Assistance Programme, Washington, DC, October 2002, pp. 1, 8; share of efficient refrigerators from Government of Indonesia and UNEP, op. cit. note 35, p. 17; Brazil from Steven Nadel, *Appliance Energy Efficiency: Opportunities, Barriers, and Policy Solutions* (Washington, DC: ACEEE, October 1997), pp. 10–11. In Brazil, if targets are not met the standards become mandatory.

38. Donald W. Aitkin, *Putting it Together: Whole Buildings and a Whole Buildings Policy*, Research Report No. 5 (Washington, DC: Renewable Energy Policy Project, September 1998), p. 6; EU, "Buildings Overview—Market Overview and Trends," ATLAS Project, *European Commission Online*, at europa.eu.int/comm/energy_trans port/atlas/htmlu/bomarover.html, viewed 16 May 2003; U.S. new homes from U.S. Census Bureau, "New Residential Construction in April 2003," fact sheet (Suitland, MD: 16 May 2003).

39. Potential savings from Ayres, op. cit. note 23, p. 35; California from David Goldstein, Natural Resources Defense Council, San Francisco, CA, discussion with author, 26 September 2003.

40. Ken Gewertz, "Pushing the Envelope: The Skin's the Thing for Conserving a Building's Energy," *Harvard Gazette*, 19 July 2001.

41. Lee S. Windheim et al., "Case Study: Lockheed Building 157—An Innovative Deep Daylighting Design for Reducing Energy Consumption," cited in Aitkin, op. cit. note 38, p. 8; 34 percent lighting and RMI from Ellen Pfeifer, "Light: The Future is Green," *Winslow Environmental News*, July 2000, p. 1; money and health savings from ibid., p. 5, and from Windheim et al., op. cit. this note, p. 8.

42. "First 'Green' High-Rise Residential Building in the World," *Real Estate Weekly*, 2 October 2002; "Governor Pataki Unveils the Solaire, First 'Green' Residential Tower in U.S.," *Silicon Valley Biz Ink*, 5 September 2003; "Powerlight/Toyota Achieve Gold Building Standard," SolarAccess.com, 23 April 2003.

43. Japan's PV installations from Paul Maycock, *PV News*, May 2003, p. 5; losses from World Bank, *World Development Report 1997* (New York: Oxford University Press, 1997), and from Indian Planning Commission, *Annual Report on the World of State Electricity Boards and Electricity Departments*, cited in M. S. Bhalla, "Transmission and Distribution Losses (Power)," in *Proceedings of the National Conference on Regulation in Infrastructure Services: Progress and Way Forward* (New Delhi: The Energy and Resources Institute, November 2000).

44. Energy savings from Alexis Karolides, "An Introduction to Green Building. Part 3: Other Green Building Considerations," *RMI Solutions Newsletter*, summer 2003, p. 13; Thor Magnusson, *A Showcase of Icelandic Treasures* (Reykjavik: Icelandic Review, 1987), cited in "Exploring the Ecology of Organic Greenroof Architecture: History," at www.greenroofs.com/history.htm; Germany from ibid.; Chicago and Amsterdam from "North American Case Studies," at www.green roofs.com/north_american_cases.htm; Ford Motor Company, "Ford Installs World's Largest Living Roof on New Truck Plant," press release (Dearborn, MI: 3 June 2003).

45. Largest share of global consumption from IEA, op. cit. note 10, p. 28; growth in sectors and energy use from Lee Schipper et al., "Energy Use in Manufacturing in Thirteen OECD Countries: Long Term Trends Through 1995," Lawrence Berkeley National Laboratory, working draft, 13 August 1998, p. 3.

46. Increasing intensity from Stuart Baird, "Heavy Industry," at www.iclei.org/EFACTS/ HEAVY.HTM, and from Ted Trainer, "The 'Dematerialisation' Myth," *Technology in Society*, vol. 23 (2001), pp. 505–14; Australia from Ted Trainer, *A Critical Discussion of Future Dilemmas* (Victoria, Australia: Commonwealth Scientific & Industrial Research Organisation, May 2002).

47. Tracy Mumma, "Reducing the Embodied Energy of Buildings," *Home Energy Magazine Online*, January/February 1995.

48. Energy used to manufacture vehicle from "Household Greenhouse Gas Emissions Questionnaire," *Alternatives Journal*, spring 2000; total energy associated with cars from R. A. Herendeen, *Ecological Numeracy* (New York: John Wiley & Sons, 1998), and from M. Wackernagel and W. Rees, *Our Ecological Footprint* (Gabriola Island, BC, Canada: New Society Publishers, 1996); Redefining Progress, *Ecological Footprint Accounts* (Oakland, CA: undated).

49. DOE, op. cit. note 7, p. 49.

50. Energy to collect foods from EPC, op. cit. note 16, pp. 30, 37; energy needs for long-distance travel from Brian Halweil, *Home Grown: The Case for Local Food in a Global Market*, Worldwatch Paper 163 (Washington, DC: Worldwatch Institute, November 2002), p. 15; growing source of

emissions from Policy Commission on the Future of Farming and Food, *Food & Farming: A Sustainable Future* (London: January 2002), p. 92.

51. Use by global food system calculated by Worldwatch with data from David Pimentel, Cornell University, e-mail to author, 23 July 2003, and from BP, op. cit. note 3; Pimentel, op. cit. this note.

52. William Moomaw, Director, International Environment and Resource Policy Program, Fletcher School of Law and Diplomacy, Tufts University, discussion with author, 5 August 2003.

53. Differences in energy prices from Baird, op. cit. note 46; efficiencies of manufacturing sectors and leapfrogging from IPCC, op. cit. note 12, p. 112.

54. Aluminum from Baird, op. cit. note 46.

55. Prices as fundamental factor from WEC, "Efficiency of Energy Supply and Use," in *Energy for Tomorrow's World—Acting Now!* (London: 2000), p. 3; impact of prices on energy intensity from Schipper et al., op. cit. note 45, p. 12, and from Anderson, op. cit. note 15.

56. Role of price from Schipper et al., op. cit. note 4, pp. 205–06.

57. Impact of taxes from Lee Schipper, "Lifestyles and the Environment: The Case of Energy," in *Technological Trajectories and the Human Environment* (Washington, DC: National Academy Press, 1997), p. 100; Denmark from Eurostat, cited in Consultores em Transportes Inovação e Sistemas et al., "Study on Vehicle Taxation in the Member States of the European Union," Final Report, prepared for the European Commission, DG Taxation and Customs Union, January 2002, p. 11.

58. U.S. tax deduction from W. Gentry, "Residential Energy Demand and the Taxation of Housing," cited in Schipper, Murtishaw, and Unander, op. cit. note 21; Sweden from Schipper, op. cit. note 57, pp. 100–01; "Annex I—Case Studies on Energy Efficiency Policy Measures: Japan," in

WEC, op. cit. note 32.

59. Eric Lombardi, *Take It Back!* (Boulder, CO: Eco-cycle, 2000).

60. Mid-1990s subsidies from UNDP, *World Energy Assessment*, 2000, cited in Group of Eight (G8) Renewable Energy Task Force, *G8 Renewable Energy Task Force Final Report 2001* (London: 2001), p. 34; reductions in 1990s from Geller, op. cit. note 5, p. 2; infrastructure and manufacturing from Alan Durning, *How Much Is Enough?* (New York: W.W. Norton & Company, 1992), p. 110.

61. Energy use in South Korea from Dzioubinski and Chipman, op. cit. note 10, pp. 7–8; subsidies to those who do not need them from UN HABITAT, op. cit. note 25, pp. 138, 142; Mark Ashurst, "Nigeria Tackles Fuel Subsidies," *BBC News*, 18 July 2003.

62. Daphne Wysham, *Sustainable Development South and North: Climate Change Policy Coherence in Global Trade and Financial Flows* (Washington, DC: Institute for Policy Studies, March 2003), pp. 3–4.

63. The target is to reduce emissions 25 percent below 1990 levels by 2005; German Embassy, *Germany's Ecological Tax Reform*, Background Paper (Washington, DC: undated).

64. Germany and Denmark from Janet L. Sawin, "Charting a New Energy Future," in Worldwatch Institute, *State of the World 2003* (New York: W.W. Norton & Company, 2003), p. 104; green power demand by end of 2002 from Lori Bird and Blair Swezey, *Estimates of Renewable Energy Developed to Serve Green Power Markets in the United States* (Golden, CO: National Renewable Energy Laboratory, February 2003); California from "Students, Activists Win Clean Energy Campaign," SolarAccess.com, 21 July 2003, and from "Berkeley, CA, USA: University of California Approves Clean-Energy and Green-Building Policy," SolarBuzz.com, 18 July 2003.

65. Energie-Cités, "Brussels Declaration for a Sustainable Energy Policy in Cities," at www.energie-cites.org/PDF/avis_bruxelles_en

.pdf; Dutch municipalities from "Groningen, The Car-Free City for Bikes," at www.globalideas bank.org/socinv/SIC-100.HTML; bicycle parking from "Bicycles and Transit: Europe and Japan," in Charles Komanoff, ed., *Bicycle Blueprint* (New York: Transportation Alternatives, 1999); Germans and Swiss from "Reduction of Air Pollution," CarSharing Co-op of Edmonton, at www.web.net; communities and members from "More Car Sharing in Europe," *Passenger Transport* (American Public Transportation Association), 22 July 2002; North America from David Steinhart, "Car Sharing: An Idea Whose Time Has Come?" *Financial Post*, 4 May 2002; U.S. cities and counties from CarSharing Network at www.carsharing.net/where.html.

66. Ecovillages from gen.ecovillage.org; California from Charles A. Goldman, Joseph H. Eto, and Galen L. Barbose, *California Customer Load Reductions During the Electricity Crisis: Did They Help to Keep the Lights On?* (Berkeley, CA: Lawrence Berkeley National Laboratory, 2002), pp. 23–25; Zero Emissions Development and quote from Greg Rasmussen, "The Kyoto Protocol: The British Approach," *CBC Radio*, July 2002.

67. Empirical relationship from Amulya Reddy, "Energy Technologies and Policies for Rural Development," in Thomas B. Johansson and José Goldemberg, eds., *Energy for Sustainable Development: A Policy Agenda* (New York: UNEP, 2002), pp. 117–19; Carlos Suárez, "Energy Needs for Sustainable Human Development," in José Goldemberg and Thomas B. Johansson, eds., *Energy as an Instrument for Socio-Economic Development* (New York: UNDP, 1995) (using UNDP 1991–92 data for 100 industrial and developing nations); national consumption levels from World Bank, op. cit. note 8, pp. 144–46.

68. Robert Prescott-Allen, *The Wellbeing of Nations: A Country-by-Country Index of Quality of Life and the Environment* (Washington, DC: Island Press/IDRC, 2001), pp. 267–68. Table 2–5 from the following: well-being rankings from ibid., pp. 267–68; energy use ranking and share of Sweden calculated by Worldwatch with 1999 data from IEA, cited in World Resources Institute et al.,

World Resources 2002–2004 (Washington, DC: 2003), pp. 262–63.

69. Degrading quality of life from Centre for Ecological Sciences, *Energy Efficiency and Sustainability: New Paradigms* (New Delhi: Indian Institute of Science, undated); Shanghai registrations from "Car Sales Booming in China," op. cit. note 10.

70. Share of population in developing countries from DOE, EIA, *International Energy Outlook 2002* (Washington, DC: 2002), pp. x–xi.

71. Population from U.S. Bureau of the Census, *International Data Base*, electronic database (Suitland, MD: updated 10 October 2003), and from United Nations, *World Population Prospects: The 2002 Revision* (New York: 2003), p. 1; consumption in 2050 calculated by Worldwatch with data from DOE, op. cit. note 70, pp. x–xi; IEA, cited in World Bank, op. cit. note 8, p. 146. This assumes the developing world's share of global population remains constant.

72. Donald J. Johnston, Secretary-General, OECD, speech given at Informal Ministerial Seminar on Energy, Pamplona, Spain, 26–28 April 2002.

Computers

1. Eric D. Williams, Robert U. Ayres, and Miriam Heller, "The 1.7 Kilogram Microchip: Energy and Material Use in the Production of Semiconductor Devices," *Environmental Science and Technology*, 15 December 2002, pp. 5504–10; Curtis Runyan, "Microchips Are Tiny, But Their Environmental Footprint Is Heavy," *World Watch*, March/April 2003, p. 8.

2. Ann Hwang, "Semiconductors Have Hidden Costs," in Worldwatch Institute, *Vital Signs 2002* (New York: W.W. Norton & Company, 2002), pp. 110–11; Silicon Valley Toxics Coalition, "Four Case Studies of High-Tech Water Exploitation and Corporate Welfare in the Southwest," at svtc.igc.org/resource/pubs/execsum.htm, viewed 11 September 2003; "toxic waste sites" refers to U.S. Environmental Protection Agency Super-

fund sites, see Hwang, op. cit. this note.

3. Data for 1988 from International Telecommunication Union (ITU) in World Bank, *World Development Indicators Database*, at media.worldbank.org/secure/data/qquery.php, viewed 12 August 2003; 2002 data from idem, *Internet Indicators: Hosts, Users, and Numbers of PCs*, at www.itu.int/ITU-D/ict/statistics, viewed 15 July 2003; Silicon Valley Toxics Coalition (SVTC), "Toxics in a Computer," fact sheet, at www.svtc.org/cleancc/pubs/computertoxics.pdf, viewed 3 July 2003; Ian Fried, "Recycling Not Easy for PC Makers," *CNET News.com*, 22 April 2003; SVTC et al., *Poison PCs and Toxic TVs: California's Biggest Environmental Crisis That You've Never Heard Of* (San Jose, CA: June 2001), p. 13.

4. Basel Action Network (BAN) and SVTC, *Exporting Harm: The High-Tech Trashing of Asia* (Seattle, WA, and San Jose, CA: February 2002), pp. 5–6; SVTC et al., op. cit. note 3, p. 6; National Safety Council, *Electronic Product Recovery and Recycling Baseline Report: Recycling of Selected Electronic Products in the United States* (Washington, DC: May 1999), p. 13.

5. BAN and SVTC, op. cit. note 4, pp. 6–7, 9, 18; SVTC et al., op. cit. note 3, p. 15.

6. BAN and SVTC, op. cit. note 4, pp. 2–3, 7–8, 12–13, 28.

7. Ibid., pp. 16, 18–19.

8. Ibid., pp. 17–21, 30–32.

9. Ibid., pp. 8, 42.

Chapter 3. Boosting Water Productivity

1. For a review of the status of freshwater ecosystems and biodiversity, see Sandra Postel and Brian Richter, *Rivers for Life: Managing Water for People and Nature* (Washington, DC: Island Press, 2003).

2. Igor A. Shiklomanov, *Assessment of Water Resources and Water Availability in the World* (St.

Petersburg, Russia: State Hydrological Institute, 1996); World Commission on Dams (WCD), *Dams and Development* (London: Earthscan, 2000).

3.   Sectoral shares from World Bank, *World Development Indicators* (Washington, DC: 2001).

4.   Robert Costanza et al., "The Value of the World's Ecosystem Services and Natural Capital," *Nature,* 15 May 1997, pp. 254–60.

5.   U.N. Environment Programme, *Afghanistan: Post-Conflict Environmental Assessment* (Nairobi: 2003), p. 62.

6.   Postel and Richter, op. cit. note 1.

7.   Murray-Darling from "Drying Out," *The Economist,* 12 July 2003, p. 38; Australia, South Africa, and Hawaii examples from Postel and Richter, op. cit. note 1.

8.   Water productivity defined as real gross domestic product (GDP) in 2000, expressed in 2000 U.S. dollars, divided by estimated total water withdrawals in 2000. Figure 3–1 from the following: water withdrawal data for all countries except the United States from U.N. Food and Agriculture Organization (FAO), *AQUASTAT database,* at www.fao.org/ag/agl/aglw/aquastat/dbase/index2.jsp; U.S. water withdrawals extrapolated from time-series data in U.S. Geological Survey (USGS), *Estimated Use of Water in the United States in 1995* (Reston, VA: 1997); real GDP from Angus Maddison, *The World Economy: A Millennial Perspective* (Paris: Organisation for Economic Co-operation and Development, 2001), converted to 2000 dollars using U.S. GDP implicit price deflator from Bureau of Economic Analysis, at www.bea.doc.gov/bea/dn/nipaweb/NIPATableIndex.htm.

9.   Water productivity defined as real GDP, expressed in 2000 dollars, divided by estimated total water withdrawals for each year. Figure 3–2 from the following: water withdrawal data for 1950–95 from USGS, op. cit. note 8; real GDP from Maddison, op. cit. note 8, converted to 2000 dollars using U.S. GDP implicit price defla-

tor from Bureau of Economic Analysis, op. cit. note 8; the 2000 figure is an extrapolation based on 1995 withdrawals and the same rate of change in withdrawals as occurred from 1990 to 1995, combined with actual GDP.

10.   FAO, *Review of World Water Resources by Country* (Rome: 2003). To avoid double-counting, we use the measure of internal renewable water resources, which excludes water flowing in from neighboring countries.

11.   National renewable water supply figures from FAO, op. cit. note 10; North China Plain figures and quote from Jeremy Berkoff, "China: The South-North Water Transfer Project—Is It Justified?" *Water Policy,* February 2003, pp. 1–28; Yellow River flow and groundwater trends from Sandra Postel, *Pillar of Sand* (New York: W.W. Norton & Company, 1999), pp. 67–69, 76.

12.   Water withdrawals are estimates, not measured quantities, and their accuracy can vary greatly from one country to another. Table 3–1 from the following: per capita water withdrawals for all countries except the United States from FAO, op. cit. note 8; U.S. per capita water withdrawals extrapolated from time-series data in USGS, op. cit. note 8, and from population estimate in FAO, op. cit. note 8; share of cropland that is irrigated from Postel, op. cit. note 11, p. 42.

13.   Phoenix precipitation from National Climatic Data Center, National Oceanic and Atmospheric Administration, *U.S. Climate at a Glance database,* at www.ncdc.noaa.gov/oa/climate/research/cag3/cag3.html, viewed 14 August 2003; "The Ethiopian Famine," *New York Times,* 28 July 2003.

14.   United Nations, *Water for People, Water for Life: The United Nations World Water Development Report* (Paris: UNESCO Publishing and Berghahn Books, 2003), p. 113; Indonesia's natural water supply from FAO, op. cit. note 10, p. 80.

15.   United Nations, op. cit. note 14, pp. 110–12; Table 3–2 from World Health Organization (WHO)/UNICEF Joint Monitoring Programme, *Global Water Supply and Sanitation Assessment*

2000 Report, available at www.who.int/water_san itation_health/Globassessment/GlobalTOC.htm.

16. WHO and UNICEF, op. cit. note 15; D. Kirk Nordstrom, "Worldwide Occurrences of Arsenic in Ground Water," *Science*, 21 June 2002, pp. 2143–45.

17. "Water Affairs and Forestry," *South Africa Yearbook 2002/3*, at www.gcis.gov.za/docs/pub lications/yearbook/ch23.pdf, viewed 11 June 2003; 6 million figure from "South Africa: A Water Success Story," at www-dwaf.pwv.gov .za/Communications/Articles/Minister/2003/ South%20Africa%20a%20water%20success%20story .doc, viewed 9 June 2003.

18. Two-tier pricing from John Peet, "Priceless: A Survey of Water," *The Economist*, 19 July 2003, pp. 3–16; Johannesburg example from Ginger Thompson, "Water Tap Often Shut to South Africa Poor," *New York Times*, 29 May 2003.

19. Agriculture's share from World Bank, op. cit. note 3; projections from Mark W. Rosegrant, Ximing Cai, and Sarah A. Cline, *World Water and Food to 2025* (Washington, DC: International Food Policy Research Institute, 2002).

20. "Irrigation Options," WCD Thematic Review IV.2, as reported in WCD, op. cit. note 2, p. 46.

21. Stuart Styles, Irrigation Training and Research Center, California Polytechnic State University, San Luis Obispo, CA, e-mail to authors, 30 June 2003.

22. Postel, op. cit. note 11, pp. 172, 174.

23. Figure of 3.2 million hectares from Hervé Plusquellec and Walter Ochs, *Water Conservation: Irrigation*, Water Resources and Environment Technical Note F.2 (Washington, DC: World Bank, 2003), p. 20. Table 3–3 from the following: 1991 areas from Sandra Postel, *Last Oasis* (New York: W.W. Norton & Company, 1992), p. 105, which were compiled from a variety of sources; circa 2000 areas from International Commission on Irrigation and Drainage, at www.icid.org/

index_e.html, viewed 30 June 2003, except for Cyprus, Brazil, Chile, and Mexico, with circa 2000 figures from FAO, op. cit. note 8, for Israel, with 2000 estimates from the Israeli Ministry of Agriculture, according to Saul Arlosoroff, Director, National Water Corporation-Mekorot, Tel Aviv, Israel, e-mail to authors, 6 August 2003, and for the United States, with a 1998 estimate from U.S. Department of Agriculture, *1998 Farm & Ranch Irrigation Survey*, Census of Agriculture, at www.nass.usda.gov/census/census97/fris/tbl04 .txt>http://www.nass.usda.gov/census/census97/ fris/tbl04.txt.

24. L. C. Guerra et al., *Producing More Rice with Less Water from Irrigated Systems* (Colombo, Sri Lanka: International Water Management Institute (IWMI), 1998), p. 11; R. Barker, Y. H. Li, and T. P. Tuong, eds., *Water-Saving Irrigation for Rice*, Proceedings of an International Workshop in Wuhan, China, 23–25 March 2001 (Colombo, Sri Lanka: IWMI, 2001).

25. Plusquellec and Ochs, op. cit. note 23, p. 22.

26. FAO, *The State of Food Insecurity in the World* (Rome: 1999); Sandra Postel et al., "Drip Irrigation for Small Farmers: A New Initiative to Alleviate Hunger and Poverty," *Water International*, March 2001, pp. 3–13.

27. P. Polak, B. Nanes, and J. Sample, "Opening Access to Affordable Micro-Plot Irrigation for Small Farmers," presented at the Irrigation Association Symposium on Small-holder Irrigation, Orlando, FL, 1999; International Development Enterprises (IDE), discussion with Sandra Postel, June 2003; see also IDE web site at www .ide-international.org.

28. Alwar district from Himanshu Thakkar, "Assessment of Irrigation in India," 8 November 1999, Contributing Paper for "Irrigation Options," WCD Thematic Review IV.2, at www.dams.org/docs/kbase; see also Anil Agarwal and Sunita Narain, eds., *Dying Wisdom* (New Delhi, India: Centre for Science and Environment, 1997), and the Centre for Science and Environment at www.cseindia.org.

29. Table 3–4 based on data in D. Renault and W. W. Wallender, "Nutritional Water Productivity and Diets," *Agricultural Water Management*, August 2000, pp. 275–96.

30. Dietary water requirements from ibid.; the medium variant U.S. population projection for 2025 is 358 million, according to United Nations, Population Division, *World Population Prospects: The 2002 Revision* and *World Urbanization Prospects: The 2001 Revision* (New York: 2003 and 2002).

31. United Nations, *World Population Prospects*, op. cit. note 30.

32. United Nations, op. cit. note 14, pp. 160–61; cities claiming less than 10 percent from World Bank, op. cit. note 3. Box 3–1 from the following: desalination capacity and growth rate from Aqua Resources International, *Desalination Market Analysis 2003* (Evergreen, CO: August 2003); Israel's plans from "Eshkol, Nizzana and Ashkelon Desalination Plants, Israel," Web site for the Water Industry, at www.water-technology.net/projects/israel/, viewed 8 September 2003; drop in energy and cost requirements from Allerd Stikker, "Desal Technology Can Help Quench the World's Thirst," *Water Policy*, February 2002, pp. 47–55; 10–25 percent figures from Amy Vickers, presentation at 2003 Community Water Leadership Program, Florida Institute of Government, St. Petersburg, FL, 14 March 2003.

33. United Nations, op. cit. note 14, pp. 160–61.

34. Kofi Annan, U.N. Secretary-General, "Towards A Sustainable Future," presented at the American Museum of Natural History's Annual Environmental Lecture, New York, 14 May 2002; "Water and the Fight Against Poverty," U.N. Environment Programme (UNEP) Executive Director's address on Water and the Fight Against Poverty—World Environment Day seminar, Beirut, Lebanon, 5 June 2003; "Leakage Control and the Reduction of Unaccounted-for Water (UFW)," Proceedings of the International Symposium on Efficient Water Use in Urban Areas—Innovative Ways of Finding Water for Cities, 8–10 June 1999; Arabian Gulf city from UNEP, "UNEP Urges

Action to Better Manage the Globe's Groundwaters," press release (Nairobi: 5 June 2003); "Taiwan Leaky Pipes Losing 1.97 Million Cubic Meters of Water Per Day," 15 April 2002, at fpeng.peopledaily.com.cn/200204/15/eng20020 415_94050.shtml.

35. Table 3–5 from the following: Albania from European Environment Agency (EEA), "EEA Report Highlights Measures to Promote Sustainable Water Use," press release (Copenhagen: 5 April 2001); Canada from Bill Hutchins, "One-third of City's Water Supply Disappears, But Where?" *Kingston This Week*, 10 June 2003; Czech Republic, France, and Spain from EEA, *Sustainable Water Use in Europe, Part 2: Demand Management (Environmental Issue Report No. 19)* (Copenhagen: 2001), p. 22; Denmark from "WATERSAVE at Loughborough," *Demand Management Bulletin* (National Water Demand Management Centre, Environment Agency, Birmingham, U.K.), February 2003, p. 3; Japan from United Nations, Best Practices and Local Leadership Programme, "Water Conservation Conscious Fukuoka, Japan," at bestpractices.org/bpbriefs/ Environment.html, viewed September 2003; Jordan from "Seminar Addresses Water Losses Plaguing the Country's Scarce Resource," *Jordan Times*, 6 March 2003; Kenya from Anna Peltola, "Simple Fixes Could Bring Water to Millions–Experts," *Reuters*, 13 August 2002; Singapore from Ng Han Tong, Public Utilities Board, Singapore, e-mail to authors, 10 September 2003; South Africa from "Growing Concern over Cities' Water Losses," *South African Broadcasting Corporation*, 19 May 2003; Taiwan from "Water Price Hike Necessary to Fix Old Pipes, Says Minister," *Taiwan News*, 11 March 2003; United States from Jon Maker and Nathalie Chagnon, "Inspecting Systems for Leaks, Pits, and Corrosion," *Journal of the American Water Works Association*, July 1999, p. 40, and from Steve Wyatt, "The Economics of Water Loss: What is Unaccounted for Water?" *On Tap* (National Drinking Water Clearinghouse, West Virginia University, Morgantown, WV), fall 2002, pp. 32–35. Box 3–2 from "Office of Water Services (OFWAT): Leakage And Water Efficiency," Eighth Report, Select Committee on Public Accounts, House of Commons, London, 4 January 2002, from OFWAT, "Ofwat Sets Leakage Targets For London," press release (Birmingham, U.K.: 26 March 2003), and from "Total Leakage Increases in Severn Trent Water and is Still Rising in Thames Water," Office of Water Services, 29 July 2003.

36. Janice A. Beecher, "Survey of State Water Loss Reporting Practices," Beecher Policy Research, Inc., Final Report to the American Water Works Association, Denver, CO, January 2002, p. 13; Wyatt, op. cit. note 35; The Kansas Water Office, "Assessment of Unaccounted for Water, Kansas 1992–2010," 2001, at www.kwo.org/Reports/2010%20Assessments/UFW%20assessment/ks_ufw_text.htm; idem, "The Kansas Water Plan, Fiscal Year 2003," *Water Conservation*, July 2001, p. 1.

37. Copenhagen from "WATERSAVE at Loughborough," op. cit. note 35; Lis Napstjert, Copenhagen Energy, e-mail to authors, 17 September, 2003. Box 3–3 from the following: Ng, op. cit. note 35; Singapore usage in 1995 from Alliance to Save Energy, *Watergy: Taking Advantage of Untapped Energy and Water Efficiency Opportunities in Municipal Water Systems* (Washington, DC: 2002), pp. 80–82; usage in 2003 from Ng Han Tong, Public Utilities Board, Singapore, e-mail to authors, 16 August 2003; Teruyoshi Shinoda, "Integrated Approaches for Efficient Water Use in Fukuoka," Proceedings of the International Symposium on Efficient Water Use in Urban Areas—Innovative Ways of Finding Water for Cities, 8–10 June 1999; United Nations, op. cit. note 35; "Final Report: Water Conservation Planning USA Case Studies Project," Environment Agency, Demand Management Centre, Birmingham, U.K., prepared by Amy Vickers & Associates, Inc., Amherst, MA, June 1996.

38. Robert Wilkinson, *Methodology for Analysis of the Energy Intensity of California's Water Systems, and an Assessment of Multiple Potential Benefits Through Integrated Water-Energy Efficiency Measures* (Berkeley, CA: Ernest Orlando Lawrence Berkeley Laboratory, California Institute for Energy Efficiency, January 2000), p. 6; QEI, Inc., *Electricity Efficiency Through Water Efficiency*, Report for the Southern California Edison Company (Springfield, NJ: 1992), pp. 23–24.

39. "Save Water to Save Energy," *Environmental Building News*, October 2002, pp. 3–4.

40. Figure 3–3 from the following: Kenya, Uganda, and Tanzania from John Thompson et al., *Drawers of Water II: Thirty Years of Change in Domestic Water Use and Environmental Health in East Africa* (London: International Institute for Environment and Development, 2001); Denmark from "WATERSAVE at Loughborough," op. cit. note 35; United Kingdom from "Water Indicators 2001/02," *Demand Management Bulletin* (National Water Demand Management Centre, Environment Agency, Birmingham, U.K.), June 2003, p. 7; Singapore from Ng, op. cit. note 37; Manila from Manila Water Company, www.manilawateronline.com/faq.htm, viewed 11 August 2003; Waterloo from Steve Gombos, Regional Municipality of Waterloo, e-mail to authors, 7 August 2003; Melbourne and Sydney from Water Services Association of Australia, *WSAA facts 2001* (Melbourne: 2001); U.S. examples from Peter W. Mayer et al., *Residential End Uses of Water* (Denver, CO: AWWA Research Foundation and American Water Works Association, 1999), pp. 91, 114. Other data from Amy Vickers, *Handbook of Water Use and Conservation: Homes, Landscapes, Businesses, Industries, Farms* (Amherst, MA: WaterPlow Press, 2001), pp. 17–19, 20.

41. U.S. General Accounting Office, *Water Infrastructure: Water-Efficient Plumbing Fixtures Reduce Water Consumption and Wastewater Flows* (Washington, DC: August 2000), p. 4; Amy Vickers, "Technical Issues and Recommendations on the Implementation of the U.S. Energy Policy Act," report prepared by Amy Vickers & Associates, Inc. for the American Water Works Association, Washington, DC, 25 October 1995, p. 4.

42. United Nations, op. cit. note 14, p. 341.

43. Figure of 30 billion liters from Vickers, op. cit. note 40, p. 140; Debbie Salamone, "Florida's Water Crisis: Chapter 2, The Human Thirst," *Orlando Sentinel*, 7 April 2002.

44. Charles Fenyvesi, "His Whole World Is Grass: Lawn Guru Reed Funk Speaks to the Tough Little Cultivar Inside Us All," *U.S. News & World*

Report, 28 October 1996, pp. 61–62; Thomas Far-ragher, "For Some Neighbors, It's a Turf War," *Boston Globe,* 5 April 1998.

45. F. Herbert Bormann et al., *Redesigning the American Lawn: A Search for Environmental Har-mony* (New Haven, CT: Yale University Press, 1993), pp. 56, 77; U.S. Senate, *The Use and Reg-ulation of Lawn Care Chemicals: Hearing Before the Subcommittee on Toxic Substances, Environ-mental Oversight, Research and Development of the Committee on Environment and Public Works, U.S. Senate, March 28, 1990,* Senate Hearing 101-685 (Washington, DC: U.S. Government Printing Office, 1990), p. 1. Box 3–4 from the following: Christian G. Daughton, "Environmental Stew-ardship and Drugs As Pollutants," *The Lancet,* 5 October 2002, pp. 1035–36; USGS from Dana W. Kolpin et al., "Pharmaceuticals, Hormones, and Other Organic Wastewater Contaminants in U.S. Streams, 1999–2000: A National Recon-naissance," *Environmental Science and Technology,* 15 March 2002, pp. 1202–11; D. Donaldson, T. Kiely, and A Grube, *Pesticide Industry Sales and Usage: 1998 and 1999 Market Estimates* (Wash-ington, DC: U.S. Environmental Protection Agency (EPA), August 2002); National Associa-tion of Chain Drug Stores, Industry Facts and Resources, Industry Statistics, "Total Retail Sales 2002—Traditional Drug Stores," at www.nacds .org/wmspage.cfm?parm1=507, viewed Septem-ber 2003; studies outside the United States from Kenneth Green, "When Pharmaceuticals Arrive at the Tap," *Environmental Science & Engineering,* March 2003; EPA, "PPCPs as Environmental Pol-lutants," at www.epa.gov/nerlesd1/chemistry/ pharma/teaching.htm, viewed August 2003; Colin Nickerson, "A Grass Roots Drive for Purity," *Boston Globe,* 3 September 2001.

46. Vickers, op. cit. note 40, pp. 147–49, 180.

47. F. Herbert Bormann, Diana Balmori, and Gordon T. Geballe, *Redesigning the American Lawn,* 2nd ed. (New Haven, CT: Yale University Press, 2001), p. 129.

48. United Nations, op. cit. note 14, p. 228.

49. Major water-using industries from webworld

.unesco.org/water/ihp/publications/waterway/ webpc/pag18.html, viewed August 2003.

50. Vickers, op. cit. note 40, p. 239; Unilever Co., *Environmental Performance Report 2003,* p. 10, at www.unilever.com/environmentsociety/ environmentalreporting/environmentalreport, viewed August 2003. Table 3–6 from the follow-ing: dairy, home construction, and produce from Environment Agency, *2003 Water Efficiency Awards: Inspirational Case Studies Demonstrating Good Practice Across All Sectors* (Birmingham, U.K.: 2003), pp. 8, 15, 21; computers from "Companies Who Show How It's Done," *Demand Management Bulletin* (National Water Demand Management Centre, Environment Agency, Birmingham, U.K.), August 2002, p. 3; steel from "Columbia Steel Casting: Using Water Wisely," Office of Sustainable Development, Port-land, OR, at www.sustainableportland.org/energy _com_best.html, viewed August 2003; pharma-ceuticals from Steve Dark, *Process Waste Water Recycling* (Jaffrey, NH: Millipore Corporation, November 2002); chocolate from *EBMUD Reports: East Bay Water Issues,* March 2001, p. 2; beer from Global Environment Management Ini-tiative, *Connecting the Drops Toward Creative Water Strategies* (Washington, DC: June 2002), p. 36.

51. Public Utilities Board, Water Authority of Singapore, at www.pub.gov.sg/water_reclamation .htm, viewed September 2003.

52. United Nations, op. cit. note 14, pp. 227–33.

53. Patrick Smith, "The Great Water Divide," *The Irish Times,* 22 March 2003; Aly Shady, Cana-dian International Development Agency, Jan-uary 2002, at www.expressnews.ualberta.ca/ expressnews/articles/news.cfm?p_ID=1829&s=a.

54. Jennifer Gitlitz, *Trashed Cans: The Global Environmental Impacts of Aluminum Can Wast-ing in America* (Arlington VA: Container Recy-cling Institute, 2002).

55. Postel and Richter, op. cit. note 1.

56. Tushaar Shah et al., "Sustaining Asia's

Groundwater Boom: An Overview of Issues and Evidence," *Natural Resources Forum*, May 2003, pp. 130–41.

57. High Plains Underground Water Conservation District, *The Cross Section*, various issues, Lubbock, TX; Shah et al., op. cit. note 56.

58. Study and Bangalore example reported in Dale Whittington, "Municipal Water Pricing and Tariff Design: A Reform Agenda for South Asia," *Water Policy*, February 2003, pp. 61–76; comparison to U.S. use is authors' calculation.

59. Effect of groundwater pumping on Ipswich flows from USGS, *Concepts for National Assessment of Water Availability and Use*, Circular 1223 (Reston, VA: 2002); American Rivers, *America's Most Endangered Rivers 2003* (Washington, DC: 2003); Massachusetts Department of Environmental Protection, *State Strikes Balance with Water Withdrawal Permits for Ipswich River Basin Communities* (Boston: 20 May 2003).

Antibacterial Soap

1. Infectious Diseases Society of America (IDSA), "Antibacterial Soap No Better Than Regular Soap, NIH-Funded Study Shows," press release (Alexandria, VA: 24 October 2002).

2. Philip M. Parker, *The 2003–2008 World Outlook for Bar Soap* (Paris: ICON Group Ltd., 2003), pp. 18–19; Euromonitor International, "Asia-Pacific Bath and Shower Products: Small Growth, Big Potential," 25 October 2002, at www.euromonitor.com/asiabath.

3. Karen C. Timberlake, *Chemistry: An Introduction to General, Organic, and Biological Chemistry* (New York: Harper Collins, 1996), pp. 524–26.

4. History and ingredients from Dru Wilson, "Antibacterial Products May Be Case of Overkill," *The Gaston Gazette* (Freedom News Service), 21 August 2002, and from Soap and Detergent Association, "Cleaning Products Overview: History," at www.sdahq.org/cleaning/history/; Diane di Costanzo, "Taking Personal Care," *The Green Guide*, January/February 2003.

5. International Network for Environmental Management, "Pollution Prevention in a Tunisian Oil Extraction and Soap Manufacturing Facility," *Case Studies in Environmental Management in Small and Medium-Sized Enterprises* (Hamburg: March 1999).

6. U.S. Geological Survey, U.S. Department of the Interior, "What's in That Water?" press release (Reston, VA: 13 March 2002).

7. J. Menoutis et al., "Triclosan and Its Impurities," *Quantex Laboratories Technology Review* (Edison, NJ: Quantex Laboratories, Inc., 1998–2001); Kristin Ebbert, "Anti-bacterial Soaps," *The Green Guide*, March 1997.

8. "Keeping Medicines Effective: Resisting Resistance," *American Medical News* (American Medical Association), 15 October 2001; Helen Phillips, "Too Much Triclosan?" *Nature*, 13 August 1998.

9. Ed Susman, "Too Clean for Comfort," *Environmental Health Perspectives*, January 2001.

10. P. Ernst and Y. Cormier, "Relative Scarcity of Asthma and Atopy Among Rural Adolescents Raised on a Farm," *American Journal of Respiratory and Critical Care Medicine*, May 2000, pp. 1563–66.

11. Susman, op. cit. note 9; Alliance for Prudent Use of Antibiotics, at www.tufts.edu/med/apua; "CDC's Advice to Doctors: Clean Your Hands," *American Medical News* (American Medical Association), 25 November 2002.

Chapter 4. Watching What We Eat

1. Rita Oppenhuizen, Public Relations, Max Havelaar Foundation, discussion with Brian Halweil, 11 August 2003; Max Havelaar Foundation, "Max Havelaar—A Fair Trade Label," background history (Amsterdam: October 1998).

2. Max Havelaar Foundation, op. cit. note 1.

3. International Federation for Alternative Trade, "A Brief History of the Alternative Trading Movement," at www.ifat.org/dwr/resource3 .html, viewed 3 August 2003; A. M. Gutiérrez, "NGOs and Fairtrade, the Perspectives of Some Fairtrade Organisations," submitted to the School of Economic and Social Studies of the University of East Anglia, International Relations and Development Studies, Norfolk, U.K., 2 September 1996.

4. Oppenhuizen, op. cit. note 1; Max Havelaar Foundation, op. cit. note 1.

5. Jim Hightower, "As American as Apple Pie: Ethical Consumerism," *The Nation*, 30 September 2002; United Farm Workers, "Union to Launch 'Fair Trade Apple' Campaign," press release, 8 August 2001; Helen Taylor, Soil Association, Technical Services & Business Development Director (SA Cert), discussion with Brian Halweil, 7 June 2003.

6. For a discussion of the rise of long-distance food shipping, see Brian Halweil, *Home Grown: The Case for Local Food in a Global Market*, Worldwatch Paper 163 (Washington, DC: November 2002).

7. Surface area from U.N. Food and Agriculture Organization (FAO), *FAOSTAT Statistical Database*, at apps.fao.org, updated 22 August 2003; Michael Brower and Warren Leon, *The Consumer's Guide to Effective Environmental Choices: Practical Advice from the Union of Concerned Scientists* (New York: Three Rivers Press, 1999), p. 58; Organisation for Economic Co-operation and Development (OECD), Working Party on National Environmental Policy, *Sustainable Consumption: Sector Case Study Series, Household Food Consumption: Trends, Environmental Impacts and Policy Response* (Paris: December 2001), p. 44; Niels Jungbluth and Rolf Frischknecht, "Indicators for Monitoring Environmental Relevant Trends of Food Consumption," ESU-services, Switzerland, at www.esu-services.ch/download/ SETAC-food consumption-indicator.pdf; Annika Carlsson-Kanyama, "Climate Change and Dietary Choices: How Can Emissions of Greenhouse Gases from Food Consumption Be Reduced," *Food Policy*, fall/winter 1998, pp. 288–89. This does not mean, however, that vegetarian food is always less polluting than animal food. A vegetarian meal made with ingredients shipped from around the world or raised in heated greenhouses may be much more polluting than meals with meat that is sourced locally or raised in a way to minimize greenhouse gas emissions.

8. Box 4–1 from the following: Ransom Myers and Boris Worm, "Rapid Worldwide Depletion of Predatory Fish Communities," *Nature*, 15 May 2003, pp. 280–83; Andrew Revkin, "Commercial Fleets Reduced Big Fish by 90 Percent, Study Says," *New York Times*, 15 May 2003; Seaweb, "Cover Study of *Nature* Provides Startling New Evidence that Only 10 Percent of All Large Fish are Left in the Ocean," press release (Washington, DC: 14 May 2003); for marine conservation organizations, see, for instance, the Audubon Society (magazine.audubon.org/seafood), the Monterey Bay Aquarium (www.mbayaq.org), and the Marine Stewardship Council (www.msc.org); Seafood Choices Alliance, at www.seafood choices.com. Box 4–2 from the following: Compassion in World Farming, Foie Gras Campaign fact sheet, at www.ciwf.co.uk/Camp/Main/Foie/ foie_gras_campaign.htm, viewed 23 September 2003; World Society for the Protection of Animals, "Forced Feeding—The Facts Behind Foie Gras Production," at www.wspa.org.uk/index.php ?page=179, viewed 23 September 2003; caviar trends from Douglas F. Williamson, *Caviar and Conservation Status, Management, and Trade of North American Sturgeon and Paddlefish* (Cambridge, U.K.: World Wildlife Fund, May 2003), p. 148; TRAFFIC, *Sturgeons of the Caspian Sea and the International Trade in Caviar* (Cambridge, U.K.: November 1996); International Trade Agency reports that the United States imported 39, 289 kilos (84,411 pounds) of caviar for consumption in 2002, at www.ita.doc.gov/ td/industry/otea/Trade-Detail/Latest-Decem ber/Imports/16/160430.html; "Shark! Danger Exaggerated, Says US Experts," *Reuters*, 17 June 2002; Brian Handwerk, "Asian Shark-Fin Trade May be Larger than Expected," *National Geographic News*, 28 April 2003; Patricia Brown, "Foie Gras Fracas: Haute Cuisine Meets the Duck Liberators," *New York Times*, 24 September 2003;

Caviar Emptor Campaign, *Roe to Ruin: Executive Summary* (Washington, DC: Natural Resources Defense Council, December 2002).

9. Peter Pringle, *Food, Inc.: Mendel to Monsanto—The Promises and Perils of the Biotech Harvest* (New York: Simon & Schuster, 2003).

10. Marion Nestle, *Food Politics: How the Food Industry Influences Nutrition and Health* (Berkeley, CA: University of California Press, 2002).

11. Germany, the United States, and the United Kingdom from J. N. Pretty et al., "An Assessment of the Total External Costs of UK Agriculture," *Agricultural Systems*, August 2000, pp. 113–36, and from Jules Pretty et al., "Policy Challenges and Priorities for Internalising the Externalities of Modern Agriculture," *Journal of Environmental Planning and Management*, March 2001, pp. 263–83.

12. Southwest England from Foundation for Local Food Initiatives, *Shopping Basket Survey for South West Local Food Partnership* (London: 2002).

13. McDonald's from David Barboza with Sherri Day, "McDonald's Seeking Cut in Antibiotics in Its Meat," *New York Times*, 20 June 2003; Kraft from David Barboza, "Kraft Plans to Rethink Some Products to Fight Obesity," *New York Times*, 2 July 2003; Bill Vorley and Julio Berdegué, "The Chains of Agriculture," World Summit on Sustainable Development Opinion (London: International Institute for Environment and Development (IIED), May 2001); William Vorley, *Agribusiness and Power Relations in the Agri-Food Chain*, background paper for Policies That Work for Sustainable Agriculture and Regenerating Rural Economies (London: IIED, June 2000), p. 21; Race to the Top project, see www.racetothetop.org.

14. Iowa Agricultural Statistics Service, *December Hogs and Pigs—Iowa* (Washington, DC: U.S. Department of Agriculture (USDA), 30 December 2002); Paul Willis, Manager, Niman Ranch Pork Company, discussion with Danielle Nierenberg, 12 June 2003.

15. Willis, op. cit. note 14; Niman Ranch Web site, at www.nimanranch.com, viewed 15 June 2003.

16. Cees de Haan et al., "Livestock and the Environment: Finding a Balance," report of a study coordinated by FAO, U.S. Agency for International Development, and World Bank (Brussels: 1997), p. 53; FAO "Meat and Meat Products," *FAO Food Outlook No. 4*, October 2002, p. 11. Figure 4–1 from the following: feed from Vaclav Smil, Department of Geography, University of Manitoba, discussion with Brian Halweil, October 2002, and from Erik Millstone and Tim Lang, *The Penguin Atlas of Food: Who Eats What, Where, and Why* (London: Penguin Books, 2003), p. 34; water from ibid., p. 35; additives from Margaret Mellon, Charles Benbrook, and Karen Lutz Benbrook, *Hogging It! Estimates of Antimicrobial Abuse in Livestock* (Washington, DC: Union of Concerned Scientists, 2001); fossil fuels from Millstone and Lang, op. cit. this note, pp. 35, 62; methane from de Haan et al., op. cit. this note; diseases from Neil Barnard, Andrew Nicholson, and Jo Lil Howard, "The Medical Costs Attributed to Meat Consumption," *Preventive Medicine*, November 1995.

17. Christopher Delgado, Claude Courbois, and Mark Rosegrant, *Global Food Demand and the Contribution of Livestock As We Enter the New Millennium*, MSSD Discussion Paper No. 21 (Washington, DC: International Food Policy Research Institute, 1998), p. 6.

18. Jo Robinson, *Why Grassfed is Best! The Surprising Benefits of Grassfed Meat, Eggs, and Dairy Products* (Vashon, WA: Vashon Island Press, 2000), p. 8; "Hay! What a Way to Fight E. coli," *Science News Online*, 19 September 1998, at www.sciencenews.org/sn_arc98/9_19_98/Food.htm.

19. World Health Organization (WHO) and FAO, *Antimicrobial Resistance*, Fact Sheet No. 194 (Geneva: WHO, January 2003); Mellon, Benbrook, and Benbrook, op. cit. note 16.

20. Ian Langford cited in Nick Tattersall, "Stressed Farm Animals Contribute to Food Poisoning: U.K. Study," *Manitoba Co-operator's*, 15 March 2001.

21. U.S. Department of Health and Human Services, Food and Drug Administration, Center for Food Safety and Applied Nutrition, *Consumer Questions and Answers about BSE* (Washington, DC: May 2003).

22. Dr. Gary Smith, University of Pennsylvania Veterinary School, discussion with Danielle Nierenberg, February 2003; "Transporting Animals in Europe," in Millstone and Lang, op. cit. note 16, p. 64.

23. I. Koizumi et al., "Studies on the Fatty Acid Composition of Intramuscular Lipids of Cattle, Pigs, and Birds," *Journal of Nutritional Science Vitaminol* (Tokyo), vol. 37, no. 6 (1991), pp. 545–54; Robinson, op. cit. note 18, pp. 12–15; see also Eat Wild Web site at www.eatwild.com.

24. Smil, op. cit. note 16.

25. Robinson, op. cit. note 18, p. 7; demand for meat from FAO, op. cit. note 7, updated 9 January 2003; idem, op. cit. note 17, p. 11; David Brubaker, former CEO, PennAg Industries, discussion with Danielle Nierenberg, September 2002.

26. Payal Sampat, "Groundwater Shock: The Polluting of the World's Major Freshwater Stores," *World Watch*, January/February 2000, p. 14; idem, "Uncovering Groundwater Pollution," in Worldwatch Institute, *State of the World 2001* (New York: W.W. Norton & Company, 2001), p. 27; Michael Mallin et al., "Impacts and Recovery from Multiple Hurricanes in a Piedmont-Coastal Plain River System," *Bioscience*, November 2002, p. 999.

27. Bobby Inocencio, Teresa Farms, Philippines, discussion with Danielle Nierenberg, August 2002; Teresa Farms, *Management Guide SASSO Free-Range Colored Chickens* (Rizal, Philippines: undated).

28. Steve Tarter, "Megahog Opponent Tours Poland," *Peoria Star Journal*, 12 June 2001; Tom Garrett, "Polish Delegation Investigates American Agribusiness, Repudiates Factory Farming," *AWI Quarterly* (Animal Welfare Institute, Washington, DC), fall/winter 1999–2000, p. 7.

29. Eggs and laying hens from Elizabeth Becker, "Advocates for Animals Turn Attention to Chickens," *New York Times*, 4 December 2002, and from "McDonald's USA Animal Welfare Laying Hens Guidelines," at www.mcdonalds.com/countries/usa/community/welfare/laying_hen/index.html; antibiotics from David Barboza with Sherri Day, "McDonald's Seeking Cut in Antibiotics in Its Meat," *New York Times*, 20 June 2003, from "Fast Food, Not Fast Antibiotics," *New York Times*, 22 June 2003, and from David Barboza, "Animal Welfare's Unexpected Allies," *New York Times*, 25 June 2003.

30. "Fast Food, Not Fast Antibiotics," op. cit. note 29; Barboza, op. cit. note 29; Cornelius de Haan et al, *Livestock Development: Implications for Rural Poverty, the Environment, and Global Food Security* (Washington, DC: World Bank, 2001), pp. xii–xiii.

31. Nadia El-Hage Scialabba and Caroline Hattam, eds., *Organic Agriculture, Environment and Food Security*, Environment and Natural Resources Series, No. 4 (Rome: FAO, 2002), p. 29.

32. Ibid.; AGRIPO Agriculture and Pollution, "Policy Options for Pollution Control—Sustainable Farming Systems," course modules prepared by Lithuanian University of Agriculture, 2001, p. 20, at distance.ktu.lt/agripo/8-sust-farming.pdf.

33. Internationale Fachmesse für Abwassertechnik, "Farmers as Water Managers? Cooperation Initiatives between Water Supply Companies and Agriculture: One of the Themes at IFAT 2002," brief produced for the International Trade Fair for Waste Water Technology No. 7, September 2001, Munich, Germany.

34. University of Essex from Pretty et al., "An Assessment of the Total External Costs," op. cit. note 11; Philippines from P. L. Pingali et al., "The Impact of Pesticides on Farmer Health: A Medical and Economic Analysis in the Philippines," in P. L. Pingali and P. A. Roger, eds., *Impact of Pesticides on Farmer Health and the Rice Environment* (Norwell, MA: Kluwer Academic Publishers,

1995), pp. 343–60.

35. Organic farms and biodiversity from El-Hage Scialabba and Hattam, op. cit. note 31, pp. 38–53.

36. "The Global Market for Organic Food & Drink," *Organic Monitor* (London), July 2003. A similar estimate of $23–25 billion in 2003 comes from Minou Yussefi and Helga Willer, eds., *The World of Organic Agriculture 2003—Statistics and Future Prospects* (Tholey-Theley, Germany: International Federation of Organic Agriculture Movements (IFOAM), 2003), p. 24; 23 million hectares and Figures 4–2 and 4–3 from ibid., p. 13.

37. Nicholas Parrott and Terry Marsden, *The Real Green Revolution* (London: Greenpeace Environmental Trust, 2002), p. 61; Rick Welsh, *The Economics of Organic Grain and Soybean Production in the Midwestern United States*, Henry A. Wallace Institute for Alternative Agriculture, Policy Studies Report No. 13 (Greenbelt, MD: May 1999); P. Mader et al., "Soil Fertility and Biodiversity in Organic Farming," *Science*, 31 May 2002.

38. Bill Liebhardt, *Get the Facts Straight: Organic Agriculture Yields are Good*, Information Bulletin No. 10 (Santa Cruz, CA: Organic Farming Research Foundation, summer 2001).

39. Peter Tschannen, Project Director, Remei AG Biore, Lettenstrasse, Switzerland, discussion with Brian Halweil, 4 May 2003; Tadeu Caldas, "A Glimpse into the Largest Organic Project in Asia," *Ecology and Farming* (IFOAM), March 2000.

40. Jean-Marie Diop et al., *On-Farm Agro-Economic Comparison of Organic and Conventional Techniques in High and Medium Potential Areas of Kenya*, joint publication of ETC (Educational Training Consultants) Netherlands and Kenya Institute of Organic Farming (Leusden and Nairobi: March 1998).

41. El-Hage Scialabba and Hattam, op. cit. note 31, pp. 8–10; Parrott and Marsden, op. cit. note 37, p. 62; Nicholas Parrott, School of City and Regional Planning, Cardiff University, U.K., discussion with Brian Halweil, 12 September 2003.

42. Brian P. Baker et al., "Pesticide Residues in Conventional, IPM-grown and Organic Foods: Insights from Three U.S. Data Sets," *Food Additives and Contaminants*, May 2002, pp. 427–46; Cynthia L. Curl et al., "Organophosphorus Pesticide Exposure of Urban and Suburban Preschool Children with Organic and Conventional Diets," *Environmental Health Perspectives*, March 2003, pp. 377–82.

43. Edward Groth, Senior Scientist, Consumers Union, discussion with Brian Halweil, 22 September 2003; National Research Council, *Pesticides in the Diets of Infants and Children* (Washington, DC: National Academy Press, 1993); Baker et al., op. cit. note 42; Misa Kishi and Joseph Ladou, "International Pesticide Use," *International Journal of Occupational and Environmental Health*, October/December 2001; Dina M. Schreinmachers et al., "Cancer Mortality in Agricultural Regions of Minnesota," *Environmental Health Perspectives*, March 1999; S. H. Swan et al., "Semen Quality in Relation to Biomarkers of Pesticide Exposure," *Environmental Health Perspectives*, September 2003, pp. 1478–84; M. F. Cavieres et al., "Developmental Toxicity of a Commercial Herbicide Mixture in Mice: I. Effects on Embryo Implantation and Litter Size," *Environmental Health Perspectives*, November 2002, pp. 1081–85.

44. Danny K. Asami et al., "Comparison of the Total Phenolic and Ascorbic Acid Content of Freeze-Dried and Air-Dried Marionberry, Strawberry, and Corn Grown Using Conventional, Organic, and Sustainable Agricultural Practices," *Journal of Agricultural and Food Chemistry*, February 2003, pp. 1237–41.

45. "Stranglehold, Pesticides: What's the Way Out?" *Down to Earth*, 15 June 2003.

46. Asami et al., op. cit. note 44.

47. Slow Food membership from Ilaria Morra, press office, Slow Food, e-mail to Brian Halweil, 14 July 2003; Carlo Petrini, President, Slow Food, e-mail to Brian Halweil, 17 July 2002.

48. Alejandro Argumedo, ANDES, Cuzco, Peru,

e-mail to Brian Halweil, 15 March 2002.

49. Trade and volume from FAO, op. cit. note 7, updated 24 December 2002; U.S. surveys from Matthew Hora and Jody Tick, *From Farm to Table: Making the Connection in the Mid-Atlantic Food System* (Washington, DC: Capital Area Food Bank, 2001), and from Rich Pirog et al., *Food, Fuel, and Freeways: An Iowa Perspective on How Far Food Travels, Fuel Usage, and Greenhouse Gas Emissions* (Ames, IA: Leopold Center for Sustainable Agriculture, Iowa State University, 2001), pp. 1, 2; U.K. data and Figure 4–4 from Andy Jones, *Eating Oil: Food Supply in a Changing Climate* (London: Sustain, 2001).

50. New Economics Foundation, "Local Food Better for Rural Economy than Supermarket Shopping," press release (London: 7 August 2001); Christopher Delgado et al., *Agricultural Growth Linkages in Sub-Saharan Africa*, Research Report 107 (Washington, DC: International Food Policy Research Institute, December 1998).

51. Vandana Shiva, Research Foundation for Science, Technology and Ecology, e-mail to Brian Halweil, 18 July 2003; for Navdanya movement, see <www.navdanya.org>; Francis Moore Lappé and Anna Lappé, *Hope's Edge: The Next Diet for a Small Planet* (New York: Tarcher/Putnam, 2002), pp. 145–50.

52. "The Old Ones are the Best," *New Agriculturalist* on-line, 1 September 2003, at www.new-agri.co.uk/03-5/focuson/focuson5.html.

53. Norman E. Borlaug, "The Next Green Revolution," *New York Times*, 11 July 2003; Sustain: The Alliance for Better Food and Farming, "Study Visit to Italy: The Italian School Meals System," at www.sustainweb.org/chain_italy_study.asp, viewed 1 September 2002; Anne Dolamore et al., *Good Food on the Public Plate: A Manual for Sustainability in Public Sector Food and Catering* (London: Sustain and East Anglia Food Link, 2003).

54. JoAnn Jaffe, University of Regina, e-mail to Brian Halweil, 26 April 2002, based on JoAnn Jaffe and Michael Gertler, "Victual Vicissitudes: Consumer Deskilling and the Transformation of Food Systems," in M. D. Mehta, ed., *The Sociology of Biotechnology* (Toronto: University of Toronto Press, forthcoming).

55. Summaries of consolidation trends are available in William Heffernan et al., *Consolidation in the Food and Agriculture System* (Washington, DC: National Farmers Union, February 1999), in Mary Hendrickson et al., *Consolidation in Food Retailing and Dairy: Implications for Farmers and Consumers in a Global Food System* (Washington, DC: January 2001) (available at www.food circles.missouri.edu), and in Halweil, op. cit. note 6; Lappé and Lappé, op. cit. note 51, pp. 138–64.

56. More than 26 million lunches from USDA, National School Lunch Program, "FAQs," at www.fns.usda.gov/cnd/Lunch/AboutLunch/faqs .htm, updated August 2003; Swedish Food Authority from OECD, op. cit. note 7, p. 53. In Box 4–3, over $300 billion in subsidies from OECD, *Agricultural Policies in OECD Countries: Monitoring and Evaluation 2002* (Paris: 2002).

57. Historical changes in food policy from Nestle, op. cit. note 10; Stuart Laidlaw, *Secret Ingredients: The Brave New World of Industrial Farming* (Toronto: McClelland & Stewart Ltd., 2003). p. 254.

Bottled Water

1. Catherine Ferrier, *Bottled Water: Understanding a Social Phenomenon* (Washington, DC: World Wildlife Fund, April 2001), p. 13; consumption growth rate from International Year of Freshwater, "Facts and Figures: Bottled Water," at www.wateryear2003.org; Rajesh Mahapatra, "Pesticide Findings Spur Indian Government Crackdown on Bottled Water Companies," *Associated Press*, 21 February 2003; global expenditure from Brian Howard, "The World's Water Crisis," *E Magazine*, September/October 2003, p. 28.

2. Description of water and Perrier recall from Ferrier, op. cit. note 1, pp. 3, 6, 17.

3. Anne Christiansen Bullers. "Bottled Water: Better Than the Tap?" *FDA Consumer Magazine*

(U.S. Food and Drug Administration), July-August 2002.

4.   Asia and the Pacific from Ferrier, op. cit. note 1, p. 13; poor access from Howard, op. cit. note 1, p. 3; Natural Resources Defense Council (NRDC), *Bottled Water: Pure Drink or Pure Hype?* (New York: March 1999); France from Ferrier, op. cit. note 1, p. 16.

5.   Maude Barlow, *Blue Gold: The Global Water Crisis and the Commodification of the World's Water Supply*, rev. ed. (Ottawa, ON, Canada: Council of Canadians, spring 2001), pp. 46–47; Anthony DePalma, "Free Trade in Fresh Water? Canada Says No and Halts Exports," *New York Times*, 8 March 1999.

6.   Container Recycling Institute (CRI), "Plastic Soda Bottle Recycling Rate Down Again…Virgin Resin Production Outpaces Recycling," at www.container-recycling.org/plasrate/ratedown .htm; PET lifecycle data from Association of Plastic Manufacturers in Europe, cited in Baxter CVG, *The Economic and Ecological Implications of a Solid Waste Reduction Program*, at www.waste reduction.org/Baxter/Bax5.htm.

7.   Ferrier, op. cit. note 1, p. 23.

8.   Water bottles in United States from Patricia Franklin, "Letter from the Executive Director," *Container and Packaging Recycling Update* (CRI, Arlington, VA), summer/fall 2003, p. 2; Kalyan Moitra, "Recycle Onus on PET Producers, Says PCB," *Economic Times of India*, 27 June 2003; CRI, Bottle Bill Resource Guide, at www.bottlebill.org.

9.   Reason to drink bottled water from Ferrier, op. cit. note 1, p. 16; NRDC, op. cit. note 4; Centre for Science and Environment, "Pure Water or Pure Peril?" press release (New Delhi: 4 February 2003); Hansika Pal, "Debate Over Pesticide Residue Clouds Bottled Water," *Economic Times of India*, 6 February 2003.

10.   U.N. International Year of Freshwater, at www.un.org/events/water; Millennium Development Goals at www.un.org/millenniumgoals.

Chicken

1.   World Society for the Protection of Animals (WSPA), World Farmwatch, "The Facts About Our Food, Eggs," pamphlet, at www.wspa.org.uk; Paul Shapiro, Compassion Over Killing, and Bruce Freidrich, People for the Ethical Treatment of Animals, e-mails to author, 15 September 2003.

2.   WSPA, op. cit. note 1; Shapiro, op. cit. note 1; Freidrich, op. cit. note 1; Stuart Laidlaw, *Secret Ingredients: The Brave New World of Industrial Farming* (Toronto: McClelland & Stewart Ltd., 2003), pp. 31, 52.

3.   WSPA, op. cit. note 1; Laidlaw, op cit note 2, p. 33; Shapiro, op. cit. note 1; Compassion in World Farming, Laying Hen Factsheet, at www.ciwf.co.uk/Camp/Main/Battery/battery _hen_campaign.htm, viewed 23 September 2003.

4.   WSPA, op. cit. note 1; Compassion in World Farming, op. cit. note 3; National Animal Health Monitoring System, *Part II: Reference of 1999 Table Egg Layer Management in the U.S.* (Washington, DC: U.S. Department of Agriculture (USDA), 2000); Michael Appleby, Vice President, Farm Animals and Sustainable Agriculture, Humane Society of the United States, e-mail to author, 26 September 2003.

5.   Anthony Browne, "Ten Weeks to Live," *The Observer*, 10 March 2002; Shapiro, op. cit. note 1.

6.   Feed amount from Richard Reynnells, National Program Leader, Animal Production Systems, USDA, e-mail to author, 26 September 2003, and from Mack O. North, *Commercial Chicken Production Manual*, 3rd ed. (Westport, CT: AVI Publishing Company, Inc., no date), p. 374; Consumers Union, "Presence of Anti-microbial Resistant Pathogens in Retail Poultry Products: A Report by CI Members in Australia and the United States," presented by Consumers International to the Codex Committee on Residues of Veterinary Drugs in Foods, 4–7 March 2003; European Commission, "The Welfare of Chickens Kept for Meat Production (Broilers)," Report of the Scientific Committee on Animal Health and Animal Welfare, 21 March 2000, pp. 31–36,

42–43; Browne, op. cit. note 5.

7.   Browne, op. cit. note 5; Erik Millstone and Tim Lang, *The Penguin Atlas of Food: Who Eats What, Where, and Why* (London: Penguin Books, 2003), p. 38.

8.   "In Praise of Family Poultry," *Agriculture 21* (U.N. Food and Agriculture Organization), March 2002.

9.   Kimberlie Cole, Westwind Farms, e-mail to Clayton Adams, Worldwatch Institute, 14 August 2003; West Wind Farms Web site, at www.grass organic.com, viewed 12 June 2003.

Chocolate

1.   Allen M. Young, *The Chocolate Tree: A Natural History of Cacao* (Washington, DC: Smithsonian Institution Press, 1994), pp. 2–8; J. C. Motamayor and C. Lanaud, "Molecular Analysis of the Origin and Domestication of *Theobroma cacao* L.," in Johannes M. M. Engels et al., eds., *Managing Plant Genetic Diversity* (Wallingford, U.K.: CABI Publishing, 2002), pp. 77–87.

2.   U.N. Food and Agriculture Organization (FAO), *FAOSTAT Statistical Database*, at apps.fao.org, viewed 3 September 2003.

3.   Overview of cocoa-growing requirements available on the International Cocoa Organization Web site, at www.icco.org/questions/tree.htm, viewed 3 September 2003; spread of cocoa throughout the tropics from Young, op. cit. note 1, pp. 14–47; Russell A. Mittermeier, Norman Myers, and Cristina Goettsch Mittermeier, *Hotspots: Earth's Biologically Richest and Most Endangered Terrestrial Ecoregions* (Mexico City: CEMEX/Conservation International, 1999).

4.   Commercial production under 60 percent canopy from Norman D. Johns, "Conservation in Brazil's Chocolate Forest: The Unlikely Persistence of the Traditional Cocoa Agroecosystem," *Environmental Management*, January 1999, p. 35; for an overview of cocoa cultivation, see Robert A. Rice and Russell Greenberg, "Cacao Cultivation and the Conservation of Biological Diversity,"

Ambio, May 2000, pp. 167–73.

5.   Global deforestation and Côte d'Ivoire from Rice and Greenberg, op. cit. note 4, p. 169; B. Duguma, J. Gockowski, and J. Bakala, "Smallholder Cacao (*Theobroma cacao* Linn.) Cultivation in Agroforestry Systems of West and Central Africa: Challenges and Opportunities," *Agroforestry Systems*, vol. 51, no. 3 (2001), pp. 177–79; forest loss in Indonesia from François Ruf and Yoddang, "Cocoa Migrants from Boom to Bust," in François Gérard and François Ruf, eds., *Agriculture in Crisis: People, Commodities and Natural Resources in Indonesia, 1996–2001* (Richmond, Surrey, U.K.: Curzon Press, 2001), pp. 106, 131–32; Brazil from Johns, op cit. note 4, pp. 31–47; Cameroon from Jim Gockowski, International Institute of Tropical Agriculture, e-mail to author, 13 December 2001.

6.   Rice and Greenberg, op. cit. note 4, pp. 169–70 and Table 2.

7.   Global worth of cocoa retail from Trevor Datson, "Chocoholism Reached Almost Epidemic Proportions," *Reuters*, 12 February 2003, and from U.N. Conference on Trade and Development, "UN Cocoa Conference Ends 10-Day Session; Will Resume Work in February 2001," press release (Geneva: 28 November 2000). High-end estimates of farm income assume an 80-percent farmgate share of annual bean export revenues; for 2002, those revenues were estimated at $4.4 billion by multiplying global production (available in FAO, op. cit. note 2) by the average 2002 trading price (available in the International Cocoa Organization's *Quarterly Bulletin of Cocoa Statistics*). Labor abuse from "New Foundation Launched to Fight Problem in Cocoa Industry," *U.N. Wire*, 2 July 2002, from Sumana Chatterjee, "Chocolate Industry to Fight Child Slavery," *San Jose Mercury News*, 2 July 2002, and from Brooke Shelby Biggs, "Slavery Free Chocolate?" at www.alternet.org, 7 February 2002.

8.   For a full account of the fair trade system, see Fairtrade Labelling Organizations International, at www.fairtrade.net; lindane from René Philippe et al., "Rational Chemical Pest Control," in Dominique Mariau, ed., *Integrated Pest Manage-*

ment of Tropical Perennial Crops (Enfield, NH: Science Publishers, 1997), pp. 25–35.

Shrimp

1.   Egypt from Don Brothwell and Patricia Brothwell, *Food in Antiquity—A Survey of the Diet of Early Peoples* (Baltimore, MD: John Hopkins University Press, 1998), pp. 55–66; "A Brief History of Shrimp Farming," *Shrimp News International*, at www.shrimpnews.com/History.html, updated 1 October 2003.

2.   Globefish, cited in Foodmarket Exchange .com, "Shrimp Production," at www.foodmarket exchange.com/datacenter/product/seafood/ shrimp/detail/dc_pi_sf_shrimp0301_01.htm, viewed 28 August 2003.

3.   Chinese production from ibid.; top exporter from Foodmarket Exchange.com, "Shrimp Trade," at www.foodmarketexchange.com/datacenter/ product/seafood/shrimp/dc_pi_sf_shrimp04.htm, viewed 28 August 2003.

4.   U.S. imports from Helga Josupeit, "An Overview on the World Shrimp Market," September 2001, at www.globefish.org/presenta tions, viewed 28 August 2003; U.S. National Oceanic and Atmospheric Administration, "Shrimp Overtakes Canned Tuna as Top U.S. Seafood," press release (Washington, DC: 28 August 2002); Japan from Globefish, "Shrimp," *Monthly Market Report*, April 2003, at www.globefish.org/market reports/Shrimp/Shrimp4.htm, and from John M. Saulnier, "European Ban on Asian Shrimp Imports Puts Even More Pressure on Prices," *Global Seafood Magazine* (Quick Frozen Foods International), April 2002.

5.   Three quarters and $7 billion from Globefish, "Shrimp," *Commodity Update*, June 2002, at www.globefish.org/publications/commodityup date/200206/200206.htm; Elliot Norse and Les Watling, "Clearcutting the Ocean Floor," *Earth Island Journal*, summer 1999.

6.   Environmental Justice Foundation, *Squandering the Seas: How Shrimp Trawling is Threatening Ecological Integrity and Food Security*

Around the World (London: 2003), p. 2.

7.   Denise Johnston, Chris Soderquist, and Donella H. Meadows, *The Shrimp Commodity System: A Sustainability Institute Report* (Hartland, VT: The Sustainability Institute, 2000) p. 3.

8.   Isabel de la Torre, "The Unpalatable Prawn," *Earth Island Journal*, spring 2000.

9.   David Barnhizer, "The Trade, Environment and Human Rights: The Paradigm Case of Industrial Aquaculture and the Exploitation of Traditional Communities," Mangrove Action Project Web site, at www.earthisland.org/map/trd-hmn .htm; Environmental Justice Foundation, *Smash & Grab: Conflict, Corruption and Human Rights Abuses in the Shrimp Farming Industry* (London: 2003).

10.   Vandana Shiva, presentation at the Peoples Summit, St. Mary's University, Halifax, NS, Canada, 17 June 1995; quote cited in Environmental Justice Foundation, op. cit. note 6, p. 1.

11.   Consortium from Network of Aquaculture Centres in Asia-Pacific, Shrimp Farming and the Environment Web site, at www.enaca.org/ Shrimp/ShrimpAquacultureCertification.htm; Todd Steiner, Sea Turtle Restoration Project, Earth Island Institute, "Sea Turtles, Shrimp Fisheries, and the Turtle Excluder Device," presentation prepared for the Shrimp Tribunal at the United Nations, 29 April 1996, at www.earth summitwatch.org/shrimp/positions/pov2.html; World Rainforest Movement, "Local Fisherfolk Protect the Mangroves in Sri Lanka," *WRM Bulletin No. 20*, 10 February 1999.

Soda

1.   Beverage Marketing Corporation, *Global Carbonated Soft Drinks: A Worldview, 2003 Edition* (New York: 2003), according to John Rodwan, e-mail to author, 29 September 2003. In 2002, 22.3 percent of the U.S. volume of soda was packaged for fountain drink dispensers and thus did not include water or carbon dioxide (ibid.); for simplicity, the focus here is on the majority of soda, packaged directly for consumers.

2. Paul Vallely, Jon Clarke, and Liz Stuart, "Coke Adds Life? In India, Impoverished Farmers Are Fighting to Stop Drinks Giant 'Destroying Livelihoods'," (London) *The Independent*, 25 July 2003; Surendranath C., "Coca-Cola: Continuing the Battle in Kerala," *CorpWatch*, 10 July 2003; Edward Luce, "Pepsi and Coca-Cola Deny Pesticide Claims," *Financial Times*, 5 August 2003; Centre for Science and the Environment, "Government Confirms Pesticides in Soft Drinks," press release (Delhi, India: 21 August 2003).

3. About 25 percent of the U.S. volume consists of diet sodas (according to U.S. Department of Agriculture (USDA), Economic Research Service, *Per Capita Food Consumption Data System: Beverages*, at www.ers.usda.gov/Data/FoodConsumption/Spreadsheets/beverage.xls, viewed 1 August 2003), which substitute high-intensity sweeteners for sugars. While these sweeteners are essentially non-caloric, and thus do not contribute directly to obesity, there is continuing debate on how safe they are. U.S. consumption data from USDA, op. cit. this note; calcium intake from Claude Cavadini et al., "U.S. Adolescent Food Intake Trends from 1965 to 1996," *Archives of Disease in Childhood*, vol. 83, no. 1 (2000), p. 19; U.S. Department of Health and Human Services, *The Surgeon General's Call to Action to Prevent and Decrease Overweight and Obesity, 2001* (Washington, DC: 2001); David S. Ludwig et al., "Relationship Between Consumption of Sugar-Sweetened Drinks and Childhood Obesity: A Prospective, Observational Analysis," *The Lancet*, 17 February 2001, p. 507.

4. Caffeine component from Euromonitor International, *Soft Drinks: The International Market, 2001 Edition*, according to Trudy Griggs, e-mail to author, 19 December 2001; Roland R. Griffiths and Ellen M. Vernotica, "Is Caffeine a Flavoring Agent in Cola Soft Drinks?" *Archives of Family Medicine*, August 2000, p. 732; Ronald R. Watson, "Caffeine: Is it Dangerous to Health?" *American Journal of Health Promotion*, spring 1988; Roland Griffiths, Johns Hopkins University School of Medicine, e-mail to author, 22 January 2002; caffeine in Pepsi from "Caffeinated Kids," *Consumer Reports*, July 2003, p. 28.

5. Beverage Marketing Corporation, *Global Beverage Packaging, 2002 Edition* (New York: 2002), according to Rodwan, op. cit. note 1; Laurel Wentz, "Global Marketers Spend $71 Billion," *Advertising Age*, 11 November 2002, at www.adage.com/images/random/GlobalMarketers2002.pdf, viewed 28 September 2003; Steven Manning, "How Corporations Are Buying Their Way Into America's Classrooms," *The Nation*, 27 September 1999, pp. 11–18.

6. Jennifer Gitlitz, *Trashed Cans: The Global Environmental Impacts of Aluminum Can Wasting in America* (Arlington, VA: Container Recycling Institute, June 2002); number of cans from Beverage Marketing Corporation, *Beverage Packaging in the U.S., 2003 Edition* (New York: 2003), according to Rodwan, op. cit. note 1, with calculations based on Gitlitz, op. cit. this note; "Summary of the National Beverage Producer Responsibility Act of 2002: Senate Bill 2220," factsheet (Athens, GA: Grassroots Recycling Network, undated); Sweden and Michigan from Gitlitz, op. cit. this note, pp. 27, 28.

7. Erika Hayasaki, "Schools to End Soda Sales," *Los Angeles Times*, 28 August 2002; junk-food tax from Michael F. Jacobson and Kelly D. Brownell, "Small Taxes on Soft Drinks and Snack Foods to Promote Health," *American Journal of Public Health*, June 2000, pp. 854–57 (currently taxes just go into general funds); Sweden and Poland from Euromonitor International, *Marketing to Children: A World Survey, 2001 Edition*, from Trudi Griggs, Euromonitor, e-mail to author, 25 January 2002, and from Ingrid Jacobsson, "Advertising Ban and Children: Children Have the Right to Safe Zones," *Current Sweden*, June 2002; sales and projections from Beverage Marketing Corporation, op. cit. note 1.

Chapter 5. Moving Toward a Less Consumptive Economy

1. Richard B. Brightwell, "A Short Biography of King Camp Gillette," at www.creekstone.net/razors/kingcampgillette.htm, viewed 18 September 2003; "King C. Gillette Disposable-Blade Safety Razor," *Inventor of the Week Archive*, June 2000, Lemelson-MIT Program, at web.mit.edu/

invent/iow/gillette.html.

2. Eric A. Taub, "DVD's Meant for Buying but Not for Keeping," *New York Times*, 21 July 2003.

3. Edward Rothstein, "A World of Buy, Buy, Buy, from A to Z," *New York Times*, 19 July 2003.

4. Lebow quote from Vance Packard, *The Waste Makers* (New York: David Mckay, 1960).

5. Alan Durning, in *How Much Is Enough?* (New York: W.W. Norton & Company, 1992), argued that from a consumption and ecological impact perspective humanity is divided into three broad classes: the consumers, the middle, and the poor. Although the specific numbers of people in these categories and the category thresholds are constantly in flux, the distinction remains useful for conceptual purposes.

6. Juliet Schor, "The Triple Imperative: Global Ecology, Poverty and Worktime Reduction" (Boston, MA: Boston College, May 2001, unpublished); 90 percent cut from Gary Gardner and Payal Sampat, *Mind Over Matter: Recasting the Role of Materials in Our Lives*, Worldwatch Paper 144 (Washington, DC: Worldwatch Institute, December 1998), p. 25.

7. Raw material trade trends from World Bank, "Change in Commodity Production and Trade," *Global Commodities Markets Online*, at www.worldbank.org/prospects/gmconline/index.htm, April 2000, p. 10; commodity prices from International Monetary Fund, *International Financial Statistics Yearbook* (Washington, DC: annual), and from Michael Renner, "Commodity Prices Weak," in Worldwatch Institute, *Vital Signs 2001* (New York: W.W. Norton & Company, 2001), p. 122; Michael Renner, *The Anatomy of Resource Wars*, Worldwatch Paper 162 (Washington, DC: Worldwatch Institute, October 2002).

8. Robert Brenner, "The Economics of Global Turbulence: A Special Report on the World Economy, 1950–1998," *New Left Review*, No. 229 (1998), pp. 258–61; China-U.S. trade surplus from Joseph Kahn, "China Seen Ready to Con-

ciliate U.S. on Trade and Jobs," *New York Times*, 2 September 2003; China-Mexico wage competition from Juan Forero, "As China Gallops, Mexico Sees Factory Jobs Slip Away," *New York Times*, 3 September 2003.

9. Organisation for Economic Co-operation and Development (OECD), *Towards Sustainable Consumption: An Economic Conceptual Framework* (Paris: Environment Directorate, June 2002), p. 41.

10. Table 5–1 adapted from Norman Myers and Jennifer Kent, *Perverse Subsidies. How Tax Dollars Can Undercut the Environment and the Economy* (Washington, DC: Island Press, 2001), pp. 187–88.

11. Herman E. Daly, "Five Policy Recommendations for a Sustainable Economy," in Juliet B. Schor and Betsy Taylor, eds., *Sustainable Planet: Solutions for the 21st Century* (Boston: Beacon Press, 2002); OECD, *Policies to Promote Sustainable Consumption: An Overview* (Paris: Environment Directorate, July 2002), p. 17; Lorenz Jarass, "More Jobs, Less Tax Evasion, Better Environment—Towards a Rational European Tax Policy," Contribution to the Hearing at the European Parliament, Brussels, 17 October 1996.

12. OECD, op. cit. note 9, p. 17; tax revenue trends and Table 5–2 from Ulf Johansson and Claudius Schmidt-Faber, "Environmental Taxes in the European Union 1980–2001," *Eurostat Statistics in Focus*, 9/2003, pp. 1, 3, 6.

13. OECD, *Making Work Pay: Taxation, Benefits, Employment and Unemployment*, The OECD Jobs Strategy Series (Paris: 1997); idem, "Environmentally Related Taxes Database," at www.oecd.org/document/29/0,2340,en_2649_37465_1894685_1_1_1_1_1,00.html, viewed 29 August 2003; Johansson and Schmidt-Faber, op. cit. note 12, pp. 3-4.

14. Carbon dioxide emissions avoided by 2002 and jobs gained from Umweltbundesamt (German Federal Environment Agency), "Höhere Mineralölsteuer Entlastet die Umwelt und den Arbeitsmarkt," press release (Berlin: 3 January 2002);

revenue growth from BUND, "Wie Hoch Sind die Ökosteuern?" at www.oeko-steuer.de/pages/fakt_oekosteuer.phtml, viewed 13 August 2003; job projection to 2010 from Deutsches Institut für Wirtschaftsforschung (DIW), "Wirkungen der Ökologischen Steuerreform in Deutschland," *DIW-Wochenbericht*, No. 14/2001.

15. European harmonization efforts from Green Budget Germany, *Green Budget News—European Newsletter on Environmental Fiscal Reform*, 2–4/2003, at www.eco-tax.info/2newsmit/newseng/GBN2.html, and from European Environment Bureau (Brussels), *Environmental Fiscal Reform Campaign Newsletter*, April 2003 and September 2003.

16. German provisions from Stefan Bach and Michael Kohlhaas, "Nur zaghafter Einstieg in die ökologische Steuerreform," *DIW-Wochenbericht*, No. 36/1999; Reinhard Loske and Kristin Heyne, "Ökologische Steuerreform: Die Stufen 2–5," press release (Berlin: German Green Party parliamentary group, 29 June 1999); Umwelt- und Prognose-Institut, "Stellungnahme des UPI-Instituts vom 23. Januar, incl. Nachträge vom 8.2. und 25.2.99," at www.upi-institut.de/oes199.htm; rise to 60 percent from Michael Kohlhaas, *Energy Taxation and Competitiveness—Special Provisions for Business in Germany's Environmental Tax Reform*, DIW Discussion Papers, No. 349 (Berlin: DIW, May 2003); recent warning from "Trittin Droht Industrie mit Höherer Ökosteuer," *Der Spiegel*, 13 August 2003.

17. BUND, "Eckpunkte zur Weiterentwicklung der ökologischen Steuerreform," at www.oeko-steuer.de/downloads/bund-oekosteuer.eckpunkte.pdf, 12 May 2003; Reinhard Loske and Frank Steffe, "Ökonomische Anreize in der Umweltpolitik," *Blätter für Deutsche und Internationale Politik*, No. 9 (2001), pp. 1082–83; Umwelt- und Prognose-Institut, "Kampagne der CDU/CSU gegen die Ökosteuer," updated 23 February 2003, at www.upi-institut.de/cdu-kamp.htm.

18. Lisa Mastny, *Purchasing Power: Harnessing Institutional Procurement for People and the Planet*, Worldwatch Paper 166 (Washington, DC: World-watch Institute, July 2003); Anne Berlin Blackman, Jack Luskin, and Robert Guillemin, *Programs for Promoting Sustainable Consumption in the United States* (Lowell, MA: Toxics Use Reduction Institute, University of Massachusetts, December 1999).

19. Michael Scholand, "Appliance Efficiency Takes Off," in Worldwatch Institute, *Vital Signs 2002* (New York: W.W. Norton & Company, 2002), p. 132.

20. Ibid.

21. Lisa Mastny, "Ecolabeling Gains Ground," in Worldwatch Institute, op. cit. note 19, p. 124; Blue Angel from "25 Jahre Blauer Engel: Von Höhenflügen und Turbulenzen," Umweltbundesamt (German Federal Environment Agency), 2 April 2003, at www.umweltdaten.de/uba-info-presse/hintergrund/blauer-engel-historie.pdf, and from www.blauer-engel.de.

22. Mastny, op. cit. note 21, pp. 124–25; OECD, op. cit. note 9, pp. 31–32.

23. Stephan Moll, Stefan Bringezu, and Helmut Schütz, *Resource Use in European Countries* (Copenhagen: European Topic Centre on Waste and Material Flows, in cooperation with the Directorate General Environment of the European Union, March 2003), p. 36; Emily Matthews et al., *Weight of Nations: Material Outflows from Industrial Economies* (Washington, DC: World Resources Institute, 2000), pp. 84–85, 112–13.

24. Moll, Bringezu, and Schütz, op. cit. note 23; Peter Bartelmus, "Dematerialization and Capital Maintenance: Two Sides of the Sustainability Coin," *Ecological Economics*, No. 46 (2003), p. 74; Figure 5–1 from Matthews et al., op. cit. note 23, pp. 84–85, 112–13.

25. Paul Hawken, Amory Lovins, and L. Hunter Lovins, *Natural Capitalism* (Boston: Little, Brown and Company, 1999); Ernst von Weizsäcker, Amory B. Lovins, and L. Hunter Lovins, *Factor Four: Doubling Wealth, Halving Resource Use* (London: Earthscan, 1997); Gardner and Sampat, op. cit. note 6, p. 26.

26. Moll, Bringezu, and Schütz, op. cit. note 23, pp. 9, 30, 53; Matthews et al., op. cit. note 23, p. XI.

27. Christer Sanne, "Willing Consumers—or Locked in? Policies for a Sustainable Consumption," *Ecological Economics*, No. 42 (2002), p. 275; Schor, op. cit. note 6; Gardner and Sampat, op. cit. note 6; automobile fuel efficiency and driving trends from Michael Renner, "Vehicle Production Declines Slightly," in Worldwatch Institute, op. cit. note 19, pp. 74–75, and from Michael Renner, "Vehicle Production Inches Up," in Worldwatch Institute, *Vital Signs 2003* (New York: W.W. Norton & Company, 2003), pp. 56–57; materials use from Ward's Communications, *Ward's Motor Vehicle Facts & Figures 2002* (Southfield, MI: 2002).

28. Blackman, Luskin, and Guillemin, op. cit. note 18; Clean Production Action, at www.cleanproduction.org.

29. Marquita Hill, Thomas Saviello, and Stephen Groves, "The Greening of a Pulp and Paper Mill: International Paper's Androscoggin Mill, Jay, Maine," *Journal of Industrial Ecology*, vol. 6, no. 1 (2002), pp. 107–20.

30. William McDonough and Michael Braungart, "The Extravagant Gesture: Nature, Design, and the Transformation of Human Industry," in Schor and Taylor, op. cit. note 11, pp. 16–18; Hill, Saviello, and Groves, op. cit. note 29. Box 5–1 from work undertaken by McDonough Braungart Design Chemistry and from William McDonough and Michael Braungart, *Cradle to Cradle: Remaking the Way We Make Things* (New York: North Point Press, 2002).

31. John Ehrenfeld and Marian Chertow, "Industrial Symbiosis: The Legacy of Kalundborg," in Robert Ayres and Leslie Ayres, eds., *Handbook of Industrial Ecology* (Cheltenham, U.K.: Edward Elgar, 2002).

32. Ibid.; John Ehrenfeld, Director Emeritus, MIT Technology, Business, and Environment Program, e-mail to author, 3 September 2003; Fiji from Gardner and Sampat, op. cit. note 6, pp.

37–38.

33. Clean Production Action, "What Are the Key Elements of an Extended Producer Responsibility Plan?" at www.cleanproduction.org, viewed 11 September 2003.

34. Blackman, Luskin, and Guillemin, op. cit. note 18; Pat Franklin, *Extended Producer Responsibility: A Primer* (Arlington, VA: Container Recycling Institute, November 1997).

35. BASF and Rohner from McDonough and Braungart, op. cit. note 30, pp. 19–20.

36. Germany from "Extended Producer Responsibility," *Environmental Manager*, August 1998; "Extended Product Responsibility: Designing for the Future," Business for Social Responsibility Education Fund, January 2000, at www.where-its-at.com/epr.html. Table 5–3 from the following: "Extended Product Responsibility: Designing for the Future," op. cit. this note; Carola Hanisch, "Is Extended Producer Responsibility Effective?" *Environmental Science and Technology*, April 2000, pp. 170A–75A; "Extended Producer Responsibility," op. cit. this note; Clean Production Action, "Extended Producer Responsibility," at www.cleanproduction.org/epr/EPR.htm; Sam Cole, "Zero Waste on the Move Around the World," Eco-Cycle: Boulder County's Recycling Professionals, at www.ecocycle.org/ZeroWaste/ZeroWasteonTheMove.cfm, viewed 28 August 2003; Michele Raymond, *Extended Producer Responsibility Laws: A Global Policy Analysis* (College Park, MD: Raymond Communications, Inc., undated); Raymond Communications, Inc., "Recycling & Solid Waste in Latin America: Trends and Policies," at www.raymond.com/promo_raymond-library/lacsum.html, viewed 9 September 2003; Total Environment Center (Sydney, Australia), "Toxic Product Fact Sheet, 2 July 2003, at www.tec.nccnsw.org.au/member/tec/projects/Waste/tpf.html; Beverley Thorpe and Iza Kruszewska, "Strategies to Promote Clean Production–Extended Producer Responsibility," January 1999, updated 2 April 2003, at www.grrn.org/resources/BevEPR.html; U.S. Environmental Protection Agency, "Product Stewardship," at www.epa.gov/epaoswer/non

-hw/reduce/epr/products/index.html, viewed 9 September 2003; Mitsutune Yamaguchi, "Extended Producer Responsibility in Japan," *ECP Newsletter*, February 2002, at www.jemai.or.jp/english/e-ecp/ecp_no19/19a.pdf.

37. "Extended Product Responsibility: Designing for the Future," op. cit. note 36; Eric Lombardi, "Take It Back!" Eco-Cycle: Boulder County's Recycling Professionals, at www.ecocycle.org/ZeroWaste/TakeItBack.cfm, viewed 28 August 2003.

38. INFORM, Inc., *The WEEE and RoHS Directives: Highlights and Analysis* (New York: July 2003), and idem, *European Union (EU) Electrical and Electronic Products Directives* (New York: July 2003); Michele Raymond, "U.S. Feels the Effects of European Recycling Debate," *Waste Age*, 1 March 2001; Hanisch, op. cit. note 36. The Directive covers large and small household appliances, information and telecommunications equipment (such as computers and peripherals, cell phones), consumer items (such as televisions, radios, stereos), lighting, electrical and electronic tools, toys, leisure and sports equipment, medical devices, monitoring instruments, and automatic dispensers.

39. INFORM, Inc., *EU Electrical and Electronic Products Directives*, op. cit. note 38; idem, *The WEEE and RoHS Directives*, op. cit. note 38; Silicon Valley Toxics Coalition and Clean Computer Campaign, "European Laws on Electronic Waste and Toxics Enacted," press release (San Jose, CA: 19 February 2003).

40. INFORM, Inc., *The WEEE and RoHS Directives*, op. cit. note 38; Lombardi, op. cit. note 37; IBM from "Extended Product Responsibility: Designing for the Future," op. cit. note 36.

41. Hanisch, op. cit. note 36; state and local interest and battery initiative from "Extended Producer Responsibility," op. cit. note 36.

42. Pressure on computer manufacturers from Clean Production Action, op. cit. note 33.

43. "Extended Product Responsibility," op. cit. note 36; Bette K. Fishbein, "Carpet Take-Back: EPR American Style," *Environmental Quality Management*, autumn 2000, pp. 25–36; Kodak from Cole, op. cit. note 36; Nike from "Extended Product Responsibility," op. cit. note 36.

44. Hanisch, op. cit. note 36; German difficulties from Michael Kröger, "Die Selbstüberlistung des Jürgen Trittin," *Spiegel Online*, 1 October 2003, at www.spiegel.de/wirtschaft/0,1518,267986,00.html.

45. Xerox Corporation, *Environment, Health, and Safety Progress Report 2002* (Webster, NY: 2002), p. 12; Nortel from Blackman, Luskin, and Guillemin, op. cit. note 18.

46. Tonnage, annual turnover, and Table 5–4 from Bureau of International Recycling, "About Recycling," at www.bir.org/aboutrecycling/index.asp, viewed 7 August 2003.

47. Global and U.S. statistics from Remanufacturing Institute, "Frequently Asked Questions," at www.remanufacturing.org/frfaqust.htm, viewed 28 October 1999; Walter Stahel, *From Manufacturing Industry to Service Economy, from Selling Products to Selling the Performance of Products*, Executive Summary (Geneva: Product-Life Institute, April 2000).

48. Xerox, op. cit. note 45, pp. 12, 14; Blackman, Luskin, and Guillemin, op. cit. note 18; Clean Production Action, "Companies Who Have Financially Benefited from EPR Programs," at www.cleanproduction.org/epr/ExistingPrograms.htm, viewed 11 September 2003.

49. Hawken, Lovins, and Lovins, op. cit. note 25; Allen L. White, Mark Stoughton, and Linda Feng, *Servicizing: The Quiet Transition to Extended Product Responsibility*, report prepared for U.S. Environmental Protection Agency, Office of Solid Waste (Boston: Tellus Institute, May 1999).

50. Xerox leasing from Clean Production Action, op. cit. note 48; other examples from Hawken, Lovins, and Lovins, op. cit. note 25.

51. Hawken, Lovins, and Lovins, op. cit. note 25.

52. Ibid.; Amory B. Lovins, L. Hunter Lovins, and Paul Hawken, "A Road Map for Natural Capitalism," *Harvard Business Review*, May/June 1999; Caspar Henderson, "Carpeting Takes on a 'Green' Pattern," *Financial Times*, 8 February 2000; Laurent Belsie, "Seeing Green from Being Green," *Christian Science Monitor*, 7 February 2000; reduced consumption from Interface, "Global Metrics," at www.interfacesustainability.com/metrics_mn.html, viewed 5 September 2003.

53. Hawken, Lovins, and Lovins, op. cit. note 25.

54. Rogelio Olivia and James Quinn, "Interface's Evergreen Services Agreement," Harvard Business School, Case Study N9-603-112 (Cambirdge, MA: 12 February 2003); Fishbein, op. cit. note 43, pp. 25–36; Evergreen lease dropped from Ehrenfeld, op. cit. note 32.

55. Herman E. Daly, *Steady-State Economics* (San Francisco: W.H. Freeman and Co., 1977), p. 20.

56. Gary Gardner, "Why Share?" *World Watch*, July/August 1999, pp. 10–20.

57. The term "infrastructure of consumption" is from OECD, op. cit. note 9, p. 30.

58. Sanne, op. cit. note 27, pp. 275, 279; OECD household debt trend from OECD, op. cit. note 9, p. 14; savings rate trend from World Bank, *World Development Indicators 2002* (Washington, DC: 2002); German data from Roman Pletter, "Dank Bankberater in die Schuldenfalle," *Spiegel Online*, 25 August 2003, at www.spiegel.de/wirtschaft/0,1518,262857,00.html.

59. Growth of U.S. consumer indebtedness from "Flying on One Engine," A Survey of the World Economy, *The Economist*, 20 September 2003, p. 4; total outstanding consumer credit and Figure 5–2 from Federal Reserve Board, "Consumer Credit Historical Data," at www.federalreserve.gov/releases/g19/hist/cc_hist_sa.txt, viewed 17 September 2003; credit card debts from U.S. Department of Commerce, *Statistical Abstract of the United States 2002* (Washington, DC: 2002), and from Robert D. Manning, "Perpetual Debt,

Predatory Plastic," *Southern Exposure*, summer 2003, p. 51; U.K. trend from Office for National Statistics, Annual Abstract of Statistics, Table 22.16: Consumer Credit, at www.statistics.gov.uk/STATBASE/Expodata/Spreadsheets/D4925.xls; Germany from Gundi Knies and C. Katharina Spieß, "Fast ein Viertel der Privathaushalte in Deutchland mit Konsumentenkreditverpflichtungen," *DIW-Wochenbericht*, No. 17/2003; Netherlands from Noam Neusner, "Credit Addiction Goes Global," *U.S. News and World Report*, 25 March 2002, pp. 35–36.

60. Joshua Kurlantzick, "Charging Ahead," *Washington Monthly*, May 2003, pp. 27–29; South Korean credit card charges from Neusner, op. cit. note 59, pp. 35–36.

61. Gilbert Le Gras, "Canada Earmarks C$1 Billion in Climate Change Funds," *Reuters*, 13 August 2003.

62. "Feebate," Energy Dictionary, at www.energyvortex.com/energydictionary/feebate.html, viewed 16 September 2003; "'Feebates'–Price Instrument Promoting Efficiency," European Partners for the Environment, at www.epe.be/workbooks/sourcebook/2.11.html, viewed 16 September 2003.

63. U.S. productivity trend, measuring the manufacturing and service sectors, from Eric Rauch, "Productivity and the Workweek," Massachusetts Institute of Technology, at www.swiss.ai.mit.edu/~rauch/misc/worktime/, viewed 9 August 2003; Schor, op. cit. note 6.

64. Sanne, op. cit. note 27, p. 285.

65. Figure 5–3 from Anders Hayden, "International Work-Time Trends: The Emerging Gap in Hours," *Just Labour* (Canada), spring 2003, p. 24; Schor, op. cit. note 6.

66. Sanne, op. cit. note 27, p. 280; Juliet B. Schor, "Sustainable Consumption and Worktime Reduction," presented at the University of Linz on the occasion of the Annual Kurt W. Rothschild Lecture, 7 November 2002; Table 5–5 based on Hayden, op. cit. note 65, pp. 27–28.

67. European preferences from Hayden, op. cit. note 65, p. 25; U.S. surveys and downshifting from Schor, op. cit. note 6, and from Juliet B. Schor, *The Overspent American* (New York: Harper Perennial, 1998).

68. U.S. wage trends from Economic Policy Institute, "Growth of Average Hourly Wages, Benefits, and Compensation, 1948–2000 (2001 Dollars)," at www.epinet.org/datazone/02/nipa_comp_2_2.pdf, and from idem, "Wages for all Workers by Wage Percentile, 1973–2001 (2001 Dollars)," at www.epinet.org/datazone/02/deciles_2_6r.pdf.

69. Wages lagging behind productivity growth from Brenner, op. cit. note 8, p. 235; need for wage increases from Schor, op. cit. note 6, and from Kenneth Lux, "The Failure of the Profit Motive," *Ecological Economics*, no. 44 (2003), p. 7.

70. Lux, op. cit. note 68; Sanne, op. cit. note 27, p. 282; Schor, op. cit. note 67, pp. 169–71.

71. William Greider, "Deflation: It Threatens the U.S.—and the World," *The Nation*, 30 June 2003, p. 12; auto industry from PricewaterhouseCoopers, "Light Vehicle Assembly by Region, Country, Category," *2001 Q4 Vehicle Outlook Reports*, at www.autofacts.com, viewed 16 December 2001; expected doubling of China's production capacities from Keith Bradsher, "A Heated Chinese Economy Piles up Debt," *New York Times*, 4 September 2003.

72. U.S. domestic demand growth from "Flying on One Engine," op. cit. note 59, p. 3; Global Policy Forum, "Balance on US Current Account 1960–2002," at www.globalpolicy.org/socecon/crisis/2003/curracctable.htm, viewed 23 May 2003; Larry Elliott, "American Deficit Dependency: Kill or Cure, the Fallout's Global," *The Independent*, 20 July 2003; EPI and Roach from Joshua Kurlantzick, "Charging Ahead," *Washington Monthly*, May 2003, p. 31. Box 5–2 from the following: For a more detailed discussion on clothing, see Juliet Schor, "Cleaning the Closet: Toward a New Ethic of Fashion," in Schor and Taylor, op. cit. note 10; apparel share of expenditure from Stanley Lebergott, *Pursuing Happiness*

(Princeton, NJ: Princeton University Press, 1993), p. 91, with share in 2001 from Consumer Expenditure Survey, Bureau of Labor Statistics, at www.bls.gov/cex/2001/share/age.pdf; for low share of revenue going to labor, see Nick Robins and Liz Humphrey, *Sustaining the Rag Trade: A Review of the Social and Environmental Trends in the UK Clothing Retail Sector and the Implications for Developing Country Producers* (London: International Institute for Environment and Development, 2000), p. 19; for more on sweatshops, see Ellen Rosen, *Making Sweatshops: The Globalization of the U.S. Apparel Industry* (Berkeley: University of California Press, 2002); calculation of Bangladesh as the fourth largest apparel-exporting country by Juliet Schor based on www.otexa.ita.doc.gov/msr/cat1.html; wages and working conditions in Bangladeshi factories from National Labor Committee, *Ending the Race to the Bottom*, Report on Bangladesh (New York: 2001), p. iii; data on wages paid in China and child labor from National Labor Committee Report, *Made in China: Behind the Label* (New York: 1998); for a current example of Disney withdrawing after workers made modest requests, see www.nlcnet.org; apparel prices in last decade from detailed Consumer Price Index data available at www.bls.gov; rate of increase in garment purchase from U.S. Census Bureau, *Current Industry Reports 1997, Summary* (Washington, DC); 48 new pieces a year based on idem, *Current Industry Reports, Apparel, 2001*, Tables 1 and 5; Goodwill donations from Christina Bergali, Media Relations Department, Goodwill International, discussion with Juliet Schor, February 2002; toys from China and wages and working conditions in their toy factories from National Labor Committee, *Toys of Misery* (New York: 2002), with 60 percent and 69 toys from Schor's calculations; other price data are calculated from Consumer Price Index, at www.bls.gov; World Bank and International Monetary Fund from Joseph E. Stiglitz, *Globalization and Its Discontents* (New York: W.W. Norton & Company, 2002).

73. Sudden versus gradual change from Schor, op. cit. note 67, pp. 169–71.

74. Hawken, Lovins, and Lovins, op. cit. note 25, p. 10.

Cell Phones

1. International Telecommunication Union (ITU), *World Telecommunication Development Report* (Geneva: 2002), p. 13; idem, "Cellular Subscribers," 24 April 2003 on Free Statistics Home Page, at www.itu.int/ITU-D/ict/statistics, viewed 23 June 2003.

2. "The Fight for Digital Dominance," *The Economist*, 23 November 2002; ITU, *World Telecommunication Development Report*, op. cit. note 1, p. 10; Michael Bociurkiw, "Revolution by Phone: Text Messaging Thrives in the Philippines," *Forbes*, 10 September 2001; Nick Wachira, "Wireless in Kenya Takes a Village," *Wired News*, 2 January 2003; ITU, "Cellular Subscribers," op. cit. note 1.

3. World Health Organization, "Electromagnetic Fields and Public Health: Mobile Telephones and Their Base Stations," fact sheet (Geneva: June 2000); George Carlo and Martin Schram, *Cell Phones: Invisible Hazards in the Wireless Age* (New York: Carroll and Graf Publishers, 2001), pp. 250–55; Independent Expert Group on Mobile Phones, *Mobile Phones and Health* (Oxon, U.K.: National Radiological Protection Board, 2000); Gautam Malkani, "Mobile Phone Safety Probed by 15 Studies," *Financial Times*, 26 January 2002.

4. Bette K. Fishbein, *Waste in the Wireless World: The Challenge of Cell Phones* (New York: INFORM, 2002), pp. 15–17; Charles W. Schmidt, "E-Junk Explosion," *Environmental Health Perspectives*, April 2002; size of market from ITU, "Cellular Subscribers," op. cit. note 1; other numbers from Fishbein, op. cit. this note, p. 1.

5. See, for example, Collective Good International, at www.collectivegood.com, and Wireless Foundation's Call to Protect program, at www.wirelessfoundation.org, viewed 19 September 2003; ReCellular, at www.recellular.net, viewed 19 September 2003.

6. Bette Fishbein, INFORM, New York, discussion with author, 15 October 2003.

7. Fishbein, op. cit. note 4, pp. 58–59; Ken Belson, "Mining Cellphones, Japan Finds El Dorado," *New York Times*, 28 February 2002.

8. Fishbein, op. cit. note 4, pp. 57–58; Blue Angel, at www.blauer-engel.de/englisch/navigation/body_blauer_engel.htm, viewed 6 August 2003; TCO Development, at www.tcodevelopment.com, viewed 6 August 2003.

9. "European Electroscrap Laws Enter into Force," *Environmental News Service*, 17 February 2003; U.K. Environment Agency, "NetRegs: Waste Electrical and Electronic Equipment (WEEE) Directive," at environment-agency.gov.uk, viewed 20 September 2003; European Commission, "Commission Tackles Growing Poblem of Electrical and Electronic Waste," press release (Brussels: 13 June 2000); U.K. Environment Agency, "NetRegs: The Restriction of Hazardous Substances in Electronical and Electronic Equipment (RoHS) Directive," at environment-agency.gov.uk, viewed 20 September 2003; Jeff Chappell, "Recycling Europa," *Electronic News*, 18 March 2002, pp. 1–2.

10. Nokia, "Environmental Report of Nokia Corporation 2002," at www.nokia.com, viewed 23 September 2003.

11. Ted Smith and Chad Raphael, "High Tech Goes Green," *YES! A Journal of Positive Futures*, spring 2003, pp. 28–30; Alice P. Jacobson, "Deleting E-Waste," *Waste Age*, 1 June 2003, p. 6; Lynn Schenkman, "New York Law Will Require Cell Phone Dealers to Recycle Their Products," *Waste Age*, 9 July 2003; Beverly Burmeier, "Happy Endings for Cast-Off PCs," *Christian Science Monitor*, 21 April 2003, p. 20; listing of state initiatives from Silicon Valley Toxics Coalition, at www.svtc.org, viewed 20 September 2003.

12. Fishbein, op. cit. note 6; Timo Poropoudas, "Mobile Giants Sign Up in Recycling Initiative," on Mobile CommerceNet, 14 December 2002, at www.mobile.commerce.net, viewed 15 October 2003; Jim Puckett and Ted Smith, eds., *Exporting Harm: The High-Tech Trashing of Asia* (Seattle, WA, and San Jose, CA: Basel Action Network and Silicon Valley Toxics Coalition, with Toxics Link India, Greenpeace China and SCOPE, February 2002).

Chapter 6. Purchasing for People and the Planet

1. Connecticut Energy Co-op, "Connecticut College Is 1st College to Buy Green Power from the Co-op," press release (Hartford, CT: 11 May 2001); Connecticut College, "Connecticut College Sets National 'Green Energy' Record; Purchases Wind Energy Certificates for 22 Percent of Electricity Use," press release (New London, CT: 27 January 2003).

2. A distinction is sometimes made between purchasing (buying products and materials for operational use) and procurement (buying parts and materials as inputs into manufactured products). In this chapter, however, the terms are used interchangeably. Michael Scholand, "Compact Fluorescents Set Record," in Worldwatch Institute, *Vital Signs 2002* (New York: W.W. Norton & Company, 2002), pp. 46–47; Janet Sawin, "Charting a New Energy Future," in Worldwatch Institute, *State of the World 2003* (New York: W.W. Norton & Company, 2003), pp. 85–109; Minou Yussefi and Helga Willer, eds., *The World of Organic Agriculture 2003—Statistics and Future Prospects* (Tholey-Theley, Germany: International Federation of Organic Agriculture Movements, 2003), p. 24; U.S. Department of Energy, Office of Transportation Technologies, "Hybrid Electric Vehicles in the United States," Fact of the Week #230, 19 August 2002, at www.ott.doe.gov/facts/archives/fotw230.shtml.

3. Estimates are conservative, per Natural Marketing Institute, "Understanding the LOHAS Market Report: LOHAS Market Size," at www.naturalbusiness.com/market.html, viewed 12 December 2002; gross domestic product (GDP) data from David Malin Roodman, "Economic Growth Falters," in Worldwatch Institute, *Vital Signs 2002*, op. cit. note 2, pp. 58–59.

4. Christoph Erdmenger et al., *The World Buys Green* (Freiburg, Germany: International Council for Local Environmental Initiatives (ICLEI), 2001), p. 13; Dunkerque from ICLEI, *Green Purchasing Good Practice Guide* (Freiburg, Germany: 2000), p. 21; Gerard Gleason, Associate Director, Conservatree, San Francisco, discussion with Clayton Adams, Worldwatch Institute, 7 April 2003.

5. Sawin, op. cit. note 2.

6. Alan Durning, *How Much Is Enough?* (New York: W.W. Norton & Company, 1992).

7. Up to three quarters of this spending goes for purchases of consumable goods and services, while the remainder goes to capital goods and investment spending. Figure 6–1 based on 1991 prices and purchasing power parities, from Organisation for Economic Co-Operation and Development (OECD), *Greener Public Purchasing: Issues and Practical Solutions* (Paris: 2000), p. 36; Commission of the European Communities, *Commission Interpretative Communication on the Community Law Applicable to Public Procurement and the Possibilities for Integrating Environmental Considerations Into Public Procurement* (Brussels: 4 July 2001), p. 5; estimate of 18 percent is based on a combined continental GDP of $11.7 trillion, per Chantal Line Carpentier, North American Commission for Environmental Cooperation, presentation at North American Green Purchasing Conference, Philadelphia, PA, 22–24 April 2002; U.S. government from Scot Case, Director of Procurement Strategies, Center for a New American Dream, e-mail to author, 11 April 2003.

8. K. Green, B. Morton, and S. New, *Consumption, Environment, and the Social Sciences*, cited in Adam C. Faruk et al., "Analyzing, Mapping, and Managing Environmental Impacts Along Supply Chains," *Journal of Industrial Ecology*, spring 2001, p. 15.

9. Figure of $250 billion includes the amount spent by students on books and school supplies. Kevin Lyons, Rutgers University (instructor), "Driving Sustainable Markets 'Teach-In,'" online course sponsored by the National Wildlife Federation's Campus Ecology Program and the National Association of Educational Buyers, 2002; 3 percent from U.S. Department of Commerce, Bureau of Economic Analysis, "Current Dollar and 'Real' Dollar Gross Domestic Product," at www.bea.doc.gov/bea/dn/gdplev.xls, viewed 7 May 2003; 18 economies from World Bank, *World*

Development Indicators 2001 (Washington, DC: 11 April 2001); Gary Gardner, *Invoking the Spirit: Religion and Spirituality in the Quest for a Sustainable World*, Worldwatch Paper 164 (Washington, DC: Worldwatch Institute, December 2002); United Nations Inter-Agency Procurement Services Office (IAPSO), *Annual Statistical Report 2000* (New York: July 2001).

10. Case, op. cit. note 7.

11. With some purchases, competitive bidding is not possible because only one supplier is technically qualified to do the job, as is the case with many aerospace and defense contracts. Box 6–1 from the following: Kevin Lyons, *Buying for the Future: Contract Management and the Environmental Challenge* (London: Pluto Press, 2000); U.S. Environmental Protection Agency (EPA), *The City of Santa Monica's Environmental Purchasing: A Case Study* (Washington, DC: March 1998), p. 7; White House Task Force on Recycling, *Greening the Government: A Report to the President on Federal Leadership and Progress* (Washington, DC: 22 April 2000), p. 25; Switzerland from OECD, op. cit. note 7, p. 67; Anne-Françoise Gailly, "Green Procurement and the Belgian Presidency," *Eco-Procura* (ICLEI), September 2001, p. 9.

12. TerraChoice Environmental Services, Inc., *Products and Services: The Climate Change Connection* (Ottawa: March 2002); U.S. Office of the Federal Environmental Executive, "Web Based Paper Calculator," at www.ofee.gov/recycled/calculat.htm; Environmental Defense, "Catalog Companies Are Selling Nature Short This Holiday Season," press release (New York: 13 November 2002).

13. Janitorial Products Pollution Prevention Project, "What Injuries Happen to Your Janitors?" at www.westp2net.org/Janitorial/jp4.htm, viewed 20 February 2003.

14. Compact fluorescent lamps from TerraChoice Environmental Services, Inc., op. cit. note 12, p. 20; cleaners from Alicia Culver et al., *Cleaning for Health: Products and Practices for a Safer Indoor Environment* (New York: INFORM, Inc., 2002).

15. Table 6–1 from the following: Bank of America, *Environmental Commitment 2001 Activity Highlights*, at www.bankofamerica.com/environment/index.cfm, viewed 29 April 2003; Boeing ,"EPA Names Boeing Partner of the Year," press release (St. Louis, MO: 14 April 1999); 16,000 homes from "Sustainability and Green Procurement," Pollution Prevention Northwest, Pacific Northwest Pollution Prevention Resource Center, fall 1999, at www.pprc.org/pprc/pubs/newslets/news1199.html; Canon, *Canon Environmental Report 2002*, at www.canon.com/environment/eco2002e/p22.html, viewed 8 April 2003; Federal Express, "FedEx and the Environment," at www.fedex .com/us/about/news/ontherecord/environment.html, viewed 29 April 2003; Hewlett-Packard, "Supply Chain Social and Environmental Responsibility," at www.hp .com/hpinfo/globalcitizenship/environment/supplychain/index.html, viewed 29 April 2003; IKEA International A/S, *IKEA: Environmental and Social Issues 2001* (Delft, Netherlands: November 2001), p. 13; John Zurcher, IKEA U.S., discussion with Clayton Adams, Worldwatch Institute, 8 April 2003; McDonald's from EPA, *Private Sector Pioneers: How Companies Are Incorporating Environmentally Preferable Purchasing* (Washington, DC: June 1999), pp. 20–22 and from "McDonald's Approves Earthshell Container for Big Mac," *Environment News Service*, 2 April 2001; William Hall, "Migros Commits to Buying Green Palm Oil," *Financial Times*, 24 January 2002; Riu Hotels from International Hotels Environment Initiative, "Case Studies," at www.ihei.org/HOTELIER/hotelier.nsf/content/c1b2.html, viewed 9 May 2003; Staples, Inc., "How Staples Recycles," at www.staples.com/products/centers/recycle/hsr.asp, viewed 16 April 2003; Staples, Inc., "Staples Environmental Paper Procurement Policy" (Framingham, MA: November 2002); Staples, Inc., "Staples Joins Green Power Market Development Group," press release (Framingham, MA: 13 March 2003); Starbucks, "Coffee, Tea, & Paper Sourcing," at www.starbucks.com/aboutus/sourcing.asp, viewed 29 April 2003; Toyota, "Procurement/Production/Logistics," at www.toyota.co.jp/IRweb/corp_info/eco/pro.html, viewed 14 April 2003. Stephan Schmidheiny with the Business Council for Sustainable Development, *Changing Course*

(Cambridge, MA: The MIT Press, 1992), pp. 9–10; L'Oreal from Amanda Griscom, "In Good Company," *Grist Magazine*, 31 July 2002; Anheuser-Busch and IBM from EPA, op. cit. this note, p. 22.

16. Craig R. Carter and Marianne M. Jennings, *Purchasing's Contribution to the Socially Responsible Management of the Supply Chain* (Tempe, AZ: CAPS Research, 2000), p. 11.

17. Steven A. Melnyk et al., *ISO 14000: Assessing Its Impact on Corporate Effectiveness and Efficiency* (Tempe, AZ: CAPS Research, 1999), p. 20; Heidi McCloskey, Nike Apparel, discussion with Brian Halweil, Worldwatch Institute, 18 February 2003; 3 percent from Nike, Inc., "Team Players," at www.nike.com/nikebiz/nikebiz, viewed 28 February 2003.

18. Nike, Inc., op. cit. note 17; Recycled Paper Coalition, "About Us," at www.papercoalition.org/aboutus.html, viewed 11 March 2003.

19. Jeffrey Hollender, "Changing the Nature of Commerce," in Juliet B. Schor and Betsy Taylor, eds. *Sustainable Planet: Solutions for the Twenty-first Century* (Boston: Beacon Press, 2002), p. 76.

20. See, for example, United Nations, *Agenda 21* (New York: April 1993), p. 33; OECD Council, "Recommendation of the Council on Improving the Environmental Performance of Government" (Paris: 1996); ICLEI, "Lyon Declaration: Enhancing the Framework, Enforcing the Action for Greening Government Operations," text adopted at the EcoProcura Lyon Conference, Lyon, France, 17–18 October 2000; United Nations, *Report of the World Summit on Sustainable Development* (New York: 2002), p. 21; OECD, op. cit. note 7, pp. 19, 20.

21. Table 6–2 from the following: OECD, op. cit. note 7, pp. 50–60; ICLEI, op. cit. note 4; Erdmenger et al., op. cit. note 4; Christoph Erdmenger, "Sustainable Purchasing—A Concept Emerging from the Local Level," *International Aid & Trade Review*, Conference & Exhibition 2002 Special Edition, 19–20 June 2002, pp. 124–25; OECD Trade Directorate, *Trade Issues in*

the Greening of Public Purchasing (Paris: 16 March 1999), pp. 4–5, 28; Canada from Natural Resources Canada, *Government of Canada Action Plan 2000 on Climate Change*, at www.climatechange.gc.ca/english/whats_new/pdf/gofcdaplan_eng2.pdf, and from Department of Justice, *Alternative Fuels Act 1995*, at laws.justice.gc.ca/en/A-10.7/text.html; Germany from www.beschaffung-info.de; Center for a New American Dream, "Environmental Purchasing Factoids," at www.newdream.org/procure/factoids.html, viewed 3 March 2003; Hiroyuki Sato, *Green Purchasing in Japan: Progress, Current Status, and Future Prospects* (Tokyo: Green Purchasing Network, 2003); Scot Case, "Moving Beyond 'Buy Recycled,'" *ECOS*, spring 2001, p. 1; U.S. General Accounting Office (GAO), *Federal Procurement: Better Guidance and Monitoring Needed to Assess Purchases of Environmentally Friendly Products* (Washington, DC: June 2001), p. 4; Christoph Erdmenger, Director of the European Eco-Procurement Programme and Eco-efficient Economy, ICLEI, "Overview and Recent Developments of Sustainable Procurement," presentation at International Aid and Trade 2002 Conference on Trade and Development: Building Capacity for Sustainable Markets, New York, 19–20 June 2002.

22. Erdmenger, "Overview and Recent Developments of Sustainable Procurement," op. cit. note 21; "The Hannover Call of European Municipal Leaders at the Turn of the 21st Century," at www.iclei.org/ecoprocura/info/Hann_call.pdf, viewed 24 March 2004; Bente Møller Jensen and Anders Schmidt, *Green Purchasing Status Report: Municipality of Kolding* (Freiburg, Germany: ICLEI, February 2002).

23. Dean Kubani, City of Santa Monica, CA, discussion with Clayton Adams, Worldwatch Institute, 11 April 2003; Mike Liles, Minnesota Office of Environmental Assistance, e-mail to Clayton Adams, Worldwatch Institute, 9 April 2003; EPA, *State and Local Government Pioneers* (Washington, DC: November 2000).

24. Taiwan from Public Construction Commission Executive Yuan, "Article 96," *Government Procurement Law*, at www.pcc.gov.tw/c2/c2b/c2b_3/2_b_3_10.htm, viewed 9 May 2003; U.N.

Environment Programme (UNEP) and Consumers International, *Tracking Progress: Implementing Sustainable Consumption Policies* (Nairobi: May 2002), p. 26, 54–55; UNEP, "DRAFT Mapping of Major Procurement Initiatives Worldwide," document prepared for the Interagency Group on Sustainable Procurement (Paris: 2003); Thailand from Burton Hamner, Hamner and Associates LLC, e-mail to author, 3 February 2003.

25. ICLEI from Erdmenger et al., op. cit. note 4, p. 13; White House Task Force on Recycling, op. cit. note 11, p. 25; Jim Motavalli and Josh Harkinson, "Buying Green," *E Magazine*, September/October 2002, p. 29.

26. The challenge, however, is ensuring that consumers activate the energy-saving features once they have purchased this equipment. Luke Brander and Xander Olsthoorn, *Three Scenarios for Green Public Procurement* (Amsterdam: Vrije Universiteit Institute for Environmental Studies, December 2002), p. 16; more than 1 million from Erdmenger et al., op. cit. note 4, p. 59; 7 percent from Scot Case, Director of Procurement Strategies, Center for a New American Dream, Takoma Park, MD, discussion with author, 2 December 2002; William J. Clinton, *Executive Order 12845: Requiring Agencies to Purchase Energy Efficient Computer Equipment* (Washington, DC: 21 April 1993); Maria Vargas, Climate Protection Partnerships Division, EPA, discussion with Clayton Adams, Worldwatch Institute, 25 April 2003; Japan from Erdmenger et al., op. cit. note 4, p. 47.

27. For vehicle incentives, see the Clean Cities International Program, at www.ccities.doe.gov; "Los Angeles Cathedral to Use Solar Power," *Reuters*, 19 August 2002.

28. European Commission, "Directive 2000/53/EC of the European Parliament and of the Council of 18 September 2000 on End-of-Life Vehicles," *Official Journal of the European Communities*, 21 October 2000; Chrysler Group, "The Chrysler Group Demonstrates Its 'CARE' for the Environment by Turning Garbage Into Car Parts," press release (Auburn Hills, MI: 20 March 2002).

29. U.S. data based on a survey of 2,267 households in November 2001, per LOHAS Consumer Research, "Nearly One-Third of Americans Identified as Values-Based, Highly-Principled Consumers, New Research Shows," press release (Broomfield, CO: 19 June 2002); Deborah Doane, *Taking Flight: The Rapid Growth of Ethical Consumerism*, report for the Co-operative Bank (London: New Economics Foundation, October 2001), p. 2. Doane points out that an ethical purchase is defined as a personal purchasing decision that is aligned to human rights, animal welfare, or the environment and that gives consumers a choice between a product and an ethical alternative.

30. See Robin Broad, ed., *Global Backlash: Citizen Initiatives for a Just World Economy* (Lanham, MD: Rowman & Littlefield, 2002); EPA, op. cit. note 15, p. 7; Environics International, Ltd., *The Millennium Poll on Corporate Social Responsibility: Executive Briefing* (Toronto, ON, Canada: September 1999).

31. ICLEI, op. cit. note 4, pp. 23–24.

32. Box 6–2 from the following: Home Depot, "The Journey to Sustainable Forestry," information sheet (Atlanta, GA: January 2003); sales, stores, products, and certification numbers from Dan Morse, "Home Depot Is Expected to Deliver Report on Timber," *Wall Street Journal*, 2 January 2003; 20 percent from "Home Depot Decision Cheered," *Environmental News Network*, 30 August 1999; competitors from Jim Carlton, "Against the Grain: How Home Depot and Activists Joined to Cut Logging Abuse," *Wall Street Journal*, 26 September 2000; Jim Carlton, "Home Builders Centex and Kaufman Agree Not to Buy Endangered Wood," *Wall Street Journal*, 31 March 2000; scrambling for certification from Barrie McKenna, "U.S. Home Builders To Ban Old-Growth Wood," *Globe and Mail*, 31 March 2000; Michael Marx, ForestEthics, presentation at North American Green Purchasing Conference, Philadelphia, PA, 22–24 April 2002; criticism from Rainforest Action Network, "Rainforest Action Network Statement on Home Depot's Wood Purchasing Policy," press release (San Francisco: 2 January 2003); price increases from June Preston, "Home Depot Says It Aims to Save Ancient Forests," *Environmental News Network*,

30 August 1999.

33. Alliance for Environmental Innovation, at www.environmentaldefense.org/alliance, viewed 12 May 2003; World Wide Fund for Nature, "WWF Climate Change Programme: Business Partners," at www.panda.org/about_wwf/what _we_do/climate_change/what_we_do/business _industry/climate_savers.cfm, viewed 7 March 2003; World Resources Institute, Green Power Market Development Group, at www.thegreen powergroup.org.

34. Julia Schreiner Alves, Compañia Estatal de Saneamiento Básico y Tecnologia, São Paulo, Brazil, e-mail to author, 7 April 2003.

35. National Pollution Prevention Roundtable, Environmentally Preferable Purchasing Discussion Group, "Environmentally Preferable Purchasing," PowerPoint presentation, at www.newdream.org/procure/resources.html#ppt.

36. Developing-world problems from Asian Development Bank, *To Serve and To Preserve: Improving Public Administration In a Competitive World* (Manila: 2000), p. 334; Erdmenger, "Sustainable Purchasing," op. cit. note 21, p. 124.

37. Commission of the European Communities, op. cit. note 7.

38. As of April 2003, some progress had been made in efforts to get specifications for "bio-based" fertilizer products into the U.S. Department of Agriculture farm bill. Tom Ferguson, Perdue AgriRecycle, LLC, discussion with author, 22 April 2002, and discussion with Clayton Adams, Worldwatch Institute, 2 April 2003.

39. Luz Aída Martínez Meléndez, Programa de Administracion Sustenable, Ministry of Environment and Natural Resources, Mexico, e-mail to Clayton Adams, Worldwatch Institute, 7 April 2003.

40. Office of the Mayor, "City Selects ComEd to Provide Clean Power, Leads Nation in Building 'Green' Electricity Market," press release (Chicago, IL: 6 July 2001); Missouri from EPA, op. cit.

note 23, p. 13; cost of ownership from Case, op. cit. note 26.

41. Brander and Olsthoorn, op. cit. note 26, pp. 11–12; Peter Bühle et al., *Stuttgart Green Purchasing Status Report* (Freiburg, Germany: ICLEI, January 2002), pp. 45–46.

42. Santa Monica has since found replacements in all 17 categories, per Kubani, op. cit. note 23; EPA, op. cit. note 11, pp. 1, 8; 1.5 tons from EPA, op. cit. note 23, p. 24; Hiroyuki Sato, Green Purchasing Network, Tokyo, e-mail to Clayton Adams, Worldwatch Institute, 21 April 2003.

43. ICLEI, op. cit. note 4, p. 42.

44. U.K. Department of the Environment, Transport and the Regions, "Action To Halt Illegal Timber Imports—Meacher," press release (London: 28 July 2000); Greenpeace U.K., "Greenpeace Catches Blair Trashing Ancient Forests to Furnish the Cabinet Office," press release (London: 10 April 2002); House of Commons Environmental Audit Committee, "Buying Time for Forests: Timber Trade and Public Procurement," *Sixth Report of Session 2001–02* (London: 24 July 2002), p. 4.

45. "Executive Order 13101—Greening the Government Through Waste Prevention, Recycling, and Federal Acquisition," *Federal Register*, 16 September 1998; EPA, *Qualitative Measurement of Environmentally Preferable Purchasing (EPP) Among Federal Employees in 2000* (Washington, DC: February 2001); GAO, op. cit. note 21.

46. Julian Keniry, Director of Campus Ecology Program, National Wildlife Federation, Washington, DC, discussion with author, 28 March 2002.

47. EPA, op. cit. note 45; Vorarlburg from ICLEI, op. cit. note 4, p. 39; Marcia Deegler, Operational Services Division, Commonwealth of Massachusetts, e-mail to Clayton Adams, Worldwatch Institute, 10 April 2003.

48. GAO, op. cit. note 21; decentralization problems from Tapio Pento, "Implementation of Public Green Procurement Programmes," in Trevor Russel, ed., *Greener Purchasing: Opportunities and Innovations* (Sheffield: Greenleaf Publishing, 1998), pp. 23–30 and from OECD, op. cit. note 7, pp. 46, 82; Berny Letreille, Environment Canada, discussion with Clayton Adams, Worldwatch Institute, 14 April 2003; Tom Snyder, Argonne National Laboratory, U.S. Department of Energy, discussion with Clayton Adams, Worldwatch Institute, 9 April 2003.

49. Møller Jensen and Schmidt, op. cit. note 22, p. 12; Holly Elwood, EPP Program, EPA, Washington, DC, discussion with Clayton Adams, Worldwatch Institute, 9 April 2003; Matthew DeLuca, Green Mountain Energy Company, Burlington, VT, discussion with Clayton Adams, Worldwatch Institute, 4 April 2003.

50. Paul Brown, Steven Morris, and John Aglionby, "Rainforests Hit By Paper Trail to U.K.," (London) *Guardian*, 26 June 2001; EPA, op. cit. note 45, p. 8.

51. Jacqueline Ottman, *Green Marketing: Opportunity for Innovation* (New York: NTC-McGraw-Hill, 1998); Recycled Paper Coalition, "RPC Listening Study on Environmental Printing and Office Papers," at www.papercoalition.org/survey.html, viewed 11 March 2003; lack of standards from OECD Trade Directorate, op. cit. note 21, p. 18; innovative products from Environment Canada, *Towards Greener Government Procurement* (Hull, QC, Canada: updated May 2000).

52. Rita Schenck, "Life Cycle Assessment: the Environmental Performance Yardstick," paper prepared for Earthwise Design, Life Cycle Assessment Realities and Solutions for Sustainable Buildings conference, Antioch University, Seattle, WA, 19 January 2002; Box 6–3 from Guido Sonnemann, Division of Technology, Industry, and Economics, UNEP, e-mail to author, 29 July 2003; Volvo from EPA, op. cit. note 15, p. 9; U.S. Department of Commerce, National Institute of Standards and Technology, Office of Applied Economics, "BEES 3.0," at www.bfrl.nist.gov/oae/software/bees.html.

53. Center for a New American Dream et al., "A Common Vision for Transforming the Paper Industry: Striving for Environmental and Social Sustainability," ratified at the Environmental Paper Summit, Sonoma County, CA, 20 November 2002, at www.conservatree.com/paper/choose/commonvision.shtml, viewed 11 March 2003; Office of the Federal Environmental Executive, "Governments Agree on National Criteria for 'Green' Cleaning Products," press release, at www.ofee.gov/gp/greencleancriteria.htm, viewed 4 March 2003.

54. See, for example, Forest Stewardship Council, "Forests Certified by FSC-Accredited Certification Bodies," at www.fscoax.org/html/5-3-3.html, viewed 6 January 2003, and Green-e, at www.green-e.org.

55. Michele Ferrari, "Ferrara, on Its Way Toward Green Procurement," *Eco-Procura* (ICLEI), September 2001, p. 19; Pennsylvania from EPA, op. cit. note 23, p. 17, and from Arthur Weissman, Green Seal, Washington, DC, discussion with Clayton Adams, Worldwatch Institute, 1 April 2003; trade issues from World Trade Organization, "Government Procurement: The Plurilateral Agreement," at www.wto.org/english/tratop_e/gproc_e/gp_gpa_e.htm, viewed 22 April 2003 and from OECD Trade Directorate, op. cit. note 21.

56. Motavalli and Harkinson, op. cit. note 25, p. 29; Arthur Weissman, Green Seal, Washington, DC, discussion with author, 8 April 2003.

57. Case, op. cit. note 7.

58. An additional 11 products were proposed for EPA recommendations in 2001 but have yet to be approved, including cement and concrete products, nylon carpet and carpet backing, roofing materials, office furniture, tires, and bike racks; EPA, Comprehensive Procurement Guidelines, at www.epa.gov/cpg; ICLEI, op. cit. note 4, p. 36.

59. ICLEI, "BIG-Net: Buy-It-Green Network," at www.iclei.org/europe/ecoprocura/network/index.htm, viewed 4 April 2003; ICLEI, "RELIEF-European Research Project on Green Purchasing," at www.iclei.org/europe/eco

procura/relief/index.htm, viewed 4 April 2003; ICLEI, "Eco-Procurement: The Path to a Greener Marketplace" (Freiburg, Germany: 2002).

60. EPA, Environmentally Preferable Purchasing (EPP), at www.epa.gov/opptintr/epp/index.htm; White House, "Executive Order #12873: Federal Acquisition, Recycling and Waste Prevention," press release (Washington, DC: 20 October 1993); EPA, EPP Database, at www.epa.gov/oppt/epp/database.htm, viewed 19 February 2003.

61. Center for a New American Dream, at www.newdream.org; EPA, "Conference Helps Further Green Purchasing," *EPP Update*, August 2002, pp. 2, 3.

62. Sato, op. cit. note 21.

63. University Leaders for a Sustainable Future, "Programs (Talloires Declaration)," at www.ulsf.org/programs_talloires.html, viewed 4 March 2003; International Hotels Environment Initiative, at www.ihei.org; Benchmark Hotel, at www.benchmarkhotel.com.

64. Erdmenger et al., op. cit. note 4, p. 33; Green Purchasing Network, at eco.goo.ne.jp/gpn/index.html, viewed 9 April 2003; King County Environmental Purchasing, at www.metrokc.gov/procure/green/index.htm.

65. Erdmenger, "Sustainable Purchasing," op. cit. note 21, pp. 122–24.

66. Robert Goodland, *Ecolabeling: Opportunities for Progress Toward Sustainability* (Washington, DC: Consumer's Choice Council, April 2002), pp. 7–8; Miriam Jordan, "From the Amazon to Your Armrest," *Wall Street Journal*, 1 May 2001; DaimlerChrysler, *Environmental Report 2001* (Auburn Hills, MI: 30 July 2001).

67. IAPSO, op. cit. note 9, p. 5; UNICEF, *Supply Division Annual Report 2001* (New York: 2002), p. 5; Goodland, op. cit. note 66, pp. 9–10.

68. World Bank, "Putting Social and 'Green' Responsibility on the Corporate Agenda. World Bank Chief Says Corporate Responsibility Is Key to Sustainable Development—and Good Business," press release (Washington, DC: 21 June 2001); Dominique Brief, Environmentally and Socially Responsible Procurement Initiative, World Bank, discussion with Clayton Adams, Worldwatch Institute, 8 April 2003; Bernard Ross, "World Bank Structural Adjustment and Investment Loans: Approaches to Environmental Conditionality in Procurement" (Washington, DC: May 2000); UNEP, "Environmentally & Socially Responsible Procurement," at www.sustainableprocurement.net, viewed 17 March 2003.

69. ICLEI, op. cit. note 4, p. 31; SWAP (Surplus With a Purpose), at www.bussvc.wisc.edu/swap.

Paper

1. Sixfold and 40 percent from Janet N. Abramovitz and Ashley T. Mattoon, *Paper Cuts: Recovering the Paper Landscape*, Worldwatch Paper 149 (Washington, DC: Worldwatch Institute, December 1999), pp. 33, 7; Japan from U.N. Food and Agriculture Organization (FAO), *FAOSTAT Statistical Database*, at apps.fao.org; about half from International Institute for Environment and Development, *Towards a Sustainable Paper Cycle* (London: 1996), p. 20.

2. Paper Trading International, Inc., "The History of Paper," at www.papertrading.com/prod01.htm; hemp printings from ReThink Paper, "Hemp," at www.rethinkpaper.org/content/hemp.cfm.

3. Abramovitz and Mattoon, op. cit. note 1, p. 21.

4. Estimate of 30 percent from FAO, op. cit. note 1; developing countries, 54 percent, 30 percent, and 16 percent from Abramovitz and Mattoon, op. cit. note 1, pp. 21–22; one quarter from ForestEthics, "Nov. 12, 2002—Office Supply Superstore Staples Inc. Agrees to Historic Endangered Forest and Recycling Policy," press release (San Francisco: 12 November 2002).

5. Energy consumption from Abramovitz and Mattoon, op. cit. note 1, pp. 26–27.

225

6. California Integrated Waste Management Board, "Why Use Recycled Materials?" at www.ciwmb.ca.gov/RMDZ/WhyUse.htm.

7. Recycled Paper Coalition, "About Us," at www.papercoalition.org/aboutus.html.

8. Germany from World Resources Institute et al., *1998-99 World Resources* (New York: Oxford University Press, 1998), p. 164; "EU Parliament Passes Tough New Recycling Law," *Recycling Today*, 2 July 2003.

Chapter 7. Linking Globalization, Consumption, and Governance

1. Indigenous leaders' visit to Worldwatch Institute; Kevin Koenig, Amazon Watch, discussion with Zoë Chafe, Worldwatch Institute, 17 July 2003; hectares from Earthrights International, "BURLINGTON: What Part of NO Don't You Understand?" press release (Washington, DC: 14 May 2003).

2. Amazon Watch, "Indigenous Leaders from Ecuador and Peru Present 'Eviction Notice' to Burlington Resources of Houston, Call on Oil Company to Leave Amazonian Territories," press release (Houston, TX: 14 May 2003).

3. U.S. Environmental Protection Agency, *Light Duty Automotive Technology and Fuel Economy Trends: 1975 Through 2003* (Washington, DC: September 2001), p. 32.

4. Benjamin R. Barber, *JIHAD vs. McWorld: How Globalism and Tribalism are Reshaping the World* (New York: Times Books, 1995), p. 4.

5. Tobacco and liquor advertisements and Coca-Cola kiosks in developing world from author's observations. Table 7–1 from the following: Hennes & Mauritz from "The World of H&M" and "Short Facts," at www.hm.com, viewed 23 September 2003, with turnover converted to dollars from Swedish krona on 23 September 2003; Levi Strauss from "About LS&Co./Worldwide" and "2002 Annual Report/Financial Highlights," at www.levistrauss.com, p. 19, viewed 4 September 2003; Tata Group from "Business Sectors" and "International Connections," at www.tata.com, viewed 23 September 2003, with turnover converted to dollars from rupees on 23 September 2003; Altria Group from "Fact Book," at www.altria.com/investors, and from "Our Companies' Global Presence," at www.altria.com/about_altria/, viewed 24 September 2003; Siemens from "At a Glance," at w4.siemens.de/annual report_2002/, and from "About Us," at www.siemens.com, viewed 24 September 2003, with net sales converted to dollars from euros on 24 September 2003; Yum! Brands from "Yum! Brands Annual Report 2002," at www.yum.com, and from "Business Performance," at www.yrigfp.com, viewed 27 August 2003; McDonald's from "May 2003 Investor Fact Sheet," at www.mcdonalds.com/corporate/investor/, and from "The McDonald's History," at www.mcdonalds.com/corporate/info/history, viewed 24 September 2003; Domino's from "Pizza Particulars," at www.dominos.com, viewed 27 August 2003; Coca-Cola from "Coca-Cola Annual Report 2002," at www2.coca-cola.com/ourcompany, and from "Coca-Cola Africa," at africa.coca-cola.com, viewed 2 September 2003.

6. Figure 7–1 from David Malin Roodman, "Trade Slows," in Worldwatch Institute, *Vital Signs 2002* (New York: W.W. Norton & Company, 2002), p. 61, updated with data from International Monetary Fund, *World Economic Outlook* (Washington, DC: 2003); sectoral breakdown from World Bank, *World Development Indicators 2003* (Washington, DC: 2003), deflated to 2002 dollars using the U.S. Commerce Department's implicit GNP price deflator; foreign direct investment from UNCTAD, *Foreign Direct Investment*, electronic database, at stats.unctad.org/fdi, viewed 30 September 2003; corporate merger trend from idem, *World Investment Report 1998: Trends and Determinants* (New York: United Nations, 1998), pp. xviii–xix; "Freer Trade Cuts the Cost of Living," at www.wto.org.

7. Figure 7–2 from U.N. Food and Agriculture Organization (FAO), *FAOSTAT Statistical Database*, at apps.fao.org; forest product export data from ibid., deflated to 2002 dollars using U.S. Commerce Department's implicit GNP price deflator; forest extent from idem, *State of the World's*

Forests 2003 (Rome: 2003), p. 1; fish export data from idem, *Fisheries Commodities Production and Trade 1976–2000,* electronic database, at www.fao.org/fi/statist/fisoft/FISHPLUS.asp; sustainable fisheries data from idem, *World Agriculture: Towards 2015/2030* (Rome: 2003), p. 197.

8. Mathis Wackernagel et al., *Ecological Footprint of Nations: November 2002 Update* (Oakland, CA: Redefining Progress, 2002), p. 6; Figure 7–3 from ibid., pp. 9–11.

9. Kenny Bruno, "Philly Waste Go Home," *Multinational Monitor,* February 1998; electronic waste from Basel Action Network and Silicon Valley Toxics Coalition, *Exporting Harm: The High Tech Trashing of Asia* (Seattle, WA, and San Jose, CA: 2002), p. 15. Box 7–1 from the following: Wendell Berry, "Back to the Land," *Amicus Journal,* winter 1999, p. 37; Basel Action Network and Silicon Valley Toxics Coalition, op. cit. this note; *Banana Production,* a film by Scott Braman, Katie Milligan, and the Center for a New American Dream, 2002, available at www.newdream.org/ consumer/bananas.html; coffee examples and data from Oxfam International, *Mugged: Poverty in Your Coffee Cup* (Oxford: 2002).

10. Global consumer class from Matthew Bentley, *Sustainable Consumption: Ethics, National Indices and International Relations* (PhD dissertation, American Graduate School of International Relations and Diplomacy, Paris, 2003).

11. China, including Council for International Cooperation on Environment and Development statement, from Matthew Bentley, "Forging New Paths to Sustainable Development," Background Paper, Asia Pacific Expert Meeting on Promoting Sustainable Consumption and Production Patterns, Yogyakarta, Indonesia, 21–23 May 2003, p. 4; leapfrogging from U.N. Environment Programme (UNEP), "United Nations Environment Programme Opens China Office," press release (Nairobi: 19 September 2003).

12. Jeffrey Barber, *Production, Consumption and the World Summit for Sustainable Development* (Rockville, MD: Integrative Strategies Forum, 2003), pp. 2–4.

13. "Chapter 4: Changing Consumption Patterns," in United Nations, *Agenda 21,* available at www.un.org/esa/sustdev/documents/agenda21; U.S. way of life from Mark Valentine, "Twelve Days of UNCED," Follow-up Report on the Earth Summit, U.S. Citizens Network on the United Nations Conference on Environment and Development, 2 July 1992, p. 5, and from Philip Shabecoff, *A New Name for Peace* (Hanover, NH: University Press of New England, 1996), p. 153.

14. Barber, op. cit. note 12.

15. U.N. Guidelines on Consumer Protection available at www.uneptie.org/pc/sustain/guide lines/un-guidelines.htm; Consumers International and UNEP, *Tracking Progress: Implementing Sustainable Consumption Policies* (Nairobi and London: May 2002), pp. 11, 19–20.

16. Organisation for Economic Co-operation and Development, Working Party on National Environmental Policy, *Policies to Promote Sustainable Consumption: An Overview,* Policy Case Studies Series (Paris: July 2002).

17. UNEP, "Promoting Sustainable Consumption and Production Patterns," discussion paper presented by the Executive Director, Twenty-second session of the Governing Council/Global Ministerial Environment Forum, Nairobi, 5–7 February 2003, p. 4; idem, "New 'Life-Cycle Initiative' Launched to Help Combat Environmental Impact of Rising Consumption Patterns," press release (Nairobi: 29 April 2002); idem, "UNEP-DTIE and Sustainable Procurement," at www.uneptie.org/pc/sustain/procurement/ green-proc.htm; idem, "Shopping for a Better World," press release (Nairobi: 2 June 2003). Box 7–2 from the following: youth population from United Nations, *World Population Prospects: The 2002 Revision* (New York: 2002); labor force from International Labour Organization, "Facts on Youth Employment," fact sheet (Geneva: 21 August 2002); survey results from UNEP, *Is the Future Yours?* (Paris: 2001), pp. 8, 10–11, 44–47; youth activism examples from Isabella Marras's discussions with youth, with U.S. example from www.sustainus.org/giftguide.pdf. For more information on the UNEP Project, see www.uneptie.org/pc/sustain/youth/youthxchange.htm.

18. United Nations, *Agreement for the Implementation of the Provisions of the United Nations Convention of the Law of the Sea of 10 December 1982 Relating to the Conservation and Management of Straddling Fish Stocks and Highly Migratory Fish Stocks* (New York: 1995), pp. 5–7; Secretariat of the Convention on Biological Diversity, *Cartagena Protocol on Biosafety to the Convention on Biological Diversity: Text and Annexes* (Montreal: 2000), pp. 6–17; United Nations, *Stockholm Convention on Persistant Organic Pollutants* (Stockholm: 2001), pp. 3–8; Stockholm Convention on Persistant Organic Pollutants, "Implementation," at www.pops.int/documents/implementation; United Nations, *Kyoto Protocol to the United Nations Framework Convention on Climate Change* (Kyoto, Japan: 1997), pp. 3–6, 19; United Nations Framework Convention on Climate Change, *Caring for Climate* (Bonn, Germany: 2003), p. 25.

19. United Nations Convention on the Law of the Sea, "Status of the United Nations Convention on the Law of the Sea, of the Agreement Relating to the Implementation of Part XI of the Convention and of the Agreement for the implementation of the provisions of the Convention Relating to the Conservation and Management of Straddling Fish Stocks and Highly Migratory Fish Stocks," at www.un.org/Depts/los/reference_files/status2003.pdf, modified 19 August 2003; UNEP/Convention on Biological Diversity, "Cartagena Protocol on Biosafety Takes Effect," press release (Nairobi: 9 September 2003); Stockholm Convention on Persistent Organic Pollutants, "List of Signatories and Parties to the Stockholm Convention," at www.pops.int/documents/signature/signstatus.htm, viewed 14 October 2003; U.N. Framework Convention on Climate Change, "Status of Ratification," at unfccc.int/resource/conv/ratlist.pdf, modified on 17 February 2003.

20. Rory Van Loo, "Coming to the Grocery Shelf: Fair-trade Food," *Christian Science Monitor*, 29 September 2003; Forest Stewardship Coun-

cil (FSC), "Forests Certified by FSC-Accredited Certification Bodies," updated 6 October 2003; historical data from World Resources Institute, *Earthtrends*, database, compilation of Forest Stewardship Council data, 1998–2002; share of forests certified based on FSC, op. cit. this note, and on FAO, *Global Forest Resources Assessment 2000: Main Report*, Forestry Paper No. 140 (Rome: 2001), p. 390.

21. Caroline Woffenden, Marine Stewardship Council (MSC), e-mail to Zoë Chafe, 10 October 2003; *Fish 4 Thought* (MSC newsletter), April 2003; products certified from MSC, "Sustainable Seafood at Anuga Exhibition," press release (London: 9 October 2003).

22. U.N. Global Compact from "The Nine Principles," at www.unglobalcompact.org, and from "Global Compact Participants by Country," at www.unglobalcompact.org/content/Companies/list_pc_040903.pdf, updated 2 September 2003; "Standard Chartered Adopts the Equator Principles," press release (London: 8 October 2003); number of countries from Lynn Swarz, Equator Principles Secretariat, e-mail to Zoë Chafe, 14 October 2003.

23. Box 7–3 and other information from United Nations, *Plan of Implementation of the World Summit on Sustainable Development* (New York: 2003), pp. 7–14.

24. Ibid., pp. 18–19; U.N. Commission on Sustainable Development, "Marrakech Meeting Takes Forward Johannesburg Summit Commitments to Sustainable Production and Consumption," press release (New York: 12 June 2003).

25. U.N. Division for Sustainable Development, "Consolidated List of Partnerships for Sustainable Development as of 3 June 2003," at www.un.org/esa/sustdev/partnerships/partnerships.htm, viewed 14 October 2003; "Bicycle Refurbishing Initiative" from www.velomondial.net and from Pascal J. W. van den Noort, Velo Mondial, e-mails to Zoë Chafe, Worldwatch Institute, 5 September and 6 October 2003; Collaborative Labeling and Appliance Standards Program from www.clasponline.org. Table 7–2 from the follow-

ing, with all e-mails to Zoë Chafe, Worldwatch Institute, on dates noted: U.N. Division for Sustainable Development, op. cit. this note; Arab Civil Union for Waste Management at www.keps74.com (text in Arabic); Awareness Raising and Training on Sustainable Consumption and Production update from Bas de Leeuw, UNEP, e-mail 2 September 2003; Cement Sustainability Initiative, at www.holcim.com and at www.wbcsdcement.org; Introduction to Social Standards in Production at www.bmz.de/en; Selling of Responsible Products update from Vincent Commenne, Reseau de Consommateurs Responsables Asbl, e-mails 14 August 2003 and 6 October 2003; Youth Dialogue on Consumption from www.yomag.net and www.youthxchange.net and from Elke Salzmann, Federation of German Consumer Organisations, e-mail 16 October 2003; Certification for Sustainable Tourism at www.turismo-sostenible.co.cr.

26. Cancún collapse from International Centre for Trade and Sustainable Development (ICTSD), "Where There's No Will, There's No Way," *Bridges Daily Update*, Fifth World Trade Organization (WTO) Ministerial Conference, 15 September 2003, and from National Wildlife Federation, "U.S. Administration's Weak Showing at WTO Reflects Neglect of Sustainable Development," press release (Washington, DC: 14 September 2003).

27. For background on the debate, see Duncan Brack, ed., *Trade and Environment: Conflict or Compatability?* (London: Earthscan, 1998), pp. 78–79; on environmentalists' concerns, see Lori Wallach and Michelle Sforza, *Whose Trade Organization?* (Washington, DC: Public Citizen, 1999).

28. WTO provisions from Jeffery S. Thomas and Michael A. Meyer, *The New Rules of Global Trade: A Guide to the World Trade Organization* (Scarborough, ON, Canada: Carswell Thomson Professional Publishing, 1997), pp. 88–91, 185–96, and from Wallach and Sforza, op. cit. note 27, pp. 53–79; evolution of legal reasoning from Howard Mann and Stephen Porter, *The State of Trade and Environmental Law 2003: Implications for Doha and Beyond* (Winnipeg, MN, Canada: International Institute for Sustainable Development,

2003), pp. v–vi. Table 7–3 from the following: beef hormone from ICTSD, "Dispute Settlement Update," *Bridges Weekly Trade News Digest*, 26 February 2002, and from European Commission, "EU Complies with WTO Ruling on Hormone Beef and Calls on USA and Canada to Lift Trade Sanctions," press release (Brussels: 15 October 2003); tuna-dolphin from WTO, "Mexico etc Versus U.S.: 'Tuna-Dolphin,'" at www.wto.org/english/tratop_e/envir_e/edis04_e.htm; shrimp-turtle from idem, "India etc Versus U.S.: 'Shrimp-Turtle,'"at www.wto.org/english/tratop_e/envir_e/edis08_e.htm; swordfish from Mark Mulligan, "Chile at Loggerheads with EU over Swordfishing," *Financial Times*, 21 July 2000, and from ICTSD, "Dispute Settlement Update," *Bridges Weekly Trade News Digest*, 30 January 2001; asbestos from Laurie Kazan-Allen, "A Breath of Fresh Air," *Multinational Monitor*, September 2000, pp. 17–19, and from WTO, "European Communities—Measures Affecting Asbestos and Asbestos-Containing Products: Report of the Appellate Body," Geneva, 12 March 2001; genetically modified organisms from ICTSD, "WTO Committees Scrutinize GMO Regulations and EU Wine Labeling," *Bridges Weekly Trade News Digest*, 2 July 2002, from "EU 'Regrets' U.S. Action on GM Crops," *BBC News*, 8 August 2003, and from ICTSD, "Coalition Seeks WTO Dismissal of GMO Dispute," *Bridges Weekly Trade News Digest*, 25 September 2003.

29. For the historical backdrop to the beef hormone case, see U.S. Office of Technology Assessment, *Trade and Environment: Conflicts and Opportunities* (Washington, DC: U.S. Government Printing Office, May 1992).

30. Paul Jacobs, "U.S., Europe Lock Horns in Beef Hormone Debate," *Los Angeles Times*, 9 April 1999; WTO, "EC Measures Concerning Meat and Meat Products (Hormones)," Report of the Appellate Body, Geneva, 16 January 1998; precautionary principle from the Rio Declaration on the Environment and Development, available at www.un.org/documents/ga/conf151/aconf15126-1annex1.htm; "U.S. Imposes Sanctions in Beef Fight," *New York Times*, 19 July 1999; sanctions from European Commission, op. cit. note 28.

31. U.S. Trade Representative (USTR), "U.S. and Cooperating Countries File WTO Case Against EU Moratorium on Biotech Foods and Crops," press release (Washington, DC: 13 May 2003); Lizette Alvarez, "Europe Acts to Require Labeling of Genetically Altered Food," *New York Times*, 3 July 2003.

32. U.S. Government perspective from USTR, op. cit. note 31; European and consumer views from Neil King, Jr., "U.S., EU Battle Rages Over Modified Crops," *Wall Street Journal Online*, 15 July 2003, and from "EU Moves to Ease Transatlantic Row over Biotech Foods," *Agence France-Presse*, 2 July 2003; U.S. polling data from Gary Langer, "Behind the Label: Many Skeptical of Bio-Engineered Food," *ABC News*, 19 June 2003; European polling data from George Gaskell, Nick Allum, and Sally Stares, "Europeans and Biotechnology in 2002: Eurobarometer 58.0," a report to EC Directorate General for Research (London: Methodology Institute, London School of Economics, March 2003).

33. Fred Pearce, "GMO Import Ban Caught in Crossfire," *New Scientist*, 10 September 2003; Lissa Harris, "Better Biosafe than Sorry," *Grist Magazine*, 11 September 2003; ICTSD, "Coalition Seeks WTO Dismissal," op. cit. note 28; 181 organizations and 48 countries from www.bite-back.org/index.htm, viewed 17 October 2003.

34. Chairman's Statement from the High-Level Roundtable on Trade and the Environment, Cozumel, Mexico, 9 September 2003; trade preferences for "green consumer goods" from Scott Vaughan, *Trade Preferences and Environmental Goods*, Trade, Equity, and Development Series, Issue No. 5 (Washington, DC: Carnegie Endowment for International Peace, February 2003).

35. Doha WTO Ministerial 2001, *Ministerial Declaration*, adopted on 14 November 2001.

36. Ibid.; Brian Halweil, "Why No One Wins in the Global Food Fight," *Washington Post*, 21 September 2003; American Lands Alliance et al., *Collective Comments re: Doha Ministerial Declaration* (Washington, DC: 25 October 2002), pp. 4–5, 13.

37. "Cancun Trade Summit to Sideline Green Issues," *Environmental News Service*, 5 September 2003; Elizabeth Becker, "Poorer Countries Pull Out of Talks Over World Trade," *New York Times*, 15 September 2003; Oxfam International, "Time for the WTO to Get Back on Track," press release (Oxford, U.K.: 17 October 2003); Walden Bello, "There is Life After Cancun," *Bangkok Post*, 21 September 2003.

Cotton T-shirts

1. Scott Fresener et al., *The T-Shirt Book* (Layton, UT: Gibbs Smith Publisher, 1995); Nancy Clark, "A Brief History of the T-shirt," *Wearables Business*, 1 May 1998.

2. Best-selling fiber from Cotton Board, "Cotton's Rise Through the Years," at www.cottonboard.org; production from U.N. Food and Agriculture Organization (FAO), "Cotton Commodity Notes," June 2003, at www.fao.org/es/ESC/en/20953/22215/index.html, and from idem, *FAOSTAT Statistics Database*, at apps.fao.org, updated 10 June 2003; pesticides from Pesticide Action Network North America (PANNA), "Problems with Conventional Cotton Production," in Organic Cotton Briefing Kit, 1998, at www.panna.org/resources/cotton.html, and from Rob Bryant, Agranova, e-mail to Brian Halweil, Wordwatch Institute, 17 July 2001; World Health Organization from Pesticide Action Network United Kingdom (PAN-UK), *Organic Cotton Production in Sub-Saharan Africa: The Need for Scaling-Up* (London: August 2002).

3. Margaret Reeves et al., *Fields of Poison 2002: California Farmworkers and Pesticides* (San Francisco: Californians for Pesticide Reform, 2002), p. 5; Ayanjit Sen, "Pesticide Kills '500' Indian Farmers," *BBC News*, 31 July 2002; Benin survey cited in PAN-UK, *Pesticide Action Network UK Review 2002* (London: 2003), p. 12; Debora MacKenzie, "Fresh Evidence on Bhopal Disaster," *New Scientist*, 7 December 2002.

4. Wildlife harm from PANNA, op. cit. note 2; Philipp Thalmann and Valentin Küng, *Transgenic Cotton: Are There Benefits for Conservation?* prepared for WWF International (Gland, Switzer-

land: March 2000).

5.   Aldicarb from Nancy M. Trautmann et al., "Pesticides and Groundwater: A Guide for the Pesticide User," fact sheet (Ithaca, NY: Cornell Cooperative Extension, 8 May 1998); immunological effects from PAN-UK, "Aldicarb," at www.pan-uk.org/pestnews/actives/aldicarb.htm; E. Michael Thurman et al., "Occurrence of Cotton Pesticides in Surface Water of the Mississippi Embayment," fact sheet (Reston, VA: U.S. Geological Survey, May 1998); Aral Sea from U.N. Environment Programme, *Afghanistan: Post-Conflict Environmental Assessment* (Nairobi: 2003), p. 62.

6.   U.S. use of 11 million bales from Cotton Board, "How U.S. Cotton Finds Its Way into the U.S. Consumer Market," at www.cottonboard.org.

7.   Chemicals and finishes from John C. Ryan and Alan Thein Durning, *Stuff: The Secret Lives of Everyday Things* (Seattle, WA: Northwest Environment Watch, 1997), pp. 23–24, and from Mindy Pennybacker, "The Hidden Life of T-Shirts," *Sierra Magazine*, January 1999.

8.   Ryan and Durning, op. cit. note 7, p. 21.

9.   Production and exports from FAO, "Cotton Commodity Notes," op. cit. note 2, and from idem, *FAOSTAT Statistics Database*, op. cit. note 2; "The Fabric of Lubbock's Life," *New York Times*, 19 October 2003; Amadou Toumani Touré and Blaise Compaoré, "Your Farm Subsidies Are Strangling Us," *New York Times*, 11 July 2003.

10.   China from Global Sources.com, "World's Hub of T-shirt Production," 29 July 2003, at www.globalsources.com/am/article_id/9000000 043332/page/showarticle?action=GetArticle; U.S. consumers from Robin Merlo, Marketing Communications Director, Cotton Incorporated, e-mail to Brian Halweil, Worldwatch Institute, 16 October 2003; Ellen Israel Rosen, *Making Sweatshops: The Globalization of the U.S. Apparel Industry* (Berkeley: University of California Press, 2002).

11.   U.N. Division for Sustainable Development, "Application of Biodynamic Methods in the Egyptian Cotton Sector," *Success Stories— 2000: Integrated Planning and Management of Land Resources, Agriculture, and Forests*, at www.un.org/esa/sustdev/mgroups/success/SA RD-27.htm; Fair Trade Federation, at www.fair tradefederation.org; for a list of companies, see International Organic Cotton Directory Web site, at www.organiccottondirectory.net.

Chapter 8. Rethinking the Good Life

1.   Schools from Woodrow Wilson School of Public and International Affairs, "Former Mayor of Bogotá to Speak on Improvement Models for Third World Cities," press release (Princeton, NJ: 26 November 2001), and from Enrique Peñalosa, e-mail to Gary Gardner, 8 October 2003; parks from ibid.; murders from Curtis Runyan, "Bogotá Designs Transportation for People, Not Cars," *WRI Features* (Washington, DC: World Resources Institute, February 2003).

2.   Highway from Susan Ives, "The Politics of Happiness," *YES! A Journal of Positive Futures*, summer 2003, pp. 36–37; passengers and car owners from Peñalosa, op. cit. note 1; Washington subway from Washington Metropolitan Transit Authority, "Metrorail Peak Ridership Continues to Grow in March," press release (Washington, DC: 25 April 2002); libraries and schools from Woodrow Wilson School, op. cit. note 1.

3.   Peñalosa quoted in Ives, op. cit. note 2, p. 37.

4.   *Merriam Webster's Collegiate Dictionary, Eleventh Edition* (Springfield, MA: Merriam-Webster, 2003), p. 1416. We are indebted to Mark Anielski for calling this etymology to our attention.

5.   Organisation for Economic Co-operation and Development (OECD), *The Well-being of Nations: The Role of Human and Social Capital* (Paris: 2001); Millennium Ecosystem Assessment, *Ecosystems and Human Well-being: A Framework for Assessment* (Washington, DC: Island Press, 2003); Canada from Mike Nickerson, "Green Party Policy Is Passed in the House of Commons," *Vancouver Greens Newswire*, at vangreens .bc.ca/news/2003/06/79.php, viewed 20 October 2003.

6. Definition adapted from Millennium Ecosystem Assessment, op. cit. note 5, p. 74.

7. David G. Myers and Ed Diener, *The Science of Happiness* (Bethesda, MD: World Future Society, 1997); Figure 8–1 from David G. Myers, *The American Paradox: Spiritual Hunger in an Age of Plenty* (New Haven, CT: Yale University Press, 2000), with updates from David G. Myers, hope College, e-mail to Erik Assadourian, 20 October 2003.

8. Ronald Inglehart and Hans-Dieter Klingemann, "Genes, Culture, Democracy, and Happiness," in E. Diener and E. M. Suh, eds., *Culture and Subjective Well-Being* (Cambridge, MA: The MIT Press, 2000), p. 171.

9. Michael Bond, "The Pursuit of Happiness," *New Scientist*, 4 October 2003, pp. 40–47; Ed Diener, "Frequently Asked Questions," at www.psych.uiuc.edu/~ediener/faq.html, viewed 23 October 2003; Ed Diener, University of Illinois, discussion with authors, 8 July 2003.

10. Caroline E. Mayer, "Trade Group to Abide by No-Calls List," *Washington Post*, 29 September 2003.

11. Organic market from "The Global Market for Organic Food & Drink," *Organic Monitor*, July 2003; consumers from Datamonitor, "Organic, Natural, Ethical, & Vegetarian Consumers," report brochure (New York: February 2002).

12. Natural Marketing Institute, "Nearly One-third of Americans Identified as Values-based, Highly Principled Consumers, New Research Shows," press release (Harleysville, PA: 14 May 2002); Amy Cortese, "They Care About the World (and They Shop, Too)," *New York Times*, 20 July 2003; share of expenditures from World Bank, *World Development Indicators Database*, at media.worldbank.org/secure/data/qquery.php, viewed 10 October 2003.

13. Seikatsu Club, "Outline of the Seikatsu Club Consumers' Cooperative Union," at www.seikatsuclub.coop/english/top.html, viewed 21 October 2003; Consumer Coop International, "Action Plan 2002–2003," at www.coop.org/cci/activities/action_plan.htm, viewed 21 October 2003.

14. Global Action Plan at www.globalactionplan.com/index.html, viewed 11 October 2003; Global Action Plan UK, *Annual Report and Accounts 2001–2002* (London: December 2002); Global Action Plan, "The Sustainable Lifestyle Campaign," at www.globalactionplan.org/Files/SLC.htm, viewed 20 October 2003; Minister of the Environment and Sustainable Development, "Week of Sustainable Consumption," press release (Paris: 25 March 2003); *la famille durable* at www.familledurable.com, viewed 10 October 2003.

15. Center for a New American Dream, "Turn the Tide: Nine Actions for the Planet," at www.newdream.org/TurntheTide/default.asp, viewed 10 October 2003.

16. Cecile Andrews cited in Michael Maniates, "In Search of Consumptive Resistance," in Thomas Princen, Michael Maniates, and Ken Conca, eds., *Confronting Consumption* (Cambridge, MA: The MIT Press, 2002), p. 200.

17. Datamonitor, "Simplicity," report brochure (New York: May 2003).

18. Surveys from Michael Maniates, "In Search of Consumptive Resistance," in Princen, Maniates, and Conca, op. cit. note 16, pp. 200–01, from Juliet Schor, *The Overspent American: Why We Want What We Don't Need* (New York: Harper-Perennial, 1998), pp. 113–15, and from The Harwood Group, *Yearning for Balance: Views of Americans on Consumption, Materialism, and the Environment* (Takoma Park, MD: Merck Family Fund, 1995); media and *Affluenza* from Maniates, op. cit. this note, p. 201.

19. Michael Maniates, "Individualization: Plant a Tree, Buy a Bike, Save the World?" in Princen, Maniates, and Conca, op. cit. note 16, p. 45.

20. Information on 07-06-05 at www.07-06-05.com/765/381.htm, viewed 11 October 2003.

21. John de Graaf, Take Back Your Time Day, discussion with Erik Assadourian, 24 October

2003; Take Back Your Time Day, "Take Back Your Time Day Campaign Launch," press release (Seattle, WA: 25 March 2003).

22. De Graff, op. cit. note 21.

23. Robert D. Putnam, *Bowling Alone: The Collapse and Revival of American Community* (New York: Simon & Schuster, 2000), p. 332.

24. Putnam, op. cit. note 23, pp. 327–28; David G. Myers, "Close Relationships and Quality of Life," in D. Kahneman, E. Diener, and N. Schwarz, eds., *Well-Being: The Foundations of Hedonic Psychology* (New York: Russell Sage Foundation, 1999), p. 377.

25. B. Egolf et al., "The Roseto Effect: A 50-Year Comparison of Mortality Rates," *American Journal of Public Health*, August 1992, pp. 1089–92; Putnam. op. cit. note 23, p. 329.

26 Putnam, op. cit. note 23, p. 327.

27. World Bank from Christian Grootaert and Thierry van Bastelaer, *Understanding and Measuring Social Capital: A Synthesis of Findings and Recommendations from the Social Capital Initiative*, Social Capital Initiative Working Paper 24 (Washington, DC: World Bank, April 2001), p. iii; benefits from OECD, op. cit. note 5, pp. 52–61; trust from Paul J. Zak and Stephen Knack, "Trust and Growth," 10 September 1998 draft, at papers.ssrn.com/sol3/papers.cfm?abstract_id=136 961; Madagascar from Marcel Fafchamps and Bart Minten, *Social Capital and the Firm: Evidence from Agricultural Trade*, Social Capital Initiative Working Paper 17 (Washington, DC: World Bank, September 1999), p. 23, and from Grootaert and van Bastelaer, op. cit. this note, p. 11.

28. Quote from Ken Grimes, "To Trust Is Human," *New Scientist*, 10 May 2003, p. 37; trust and growth from Stephen Knack, *Social Capital, Growth and Poverty: A Survey of Cross-Country Evidence*, Social Capital Initiative Working Paper No. 7 (Washington, DC: World Bank, May 1999), p. 17.

29. Thierry van Bastelaer, *Does Social Capital Facilitate the Poor's Access to Credit? A Review of the Microeconomic Literature*, Social Capital Initiative Working Paper No. 8 (Washington, DC: World Bank, February 2000), pp. 8–15.

30. "Grameen at a Glance," at www.grameen -info.org/bank/GBGlance.html, viewed 13 October 2003; Microcredit Summit from Sam Daley-Harris, "State of the Microcredit Summit Campaign Report 2003," from Mya Florence, Microcredit Summit Campaign, e-mail to Gary Gardner, 27 October 2003.

31. New U.S. interest groups from Rob Sandelin, "Clearly Something Is Happening Here," *Communities: Journal of Cooperative Living*, spring 2000.

32. Nancy Hurrelbrinck, "Energy Conservation in Cohousing Communities," *Home Energy*, July/August 1997, pp. 37–41; private ownership from Graham Meltzer, "Cohousing and Sustainability: Findings and Observations," at www.bee.qut.edu.au/people/meltzer/articles/ Cohousing%20Journal/cohojourn.htm.

33. Average attendance from Graham Meltzer, "Cohousing: Verifying the Importance of Community in the Application of Environmentalism," unpublished paper, circa 1998; cooking time from Zev Paiss, discussion with Gary Gardner, March 1999.

34. Watersheds from Anirudh Krishna and Norman Uphoff, *Mapping and Measuring Social Capital*, Social Capital Initiative Working Paper No. 13 (Washington, DC: World Bank, June 1999), p. 1; garbage collection from Sheoli Pargal, Mainul Huq, and Daniel Gilligan, *Social Capital in Solid Waste Management: Evidence from Dhaka, Bangladesh*, Social Capital Initiative Working Paper No. 16 (Washington, DC: World Bank, September 1999), p. 1. Box 8–1 from the following: "Gaviotas," *Social Design Notes*, 9 August 2003; windmills from "Utopia Rises Out of the Colombian Plains," *All Things Considered*, National Public Radio, 29 August 1994; self-sufficiency from "Gaviotas," op. cit. this note; hospital from "Colombia's Model City," *In Context*, fall 1995; organic agriculture and reforestation from "Gaviotas," op. cit. this note.

35. OECD, *Towards Sustainable Consumption: An Economic Conceptual Framework* (Paris: Environment Directorate, June 2002), p. 29.

36. "Santa Monica Sustainable City Plan," 2003 update (Santa Monica, CA: 11 February 2003).

37. Box 8–2 from Robert Prescott-Allen, *The Wellbeing of Nations: A Country-by-Country Index of Quality of Life and the Environment* (Washington: Island Press, 2001).

38. Anders Hayden, "Europe's Work-Time Alternatives," in John de Graaf, ed., *Take Back Your Time* (San Francisco, CA: Berrett Koehler, 2003), pp. 202–10.

39. Jukka Savolainen et al., "Parenthood and Psychological Well-being in Finland: Does Public Policy Make a Difference?" *Journal of Comparative Family Studies*, winter 2001, pp. 61–74.

40. Roger Levett et al., *A Better Choice of Choice* (London: Fabian Society, August 2003).

41. World Bank, "Participatory Budgeting in Brazil," project commissioned by the World Bank Poverty Reduction Group, at poverty.world bank.org/files/14657_Partic-Budg-Brazil-web.pdf.

42. Commuting time from Putnam, op. cit. note 23, p. 212–13; Rob Stein, "Suburbia USA: Fat of the Land? Report Links Sprawl, Weight Gain," *Washington Post*, 29 August 2003.

43. John Pucher and Lewis Dijkstra, "Making Walking and Cycling Safer: Lessons from Europe," *Transportation Quarterly*, summer 2000, pp. 25–50.

44. "Walking in the City," *EcoNews Newsletter*, October 2002.

45. City of Austin, Matrix Program Web site, at www.ci.austin.tx.us/smartgrowth/incentives.htm, viewed 12 October 2003; Dauncey quote from "Smart Planning, Smart Growth," *EcoNews Newsletter*, December 2000.

46. Kate Zernike, "Fight Against Fat Shifts to the Workplace," *New York Times*, 12 October 2003.

47. Sweden from Ingrid Jacobsson, "Advertising Ban and Children: Children Have the Right to Safe Zones," *Current Sweden*, June 2002; U.S. Department of Agriculture, "Import Restrictions and Requirements," *Tobacco Circular*, 1998, at www.fas.usda.gov/tobacco/circular/1998/impre qmts/us.pdf, viewed 23 October 2003; "EU Health Ministers Ban Tobacco Ads," *U.N. Wire*, 3 December 2002.

48. "Smoke Free Movies" at www.smoke freemovies.ucsf.edu/, viewed 13 October 2003; Adbusters, "Reclaim the TV Airwaves," at adbusters.org/campaigns/mediacarta/toolbox/ resources/tvjam, viewed 15 October 2003; Thailand from U.N. Environment Programme (UNEP), Division of Technology, Industry, and Economics, *Report from UNEP International Expert Meeting: Advertising and Sustainable Consumption* (Paris: January 1999); Box 8–3 from Solange Montillaud-Joyel, UNEP, Nairobi.

49. Australia and Canada from New Jersey Media Literacy Project, "Plugging in to Media Education," Center for Media Studies, at www.media studies.rutgers.edu/cmsyme.html, viewed 21 October 2003; Instituto Akatu, "Akatu Institute for Conscious Consumption," at www.akatu.net/ english.asp, viewed 21 October 2003.

50. Benjamin Hunnicutt, "When We Had the Time," in de Graaf, op. cit. note 38, pp. 118–21; Jerome Segal, "A Policy Agenda for Taking Back Time," in de Graaf, op. cit. note 38, p. 214.

51. Lao-Tzu, *Te Tao Ching* (New York: Ballantine Books, 1989), p. 85.

52. Gould quoted in David Orr, "For the Love of Life," *Conservation Biology*, December 1992, p. 486.

53. Prescott-Allen, op. cit. note 37; OECD, "DAC Tables from 2000 Development Co-operation Report," at www1.oecd.org/dac/images/ ODA99per.jpg.

54. Ruub Lubbers cited in Hayden, op. cit. note 38, p. 202.

Index

Knowing the issues is just the first step in moving toward a more sustainable world.

For more than 20 years, *State of the World* has guided a global audience in identifying and understanding the pivotal environmental and social issues of our time. This year, we invite you to do more than just read the book—please join us online, where you will find a wealth of additional material about our special *State of the World* focus: The Consumer Society.

Visit our Consumption Web portal at http://www.worldwatch.org/topics/consumption/ throughout the year and you will constantly find new resources to review, including:

- Our first online-only publication, *Good Stuff? A Behind the Scenes Guide to the Things We Buy*, which traces the environmental and social impact of more than 25 everyday consumer items—from food and cars to cleaning supplies and electronics.

- Video and audio interviews with *State of the World 2004* chapter authors.

- A series of online discussions with authors and partners from *State of the World 2004* and *Good Stuff.*

Reading *State of the World 2004* is a great first step. Looking at what you can do next seals the deal. No matter where on Earth you live, and no matter where your true interests lie, take that extra step with us online and help build a sustainable world.